Fraser Darling in Africa

For

Alex Kerr,
'Ma Frien'!'
with very best wishes

from

Morton

In memory of Ian Robert Grimwood and David Leslie William Sheldrick and in salute to all those who accompanied Fraser Darling in his African travels.

Fraser Darling in Africa

A Rhino in the Whistling Thorn

edited and presented by

JOHN MORTON BOYD

J Morton Boyd
20.10.92

EDINBURGH UNIVERSITY PRESS

© Edinburgh University Press, 1992

Edinburgh University Press
22 George Square, Edinburgh

Typeset in Lasercomp Palatino
by Alden Multimedia, Northampton, and
printed in Great Britain by The University Press, Cambridge

A CIP record for this book is available from the
British Library

ISBN 0 7486 0368 9

Africa is in turmoil, and tragedy deepens . . . It is my belief that the rhinoceros is a key species in the management of African vegetation, and that its wanton and drastic reduction in half a century is one of the factors which have led to a decline of much habitat . . . and there is a tragic irony in the fact that when ecological understanding of the idiosyncrasies of the African terrain is growing, the possibility of enlightened land management is waning . . .

FRASER DARLING, 1960–2

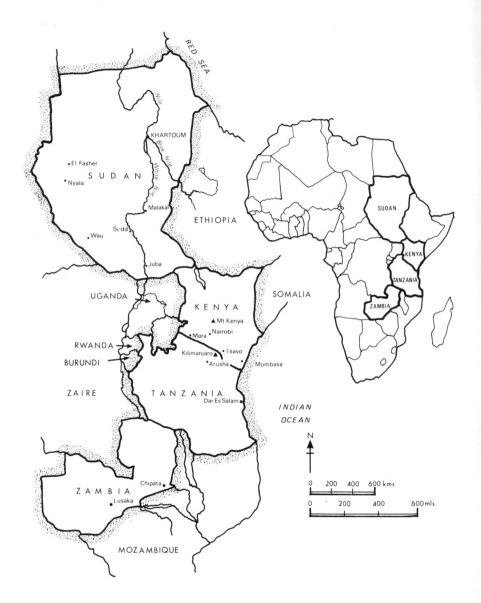

Map 1: Zambia (Northern Rhodesia), Tanzania (Tanganyika), Kenya (Kenya Colony) and Sudan: the African countries visited by Fraser Darling between 1956 and 1961.

Contents

Preface

As Sir Frank Fraser Darling's colleague and friend I promised that, after his death, I would edit and publish his African Journals, manuscripts and type-scripts, which he entrusted to me for that purpose. To prepare the ground for this, I wrote a preparatory work *Fraser Darling's Islands* (1987), which described his life and work in Scotland from 1928 – when he came from England as a research student at Edinburgh University – until 1948, when he left Scotland to live in the south of England again. His Scottish years are known to, and appreciated by, a different public from those who followed his international career. The former he described in his evocative and scholarly books about the lives of red deer, sea-birds and grey seals. These were made the more charming by their setting of family life in the remote and desolate West Highlands and Islands of Scotland.

Fraser Darling described his international years in several books and reports, but not at the personal level which marked his Scottish writings. His American experience is contained in *Wildlife in Alaska* (with A.S. Leopold, 1953), *Pelican in the Wilderness* (1956), and his African experience in *Wildlife in an African Territory* (1960), and 'An Ecological Reconnaissance of the Mara Plains in Kenya Colony' (*Wildlife Monographs*, 5, 1960). The first of these is largely drawn from his American journal, but the other two are technical works, and give little impression of the reality of bush travel in Africa in the late fifties. None of them record his travels in Sudan, nor his brief visit to Ngorongoro Crater and the Serengeti Plains in Tanzania.

The African travel studies fall into three parts – Zambia (formerly Northern Rhodesia), Kenya (formerly Kenya Colony), and Sudan:

Zambia

February to April 1956 in the Luangwa Valley, the Plateau and the Kafue Flats; August to October 1956 in the Kafue Basin and Barotseland (now the Western and North Western Provinces); July to October 1957 in the Bangweulu Swamps, Mweru Wa Ntipa, Sumbu, Nyika Plateau, and the East Luangwa.

Kenya

August 1956 in Tsavo; September to December 1958 in the Mara; January 1961 in Nairobi with visits to Ngorongoro Crater, Serengeti (both in Tanzania) and the Mara.

Sudan

January to March 1961 in Upper Nile (now Sobat and Jonglei), Equatoria, Bahr-el-Ghazal, and Dar-Fur.

The spirit of adventure in the Dark Continent, enshrined in the journals of Livingstone, Speke, Baker, Stanley and the other great explorers of the nineteenth century, lives on in these pages written by Fraser Darling almost a century later. The *ulendo* (foot safari with porters) in Zambia, chimes with Livingstone's days in bush and forest (without slave caravans). Travel by landrover and light aircraft transforms the adventure to the modern idiom of African safari, albeit in the same immense tropical countries with their dwindling panoply of wildlife, and their burgeoning numbers of African peoples.

To travel with Fraser Darling is to travel with a seeing eye for country and people. Although dealing with incidents of the day, his Journal is full of intuitive responses to the human situations in which he finds himself, revealing his preoccupation with the dignity of life, the gifts of personality, the quality of craftsmanship, the absolute beauty of nature. Above all there is his ecological sense of right and wrong in the use of country; the scourges of deforestation, overgrazing, and burning were a restatement of what he had already encountered and deplored in the Scottish Highlands and in Alaska.

I first went to Africa as a Nuffield Travelling Fellow in 1964, when I was introduced to the world-famous scene of African wildlife conservation which was my primary reason for going, and in which Fraser Darling was already a prominent figure. Having just spent two years revising his popular book *Natural History in the Highlands and Islands*, I was already his close friend. Thus, I travelled to Africa with his blessing and an introduction to many people who were the companions of his own days on safari, and whose names appear in his Journal. Until his death in 1979, I remained close to him and maintained the African connections which he made. I have made many visits to East Africa, particularly to Amboseli, Mara, Tsavo and Samburu in Kenya, and Ngorongoro Crater, the Serengeti Plains, Lake Manyara, and Tarangeri in Tanzania. I was not fortunate enough to visit Zambia or Sudan, but I have travelled in northern Uganda close to the Sudanese border where Fraser Darling had been three years previously.

Unfortunately, I was not given time to go through these Journals with Frank before he died, to obtain from him further explanation of events at the time of writing. In presenting the Journals, therefore, I have relied on his published writings and have cast my net wide in consulting his contemporaries. However, as his pupil of many years, I have also given my own interpretation of his thoughts and actions.

In 1987 my wife and I accompanied Major Ian Grimwood on a camping safari 'in the tracks of Fraser Darling' in Amboseli, Chyulu Hills, Tsavo and Mara during which Ian was able to relate to us his recollections of travelling with Fraser Darling in Zambia and Kenya. This he did with particular effect round our camp fire, breaking off occasionally to listen to noises of the African night which he knew by heart. At that time we met Colonel Mervyn Cowie, formerly Director of the National Parks of Kenya, who related to us in some detail the 'inside' story of Fraser Darling's assignment in Kenya. Daphne Sheldrick, whose husband, David, was his companion and counsellor in the Mara in 1958, gave us her recollections on the verandah of her bungalow overlooking the vast spaces of the Nairobi National Park and the Athi Plains beyond. Dr Hugh Lamprey, formerly Director of the East African College of Wildlife Management and Director of the Serengeti Research Institute, who visited Fraser Darling at his camp in the Mara and travelled with him in Ngorongoro and the Serengeti, did likewise. Noel Simon, formerly chairman of the Kenya Wildlife Society, was a prime mover in inviting Fraser Darling to Kenya, and has sent me important papers relating to the Mara reconnaissance. Dr Michael Gwynne, who also met Fraser Darling in Africa, arranged meetings for us in Nairobi with the then British High Commissioner, John R. Johnston, and Richard Leakey, then Director of the National Museum of Kenya. Michael, with his wife Maureen, had us as their guests at Karen.

Major Eustace W. Poles, formerly the Chief Game Ranger, Mpika District, Northern Rhodesia very kindly sent me a typescript copy of his own journal covering the three-week *ulendo* which he made with Fraser Darling in the Luangwa Valley in 1956 (which is now lodged within his twenty-five large field journals at the Zoological Society of London). Roelf Attwell, formerly Provincial Game Officer, Northern Rhodesia has sent me notes on his travels with Fraser Darling, a copy of his obituary of Eustace Poles from *African Wildlife* (1990), and was particularly helpful in contacting others. Frank Ansell, formerly Deputy Director of the National Parks and Wildlife Service of Zambia, has sent me a riposte to Fraser Darling's personal comments which I have included verbatim in the book. Dr Robin Pellew helped to contact his elderly uncle Lt Colonel Ronnie Critchley in Australia concerning Fraser Darling's personal comments. Because of Colonel Critchley's indisposition, he was unsuccessful in obtaining a reply, but has provided a riposte on his uncle's behalf which I have quoted. Bill Astle, formerly Chief Wildlife Research Officer in Zambia, provided maps of the Luangwa Valley, and a useful commentary on conditions in Zambia since Fraser Darling's visits. Barry Shenton and F.I. Parnell kindly sent their recollections. Norman Carr, Richard Bell, Jack Barrah, and David Cruickshank gave advice. All of those people have my special thanks; but there are others from Africa with whom I have spoken and had correspondence about these Journals, who also have my grateful thanks.

William Conway and Steve Johnson of the Zoological Society of New York, and Paige MacDonald of The Conservation Foundation, Washington DC very kindly sent me papers on Fraser Darling's visit to the Sudan. Douglas Grant kindly provided me with background on conditions at home surrounding the illness and death of Fraser Darling's wife, Averil, in 1957.

The Carnegie Trust for the Scottish Universities gave grants to carry out research in East Africa, and to help with colour illustration of the book. The Dulverton Trust also gave a grant towards the production costs – Fraser Darling was Lord Dulverton's friend and counsellor in the ecology of wildlife in the Scottish Highlands and East Africa. The Conservation Foundation, Washington DC gave a grant towards preparing Fraser Darling's African Journals for publication.

I thank the Fraser Darling family for their support and understanding. Lady Christina has given me access to Sir Frank's collection of transparencies, which provide the bulk of the illustrations in the book. Alasdair, Richard and Francesca read and commented on parts of the text. Thanks also to Miss Moira Munro who lettered my maps, and to Roger Wheater and Professor Aubrey Manning for help and encouragement. I have greatly appreciated the scholarship, skill, and perseverance of Dr I.D.L. Clark and the staff of Edinburgh University Press. Lastly, the book would not have been written without the continuous support of Winifred, my wife, throughout the work.

J. MORTON BOYD
Edinburgh

Fraser Darling: A Profile

CHAPTER ONE

The Coming of a Pioneer

FRANK FRASER DARLING was born in obscurity at a farm near Chester-field on 23rd June 1903. He was the only child of Harriet Ellse Cowley Darling, daughter of a prosperous cutler family in Sheffield, and one Frank Moss, who left Britain for East Africa about the time of the child's birth, and was subsequently killed in action on the Kenya–Tanganyika border in 1917. Born in disgrace in a stable loft, the plan was to have Frank fostered and ultimately forgotten, but the family underestimated his mother's resolve. In her loneliness and grief she refused to part with the child, an act of love which was sacred to Fraser Darling throughout his life and which, more than any other, determined the shape of his personality.

In old age he described his life as 'a hole out of which he had climbed'. Yet he saw some advantages of being illegitimate: 'that feeling I was outside gave me the freedom I might not have had, and lack of conventionality being outside class and inwardly convinced that I was tolerably well bred.'

Frank never met his father, nor ever talked of him to his friends. Yet in later life, when his travels took him to East Africa, he spoke with pride in his voice of the soldiery on that tropical frontier. Sensitive about his roots, he showed a sense of patrimony with that vast and beautiful country which held his father's remains. He was fifty-three years of age when he first set foot in Africa, but throughout his life he probably nursed a desire to go there, generated as much by curiosity as by filial attachment. On 24th August 1956 he visited the scene of the 1917 action, but makes no mention of his father in his Journal, though in a letter to his son Alasdair in 1977 he wrote: 'by the way, did I ever tell you that I got to the site of the battle in Tanganyika where he [his father] was killed. There was still a bully beef can or two in the bush.'

Before going to Africa in 1956 Fraser Darling had already made his name as an ecologist and conservationist in Britain and America. Having run away from school at the age of fifteen, and after a spell of work on a Pennine farm, he studied at the Midland Agricultural College, Sutton Bonington, and obtained diplomas in agriculture and dairying. Soon afterwards as 'Frank Darling' he married Marian Fraser ('Bobbie'), and took the style 'Frank Fraser Darling' which has stuck ever since, though he was divorced from Marian Fraser in 1948.

He disliked his first job as a Clean Milk Adviser in Buckinghamshire and longed for a post in research which would take him to the wild uplands and islands of Scotland. He had a streak of scholarship and was attracted by the work of F.A.E. Crew, professor of genetics at Edinburgh University, who offered him a studentship. This led to a PhD thesis on the fleece of Blackface sheep and took him into the Scottish hills. He was then appointed to the newly formed Imperial Bureau of Animal Genetics in Edinburgh, which was largely a bureaucratic grind from which he was soon to seek his escape (with wife and son). On shoe-string grants from the Leverhulme and Carnegie Trusts he went to study red deer in Wester Ross, and later sea-birds and grey seals in the Western Isles.

Living at Dundonnell and later in the Summer Isles, he began the work which was to mark him as a naturalist–philosopher of original turn of mind and great intellectual drive. Before the Second World War he saw himself as working in the middle ground between ecology and animal behaviour. He described the social and breeding behaviour of the red deer (*Cervus elephus*), gulls (*Larus* spp.), and the grey seal (*Halichoerus grypus*) respectively, in three academic works *A Herd of Red Deer* (1937), *Bird Flocks and the Breeding Cycle* (1938) and *A Naturalist on Rona* (1939).

His main bid for a place in the history of science strangely enough did not come in the work on deer and seals for which he is most remembered, but from that on gulls which touched fundamental levels of science. He believed that gregariousness in birds was a stimulus to breeding, providing a shorter breeding season and greater breeding success. Thus flocking behaviour is genetically maintained. He was unable to demonstrate this by his observations and the theory was discredited. However, there are those who still believe that Fraser Darling was on the right track in explaining gregarious behaviour in birds and, in some biology texts, the selective process of synchronous breeding in birds is referred to as the 'Darling Effect'.

It was through his more popular books *Island Years* (1940), and *Island Farm* (1943) that he became widely known in Britain in the war years. During this time, his locus of work moved to the middle ground between man and nature. The war had put an end to his aspirations of further work on the grey seal, and he had already purchased the ruined fishing station with adjacent land on Tanera Mor in the Summer Isles. Being at the age of thirty-six too old for active military service in 1939, he elected to farm rather than to leave the West Coast for wartime civilian work. From that point his outlook changed.

Fraser Darling returned to the precepts and disciplines of his agricultural upbringing, and began to mix these with others he had acquired as a field naturalist. He believed that man was a part of, and not apart from, nature. His idea of human ecology saw mankind in a species–habitat relationship. To him political actions were as potent ecological factors as climate. The oneness of man and nature became an imperative in his life. He set off on the path which was to lead him to detailed surveys of natural resources in different human, social, and economic orders in the wild lands of Scotland, the United States, East and Central Africa, and Sudan.

Between 1939 and 1943 Fraser Darling reclaimed derelict land to agricultural production on Tanera. In 1942 the wartime Secretary of State for Scotland, Tom Johnston, saw in Frank's writings a strong evangelism in the wise use of natural resources. A messianic figure was badly need in the Highlands and Islands, and Johnston sent for Frank in May 1942: would he run an agricultural advisory programme in the crofting areas? He agreed, and for two years he travelled, taught, and wrote agricultural articles for a syndicate of six local newspapers. The articles were afterwards published in book form, *Crofting Agriculture* (1945).

In his spare time he wrote children's books and works in popular natural history as wage earners: *Wild Country* (1938), *The Seasons and the Farmer* (1939) and *The Seasons and the Fishermen* (1941), *The Story of Scotland* (1942), *Care of Farm Animals* (1943), and *Wildlife in Britain* (1943).

Fraser Darling's growing popularity was confirmed by his appointment to the Scottish National Parks Survey Committee in 1943, and as Director of the West Highland Survey in 1944 – a study of the causes underlying depopulation and economic decline in the Western Highlands, which he sub-titled 'An Essay in Human Ecology'. As will emerge later, he had no clear definition for the term 'human ecology', using it to convey the broad idea of man's relationship to the natural order.

In 1944 he left Tanera (and Bobbie) for a new life with Averil Morley, who was later to become his second wife. He was on the threshold of recognition as a national figure and at his intellectual best. Then he produced his two memorable contributions to natural science: *Natural History in the Highlands and Islands* (1947) and *West Highland Survey* (1955). However, relations between Frank and the professional civil servants were not good – their orthodoxy clashed with his non-conformist ideology – and this, in part, resulted in a delay of five years in the publication of the *Survey*. It was not well received by the Scottish Office and, to Frank's great disappointment, was disregarded by the Government. However, the work was funded by the Development Commission, and agreement was eventually obtained for its publication privately, with a Government disclaimer. The outcome did not augur well for further encounters, which lay ahead in Africa, with the Colonial Office.

By 1949 Fraser Darling as a member of the Scottish Nature Reserves Committee, and an already established figure in the new post-war conser-

vation movement, was a hot contender for a senior post in the Nature
Conservancy. That year the Conservancy was set up in Great Britain under the
Privy Council, by the National Parks and Access to the Countryside Act. He
made a bid to become the Director for Scotland, but the post went to Dr John
Berry, formerly of the Scottish Information Office and until then Biologist of
the North of Scotland Hydro-Electric Board. Frank was appointed to the
Scottish Committee of the Conservancy, and later Director of the Conser-
vancy's Red Deer Survey. Michael Swann (later Lord Swann) offered him a
senior lectureship in ecology and conservation in the zoology department,
Edinburgh University, which he also accepted.

Frank's personal conflicts with the staff of the Scottish Office in the 1940s
probably cost him the job he desired. Nonetheless, though he outwardly
relished the other appointments which continued throughout the next decade,
his whole heart was not in them, and his interests moved elsewhere, to the
United States and Africa. The final report of the Red Deer Survey never
appeared and the senior lectureship was prematurely given an honorary status
in the School of Scottish Studies. He simply did not afford these jobs the time
they required.

CHAPTER TWO

The American Connection

JULIAN HUXLEY, the doyen of classical natural history in the late 1930s and '40s, took an interest in Frank's studies in animal behaviour and corresponded with him in Tanera. This was more than a passing respect for, in 1949, UNESCO, of which Huxley was then the first Director-General, invited Frank to be one of its representatives at the United Nations' Conference on Conservation at Lake Success on Long Island. For the first time he saw the prospect of international distinction, and he liked it.

> As UNESCO's man I was a cosmopolitan and my own countrymen seems a body of folk remote from me, as were their interests. Even our own Nature Conservancy . . . had ignored the conference.

There is a touch of self-esteem in that statement from *Pelican in the Wilderness* (1956). He had found consolation for the disappointments which he had endured at the hands of the British establishment. From that time his work in Britain lacked the vitality which had come to characterise him at home but which now became vested in his American connection: 'It was in the enormous lounge of the Lake Success building that I got to know many of the American friends I now hold dear.'

The *West Highland Survey* had made a greater impression in the United States than it had in Britain, mainly, it would appear, because of its sub-title, *An Essay in Human Ecology*. The book had found a target in the Rockefeller Foundation in New York, where Frank was hospitably received and offered a Special Research Fellowship. Ostensibly this was to enlarge his experience in human ecology by looking at America through the same eyes, and with the same analytical mind, as he had done in the Scottish Highlands and Islands.

> Human ecology, the Foundation was ready to think, might be a subject which would help to use aright the scientific advances in other fields by being able to forecast critically the consequences of application . . . At present (1956) we do not quite know what human ecology is. The geographer, the sociologist, and the student of social medicine, all think their's is the real field of human ecology. The man who says it embraces all of these and more also may be right, but he will get nowhere, because no one can span all knowledge and digest it. For myself, I think that human ecology can be

little more than an attitude of mind, and that in no way an arrogant one, till we have done a good deal more thinking. If we must give it a base line, let us say that we must deal with the human habitat, accepting the fact that we still live from the produce of the earth.

This resulted in Frank having a six-month roving commission in the United States and Mexico. This he romantically described as 'a Naturalist's Odyssey in North America', which is the sub-title of *Pelican in the Wilderness*, a book written substantially from his American Journal. The title is taken from Psalm 102, a prayer of the afflicted when overwhelmed, and pouring forth his heart to the Lord. This title conveys the unremitting sense of outrage at man's maltreatment of nature which he possessed during the *West Highland Survey* and which greatly coloured his work in America and Africa. Though different in style, the thinking in this book strikes a strong harmonic with *A Sand County Almanac* (1949) by Aldo Leopold, whom Frank greatly admired.

> Aldo was the man above all others in America who reduced an immense body of sentiment and goodwill for conservation to reasoned, documented, scientific reality . . . Aldo was that rare creature, a naturally wise man who had the intellectual capacity and the continuing simplicity to become a good deal wiser.

Aldo Leopold and Fraser Darling never met; Aldo died suddenly in 1948 at the age of fifty-eight, a year before Frank went to America. They were two of a kind and it was a great pity that they did not have an opportunity of combining their visions and philosophies. We shall never know the influence, if any, of the latter on the former, but *Pelican in the Wilderness* speaks of the influence of Leopold on Fraser Darling. However, the book did not receive the acclaim for which Frank hoped and was never reprinted.

'The Rockefeller Foundation unlocked the door, Fairfield Osborn held it open, and the American people said "Come right in".' So wrote Frank in looking back over his early times in the United States. In 1950 Rockefeller had sent him round America to look at human ecology. On his return after six months of 'looking', he was the guest of the President and Heads of Divisions of the Foundation. The top brass wished to hear about human ecology; what it was, what it could do as a science, what the Foundation might do to further its study? Whatever impression he created by saying that 'more academic work was necessary, and disciplined thinking' is not recorded, but Fairfield Osborn, the President of the Conservation Foundation, was present. He was impressed and later invited Frank to visit Alaska with A. Starker Leopold (Aldo's son) in 1952 to look at the relationships between man and wildlife.

The study, sponsored also by the New York Zoological Society, is particularly valued for its treatise on the ecology of the native caribou and eskimos, and the reindeer introduced from Siberia to sustain immigrant communities exploiting the resources of land and sea. Though different in detail, in principle the ecological processes were similar to Frank's early studies on native red deer and highlander, and the sheep introduced by lowlanders during the industrial revolution in Scotland. As we shall see, he found the same principles applying on a grand scale in Africa. The relationship of native game species to native peoples is again seen to be disrupted by the arrival of immigrant pastoralists and colonialists. Frank was alive to the common denominators of these ostensibly different and geographically distant enclaves of man's domain within the natural world order.

JOURNAL I

Zambia (Northern Rhodesia)

Introduction

IN 1952, ROELF ATTWELL of the Game and Tsetse Control Department of Northern Rhodesia came to Oxford University on a sabbatical study leave at the Bureau of Animal Population. He broached the question of the priorities of the Bureau with the Director, Charles Elton, saying broadly that the need in Africa for ecological research was altogether greater than in Britain, taking Wytham Wood, the Bureau's celebrated study area near Oxford as an example against, say, the Serengeti Plains or the Luangwa Valley. The scientific worth of the Whytham research and its relevance to training of ecologists was not in question. Attwell saw that while this classical work progressed in somewhat cloistered precincts, the magnificent wildlife of British Africa was dwindling rapidly, partly for want of ecological knowledge applied on the spot. Game wardens in Africa desperately needed expert advice to get them thinking and working on ecological lines. Why was he not assisting the Colonial Office in providing this support? Elton reacted strongly and adroitly to Attwell's temerity, giving his reasons, and suggested that Fraser Darling was the man for the job in Africa.

Attwell and Fraser Darling first met at the then recently created national nature reserve at the Beinn Eighe, Wester Ross in December 1952. The same question was put to Darling as had been put to Elton: 'Why have you not helped us in Africa?'

'Because I've never been asked; I'm waiting to be asked!' was the disarming reply to Attwell who, on his return to Northern Rhodesia, suggested to his Director, T.G.C. Vaughan-Jones, that the services of a senior ecological consultant be obtained to advise the Government on the conservation of wildlife resources. Ironically, it took three years for Attwell's idea to bear fruit in 1956. By then Vaughan-Jones had become Commissioner for Rural Development, and though he had piloted the proposal through the difficult straits of colonial affairs, it fell to his successor F.I. Parnell and his Assistant Director, Major I.R. Grimwood to make arrangements for Fraser Darling's study tours in Northern Rhodesia in 1956–57 as a senior research fellow of the Conservation Foundation, Washington DC.

He was invited to spend six months travelling in areas where large ungulates and their predators were still to be found in appreciable numbers, and to report on their status, the nature and condition of their habitats, reserve and controlled area policy, and the relation of wild-life conservation to land-use policy and African nutrition.

The study tour was done in three parts to see the country in wet and dry seasons, and from the beginning Frank was pitched into an overwhelming welter of interest which beckoned to him from all quarters all of the time. Though he had spent much time in the 1930s studying the ecology and behaviour of animal species of his choice, and was erudite in life science through reading and travel, he was not a student of detail. He did not have a classical training in zoology or botany to quicken his curiosity in the particular, or distract him from the broader ecological perceptions for which he was becoming well known. His diploma in agriculture provided him with leading lines in the interactions of water, soils, forests, pastures, and grazing stocks. When, therefore, he was confronted with such a surging mass of life in this, his first visit to tropical Africa, he realised that he had neither the time, knowledge, nor indeed the basic training to absorb other than the salient features of the country and its peoples. He knew that Northern Rhodesia had its complement of dedicated life and earth scientists with years of field work and research behind them. He defers to them frequently, and with respect.

Fraser Darling was an outsider, articulating objectively the expertise of many in a holistic and educational way, and his reports carried a political charge possessed by none of the insiders who accompanied him. Correspondence which accompanied his visit shows him to be highly respected by the expatriate staff of the Game and Tsetse Control Department in whose cause he became a champion in dealing with the wrong-headedness (as they saw it) of the Colonial Administration. However, W.L. Astle (in a letter) is of the opinion that Fraser Darling was 'most unfair to the provincial administrators and nowhere showed that he understood their problems in the period just prior to independence [from colonial rule]'.

When Frank was taken by the game men to meet the local District Commissioners, and tribal chiefs, he had the aura of a 'great white hope' in solving game conservation problems. As a man of considerable self-esteem this pleased him greatly and instilled within him a fearlessness of utterance. He delighted in speaking out to office-bound authority from the high ground of personal experience in the field, and knew that his voice would carry far to Britain and the United States. Noel Simon in his fine book *Between the Sunlight and the Thunder* states that, in the late 1950s and early '60s, Fraser Darling dominated the field of African ecology.

CHAPTER THREE

The Luangwa Valley *ulendo*, the Plateau, and the Kafue Flats

February to April 1956

IN THE FIRST JOURNEY on foot from Mpika to the Luangwa Valley and back, a journal was also written by Major Eustace Poles, a Chief Game Ranger who was in charge of the *ulendo*. Accordingly, the Fraser Darling journal is supplemented by quotations, in chevron brackets, from the Poles journal to provide additional detail and interest. Poles' account has trivial differences in timings and in the spellings of names of places and people. The double accounts of such episodes as the buffalo and elephant hunts are revealing of two distinct personalities who, despite their differences and the strains of the most trying of circumstances, came out of the trek full of admiration and respect for each other which lasted for the rest of their lives. Major Poles died in 1990.

Roelf Attwell has written (in a letter) that 'Much of his [Poles'] field touring was done on foot, with porters, in the wilderness of the Luangwa Valley . . . his whole approach to his job was meticulous – this included a daily field diary, personally cleaning his firearms (he owned some of the finest British sporting rifles and guns), and daily sick parades for his carriers. His temporary bush camps were models of neatness; he was of the opinion that any damned fool can be uncomfortable in the bush. His conscientious approach to field and administrative work provided an excellent example for aspirant game rangers . . . A considerate and thoughtful person, he nevertheless had a mischievious sense of humour associated with a touch of showmanship. He developed a special photographic technique known as "bottom bouncing", not to be practised by the faint-hearted. He took a delight in initiating the over-enthusiastic game photographer into this practice: it entailed a close approach on one's bottom to potentially dangerous animals, armed only with a camera around one's neck. If successfully applied, this method renders even medium-sized telephoto lenses superfluous; but many a shakey negative can result.' Nothing is said by Attwell about results if the technique is unsuccessfully applied!

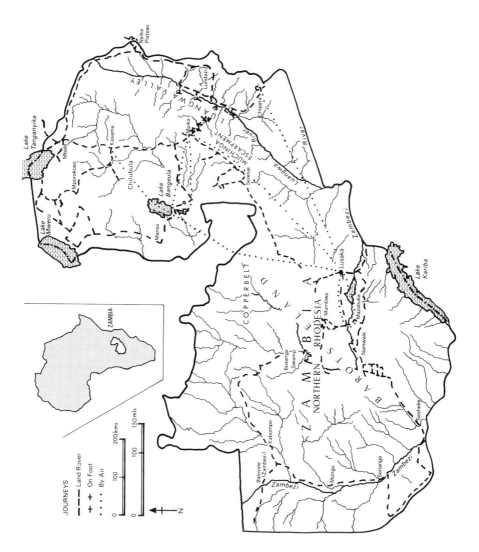

Map 2: Zambia (Northern Rhodesia) showing the routes taken by Fraser Darling by Land Rover, light aircraft, and on foot during his ecological reconnaissances in 1956 and 1957.

2 March 1956

Eustace Poles is the Chief Ranger of this Mpika area. He is a regular army officer who continues to run this out-of-the-way station on military lines. His Game Guards stand smartly to attention and salute you. I fall in with it completely and when Eustace wakes one up before it is light I jump to and eat a slice of bread and butter and drink a mug of tea. Eustace starts his day's trek in that way. Today we start [from Mpika] on our Luangwa *ulendo*, and as I look outside I see 42 Africans waiting each in his own way for the allotment of loads. Game Scouts are in charge as non-commissioned officers and one tall fine-looking Bantu is a Chief in his own right and is in effect Eustace's sergeant-major. The *ulendo* looks glum, and what a variety! The youngest are a couple of boys of 16; the rest range up to one or two with grey hair. They are dressed in the most indescribable rags. At last the Chief gets the loads allotted after the *ulendo* has surveyed the piles in front of them and tried to shuffle towards those piles that look as if they would carry well. At last all the boxes and tents and baggage is up on shoulders and a mere newcomer like myself finds it hard not to help the poor devils get their loads up. Eustace blows his little hunting horn which is his constant companion and we start forth at 8.45 a.m., Eustace walking in front, then myself, then Ian Grimwood, then our personal orderlies, then the long string of the *ulendo* and a couple of Game Scouts in the rear. It is a beautiful sunny morning with a breeze. We walk through *Brachystegia* secondary woodland for one hour and then have a rest. Then another hour and three-quarters and breakfast [at Mpelembe's village] which takes nearly an hour and a half. Eustace goes on *ulendo* in style: we have canvas armchairs, a trestle table, table cloths and table napkins. Porridge, bacon and fried bread and bread and butter and marmalade and coffee. The *ulendo* looks after itself. We have passed one or two small native villages and have been greeted in native fashion. The men cup their hands and clap them gently, murmuring. We acknowledge the greeting. The women do the same but bend their knees as if in a curtsey. The huts are circular, thatched and made of mud and wattle. Small coloury chickens wander about. We had passed some boys carrying out crates of chickens, rope net over wattles, and these they sell in the European places. It is a long trek for them. Each chicken costs six shillings to seven and six pence and it is wisest to fatten them for a week or two. We continue through more miles of secondary woodland showing signs of *quondam* gardens of *chitemene,* and then onto a range of hills which are the preliminary of the escarpment. To the ridge and just beyond where a stream comes down, and there we decide to camp [at Pregnant Twin's Camp]. Eustace, Ian

Grimwood and I sit in the shade while the Game Guards set up the camp, tents, shelters for the *ulendo* and so on. Tea comes in due course, and when the tables are set up, we write diaries and what not. Eustace and Ian have a tot of whisky and I a glass of sherry, which tastes extraordinarily good. More writing and then dinner, all well served. I forgot to say that between tea and sherry, canvas baths were prepared for us, very acceptable, and one felt fine after bath and shave. To bed under mosquito net at 9.25 p.m., Eustace and I sharing a tent as there was too much gear for the *ulendo*, which was intended to be 45, but is only 42. The cicadas and crickets are noisy. The country we have come through today is relatively lifeless. The game has been hunted out by Africans with muzzle-loading guns. Northern Rhodesia allows Africans to have these and the effect on game over large areas has been disastrous. The African is incapable of conservation sense and the Administration here is exceptionally soft towards the African, whose population is increasing rapidly and the habitat will not withstand the hunting pressure. European hunting is no serious factor now.

3 March 1956

Eustace blows his hunting horn at 5.15 a.m. I did not sleep till well after midnight and then fitfully. Got out of my sleeping bag and just laid it over me. Had no sooner got my trousers and shirt on than a pint of tea was on the table with a slice of bread and butter. Eustace then parades his Game Guards, inspects them, marches them up and down a time or two and dismisses them in Guards fashion. Each Game Guard and personal servant comes up to each of us, salutes and says 'Morny, morny', which greeting we acknowledge. We are on the march by 6.45 a.m., all carriers loaded. We walk for an hour and I begin to feel a spot on my left ankle and another in the pad of my right big toe. Eustace had ordered me to say if my feet should begin to trouble me. There is a blister forming under my right big toe, so Eustace pricks right through the pad and releases the fluid and puts on a pad of Elastoplast. The rest of the day goes well, especially as Ian gives me a pair of woollen socks to put over the nylons. Fills the boot rather but saves further friction. ⟨Frank, as Dr Fraser Dorling has insisted I call him, is most appreciative of everything that is done for him and charmed with the scenery and climate. He is a great talker and most interesting to listen to, besides being a man of easy charm. After the first stage during which we negotiated steep descents, Frank developed foot trouble and had the sense to tell me in time for me to attend to the rubs and blisters before they became too severe.⟩ We walk on till 10 o'clock and have breakfast by a stream [Dokitera] in these interminable woods. The only game seen so far is a duiker, and we have seen the tracks across our path (a former elephant path) of a zebra, a kudu and a buffalo. We come to the edge of the Muchinga Escarpment and see the vast Luangwa Valley laid out before us to the horizon. The Munyamadzi River (a tributary) is visible below with a few leafless winterthorns along its banks and then the ordinary *Mopani* woodland. It is a magnificent scene, all clear and green and a beautiful sky. We descended the steep escarpment [Mwasandema] and found it getting hotter all the time. My face is sore with sunburn, and sweat trickled off the end of my nose, but how wonderful it all is! Our camp tonight is at the foot

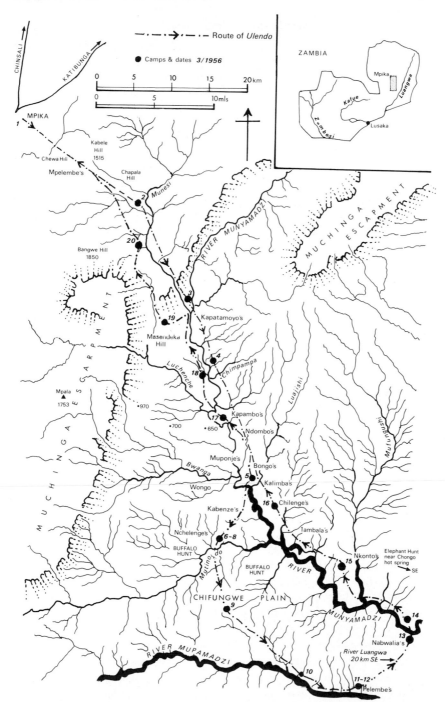

Map 3: The approximate route followed by Fraser Darling and Eustace Poles from Mpika over the Muchinga Escarpment into the Luangwa Valley, compiled by JMB from the journals of both Darling and Poles, and from maps and advice given by W.L. Astle.

of the escarpment [Katengeshi stream], in the woods. A few of the carriers come to Eustace this evening and squat silently in a row beside him. He says 'damn', gets his medicine chest and begins to dress minor cuts and sores. We sit in the warm darkness, wonderfully relaxed and comfortable. Today I saw my first honey guide bird, and two ravens were over the camp this morning, calling exactly like our own. 13 miles yesterday, and only 10 today.

4 March 1956

Eustace blows his little horn at 5.30 a.m. and the camp is moving very quickly. Our pint of tea and slice of bread are eaten and we are marching by 6.30 o'clock. We descend the last 250 feet of the escarpment foothills into the valley of the Munyamadzi. The country changes immediately to lushness and moist heat of over 100°F. To me it feels pretty good and I sweat like a bull as we march along. The grass is up to 8 feet high. There are palm trees, trailing creepers and larger trees. Soon we come to a village and Eustace curses because it is a new one creeping up the valley and in a place where formerly he could always expect to find game. Now there is none, and he is thinking of the *ulendo* which will need meat and will soon start grumbling for it by clapping of hands. On and on with the Munyamadzi River occasionally visible on our left. And then we come to it, a 50 yard river running at 3–4 knots, browny-grey, with suspended silt. Below the steep bank is a dug-out canoe, the craziest thing you ever saw. It paddles across to us by going up the river pulling on the foliage and then coming down and across the current, the paddler at the stern. I am sent across first and I kneel down and crouch as low as possible. We pull up river by the grasses and then off we go. The dug-out log rolls to one side and the river pours in, bringing two little fish the size and shape of sticklebacks. I lean the other way as quickly as possible. We continue our crazy voyage and I watch the freeboard. My servant comes next with my camera and gear, then Ian, then Eustace, and then the Game Guards. The *ulendo* is taken much farther downstream where it is possible for them to wade across waist deep. When I got over the river myself I was met by several men and boys of the next village and greeted by their cupped hand-clapping. I now reply 'Mapulene' and 'Mutande' which is the proper reply. Soon one of the Messenger Corps arrives, a magnificent Bantu, and he stands to attention and salutes me. I reply and he grins in appreciation. Then a group of women who go down on their knees and ululate at me, their tongues crossing their mouths rapidly to and fro, 'Mapulene' and they stop. Ian and Eustace get the same. It seems that I am being treated as a V.I.P. and the two messengers were sent down ahead of us by the District Commissioner to make sure the villages would turn out and help us across the river, provide mealies and cassava for the *ulendo* and generally be helpful instead of indifferent. Once across we marched through crops of kaffir corn or sorghum and were met by more ululating women who then danced before us to the village [Kapatamoyo's] singing and waggling their behinds. Eustace says 'Jesus Christ' and bears up. The sorghum is ten feet high and down on the ground are some earth-nuts and pumpkin plants. Eventually we reach the village of round huts and there is still more ululation and greetings. Less than a mile beyond the village is a grass shelter erected for us, its floor laid with grass. We rest there luxuriously and wait for the *ulendo*. The village folk gradually fade away except for a few little boys. I felt pretty tired today and my blister was hurting, but I found I could walk on the side of my foot and favour it. Once, crossing a deep *donga*, I came down full on it, and perhaps this squeezed out the rest of the fluid. We made only 8 miles today and Eustace would go no further, so I just rested and thoroughly enjoyed it. You cannot see far anywhere because of the long grass and orchard-like forest. Into bed 8.30 p.m. after a good dinner, and how happy I am [at the foot of Makondeka Hill].

5 March 1956

A good deal of rain after a tremendous lot of thunder and lightning which we had last night at dinner time. In the night I thought of the extra weight the tents would be for the porters, who are carrying between 50 and 60 lbs. As it is still raining at 5.30 a.m., Eustace lets the *ulendo* down lightly and it is 7.30 before we are away. The narrow path, scarcely wide enough for my big feet splaying east and west, is in a swamp and we slide about on the clay. The grass is long and wet and we are wet through. Because of the rain Eustace decides to follow the side of the river we are on ⟨through Ndombo's, past Muponje's (on the right bank) to Bongo's village⟩, which means an overgrown path, but saves us two crossings of the Munyamadzi, which will probably have flooded during the night. The route today is across a series of deep clayey hollows and wide ridges of *mopani*, a characteristic tree of the valley. At one point we pass a kind of fig tree of monumental proportions. It covered a quarter of an acre and had an immensely thick trunk. Once I saw a local partridge, sometimes some weaver birds and a lyra. And once a pair of Bateleur eagles. Eustace keeps us going till 12.15 before we stop for breakfast, and as the *ulendo* had got behind it was 1.15 before we got it. I don't mind, for we are lying under a large nyanza or tobacco tree, with leaves like a rowan. My feet have improved and I am travelling better today. In fact, I have turned the corner and I know now that I shall be able to keep going at whatever is needed. I feel fitter than I have felt for months. I like to think I am sweating off flesh at a pound or two a day. Indeed, I am already pleasurably noticing a tendency of my trousers to slip over my hips. We do ten miles today through this fairly stiff going and then come to a village. The women, a score or so of them, meet us half a mile outside, ululating and then singing a chant and dancing before us. The village seems *en fête* for us. Women have been coming along before dark with baskets of corn cobs and of white maize meal and cassava meal. Now as I am writing this [camping half a mile beyond Bongo's], the drums are going back in the village and I imagine there is going to be a big beer drink. In Northern Rhodesia there seems to have been a pretty thorough movement for women to clothe themselves from neck to knee, but in this village we seem to have got a more primitive Africa. Few of the women are clothed above the waist, and this afternoon as we came through and the dancing reached a peak of welcome or whatever you call it, one elderly female came alongside me and performed a most grotesque shimmy. Eustace said, 'My God! You must have made a hit, Frank,' and candidly I haven't been in this territory long enough not to feel a trifle uncomfortable. All I could do was to give the old woman a smile, and possibly walk a little faster.

Coming along through the forest today in the wet, sludgy conditions, along the elephant track (our whole journey so far has been along an elephant track) there were places where the elephants had each put their great round feet in the same places and made a series of round puddles which we had to evade. Game has been hard to see, only a 3 year old kudu bull disappearing in the long grass. Eustace went out from camp to shoot for the *ulendo* and saw some kudu and impala. ⟨Two honey guides tagged along, warning every animal of our presence.⟩ He got one kudu bull in beautiful range and the wretched cartridge did not fire. He complains bitterly about I.C.I. ammunition, which is a monopoly here as in Britain. One batch of rifle ammunition misfired 60 per cent. Shooting buffalo one day he shot once and misfired three times. I.C.I. take no notice of complaints, and when I suggested a letter in 'The Field', Eustace said they refuse to print any letter critizing such a considerable advertiser.

The firearm situation in this territory is fantastic. Any black man can have a rifle or muzzle-loader. There are 50,000 registered muzzle-loaders and 1,000 modern shotguns, and the Lord knows how many unregistered firearms. Game conservation is not prejudiced by the European but by the native, who has cleared great areas. ⟨Ian and I have given Frank something of an insight into the country's maladministration and our problems. He is appalled and not very optimistic.⟩

6 March 1956

Up at 5.30 and a pint of tea but no slice of bread this morning to start the march on. Trouble with the *ulendo* which says it feels tired and wants a day's rest. Eustace says they must go another day because there is no game near camp and unless they come another day there will be no meat. When the Game Guards and the chief Messenger are getting the *ulendo* going, one carrier gives a most polished exhibition of insolence. The Guards are furious and the carrier smiles lazily at his success. We march two miles to the river and we must cross again, this time the *ulendo* and *kutundu* (loads) as well. It is going to take all day. The dug-out is bigger this time but the current swifter – 5 knots. Eustace, Ian and I cross first and after seeing a few loads over we decide to march, in order to hunt for food near the camp site for tonight. We march fairly hard, 10 miles, reaching a deserted village [Nchelenge's] at 12 noon. One man and his wife appear, and as the young Chief whose name sounds like Stamfordham has come with us, it is arranged that we shall have a little kaffir corn to make porridge. By one o'clock we have a small bowl between the three of us of thin gluey porridge without salt or anything else. We eat it with little spatulae cut from a living sapling. It is not very tasty or very sustaining, but when you have had nothing else that day one doesn't grumble. The country we have walked through today has been lovely, rather closer forest with glades here and there and pools. Sometimes we would pass through long grass above our heads and one gets a bit fed up with it, the leaves hitting your face and you can't see a thing. We have crossed over from the Munyamadzi to the

Crossing the Munyamadzi River on 6 March 1956 in a dug-out canoe with Sandiford Kasembe paddling.

Mutinondo River, which comes in to the Munyamadzi much lower down. There are one or two villages up this river but the people are about as remote as any in Northern Rhodesia. After our thin porridge we went hunting, over country where elephants had been working. We, going along their paths, had a job to get along without going into the great wet holes where the elephants put their feet. It was a great day for me as we saw two elephants, both bulls of fair size. How truly wonderful they are – the slow bulk, the great ears and the beauty of the trunk. The elephant is like a great ship of the long grass. We were up to 50 yards range but because of the extreme heat of the afternoon I did not carry glasses or camera. There will be more and better chances when the grass has gone. I was tired, thirsty and hungry when we got back, and I was delighted to see the first carriers had made the journey from the Munyamadzi. The dug-out canoe had overturned and tipped three carriers into the river but no one was drowned and the crocodiles seemed to be out of the way. Yesterday afternoon, Ian and I bathed in the Munyamadzi, but it was a quick in and out because of the fear of crocodiles.

⟨No sooner finished [the porridge] than I took Sandiford, together with Nchelenge, and went out to hunt. Frank came too but I soon regretted suggesting he should join us. He is a grand chap, full of guts, but of course not acclimatised nor are his feet hard. In addition he had no proper food inside him and already had done a hard march . . . we came across a couple of bull elephants . . . one had tusks of about 35 lbs, the other

25 lbs. We passed them and went on through very dense undergrowth where there was plenty of signs of game, principally buffalo and waterbuck and some impala.⟩

7 March 1956

Not too good a day. The *ulendo* is resting and we are keeping to this camp. We rise as usual at 5.30 o'clock and have a pint or two of tea and a wafer of Ryvita. Eustace is going to the Chifungwe Plain across the Mutinondo River to shoot meat of some sort. Ian is going to do the same in a northwesterly direction and I come with him. The carriers are dead keen to come along, each man with his axe. Chilenji, the Mutinondo Resident who arrived from nowhere yesterday afternoon comes as guide to us and brings his assegai. We walk 4–5 miles through the *mopani* woodland and behind a bush a roan antelope appears almost on the end of Ian's rifle. Why the hell doesn't he shoot, I wonder, and imagine perhaps he cannot see the creature as well as I could. In fact, as it happened, he couldn't tell whether it was a bull or a cow, and then snort and it was away. The carriers (six of them) behind me looked pained. Then we came out on to open grassland and savannah, with pebbly hills and great shallow hollows. There below was a herd of zebras, perhaps 400 yards away, the flies now beginning to trouble them. Then a herd of 100 buffalo came into the hollow, making for a blackened area where they were apparently accustomed to stand. Ian and Chilenji and a Game Scout went down and I thought the approach a little unconsidered by Scottish stalking standards: he should have made for the bare dark place and met them coming into it and picked his beast. As it was they were more or less moving away when he shot. The herd went up close together and only very slowly moved away. From my anthill I could then see a beast break away from the herd and stumble at right angles nearer to us. She was hit through the shoulder. I saw her lie down twice and thought she was done, but no, she got up and blundered on. I could see that Ian did not know she had left the herd, the wretched long grass preventing his seeing. So, one of the carriers with me went up a tree and he got the idea that I wanted Ian and his Game Guard told. His voice yelled high and clear and we got them turned but I wished I had a rifle because I could have run down and finished her off. Now she was out of sight in the long grass and I wanted to show Ian the blood spoor where I had made careful marks. He said he preferred to give the buffalo time, as his rifle had misfired three times before he fired the shot which had hit the buffalo in the shoulder. Then began a slow following of blood spoor through the long grass, then through dense woodland and then into more 10 foot high grass. She was lying there and Ian fired – misfire – then again and this hit, but she was up and away. Another follow in long grass, which is hellishly dangerous with a wounded buffalo in it. For the last hour a little old man [Mandeze] among the carriers had taken on the job of shikar, noting the traces of blood spoor with great care. He would shin up trees as if he were a monkey and was immensely cheerful. Ian halted me at an anthill with the carriers and went on with the old man and Chilenji the guide. I heard four shots 200 yards away and then Chilenji came running for me. I hurried after him, the carriers behind me, and on coming up with the young cow buffalo lying dead in the long grass, Ian said the old man was hurt. The ammunition had played more tricks and the buffalo had charged Ian. He was standing to shoot again, aiming at the spine as she came, but suddenly the old man had crossed between the buffalo and Ian to deflect the charge and had succeeded. The buffalo was after him and he had gone into a bush. The buffalo gored the bush and Ian thought the old man was getting it but suddenly he appeared smiling round the bush and called in Bisa, 'You made a mistake, she's a cow.' Hitherto Ian had thought he had chosen a youngish bull. Then he shook hands with Chilenji and Ian and I took his trousers (shorts of most abject character) down. He had a two inch cut in the groin and another on his shin. Now his bravado was gone and I saw him trembling: so started in on his wound to pad it and tie the padding of my and Ian's handkerchiefs to his body. Poor old man! I wanted to get him back to camp before he stiffened up, so Chilenji led the way to make the shortest time. The old man did his best and kept up a good pace. I carried him over the tree trunks and *dongas* which crossed our path. Back at camp I laid him on a tarpaulin, bathed the wound with boiled water and permanganate and then stuffed it with sulphanilamide powder. Then lint and sticking plaster. I got him laid on grass in the carriers' lines and also gave him a drop of whisky and two veganins. He complained of pain all round his middle and I think he is bruised all round his pelvis, but not seriously. Then Eustace came in and gave him some morphine orally. The old boy was very brave and enduring, but once laid out in camp he got sorry for himself understandably. Eustace went out this morning on one Ryvita biscuit, did 24 miles and forded the Mutinondo up to his neck. No food till dinner tonight. So damn silly: Eustace lives on himself.

⟨Sandiford and I [Eustace] did a stalk and I got in to within about 40 yards where I fired at a youngish bull [buffalo]. It ran off with the herd and I could not tell whether my shot was a good one or not . . . I fired at another, galloping, at a range of about 80 to 100 yards. Soon these two animals stopped and the herd also halted and turned to face us. The stricken bull walked slowly into a small morass of muddy water where it fell and died; the other stood dying on its feet and I killed it with another shot. We butchered the buffalo and started homeward . . .⟩

As you [Averil] said, when he visited us at Cole park he was courting, after a much earlier divorce. Well, he married this party during that leave, and then she suddenly said she would not come to Africa. Poor Eustace was heartbroken. He was in Burma on the Chindit business from the very start and apparently thoroughly enjoyed the misery of it because he and another fellow were so generally successful.

Ian and I bathed in the Mutinondo River which is 20 yards broad but swift. We have got to cross this with carriers and loads tomorrow and I am not looking forward to it much. Three pied wagtails flirped round us as we bathed and I was much struck by the quantity of song. Ian is going to leave us tomorrow, so he will take eight carriers and the old man on a litter. Eustace and I will carry on, but we are losing time and Eustace is getting worried.

Am struck here with the absence of pernicious flies in the evenings, but tsetse were biting me most of the day. Africa seethes with ants of all kinds. They are everywhere: some bite and some don't, and you never quite know.

Saw a roan antelope out on the savannah this morning, such a handsome chestnut animal. It was nearly a mile away, but I had it in my glasses. The butterflies are wonderful.

My nose and cheeks are peeling. Very sore. Feet improving all the time. Am toughening up quite well: no stiffness, no cramp. Tasted some native beer tonight, buff-coloured made from sorghum meal and tastes as much like buttermilk as anything.

It seems to me that the game is very widely dispersed in the valley in the wet season and therefore thin on the ground. The dry season concentrates them on the permanent streams which are the Munyamadzi and the Luangwa River itself. The Munyamadzi is really the heart of the game reserve, but in 50 miles there is a thin string of 5,000 people who are extraordinarily placed for living on game. If they were conservation-minded all would be well, but they are not. This valley, I suppose, is as dark an Africa as you are likely to find, the escarpment being a very real barrier and the nature of the valley another. These Munyamadzi people, living permanently in very fertile alluvial deposits but in an unhealthy low altitude, steamy climate, are a focus of Rhodesian sleeping sickness. The healthy ones are healthy because they are well fed on game, but when there is any chirp of a doctor coming down to the valley (say once in three years) the sick are carried into the bush and hidden. The doctor doing anything at any of them is thought to make them vulnerable to bad magic. The witch doctor is well in command.

8 March 1956

A week since we started, a fortnight to go, and I don't want to miss one day of this real work, so different from interviewing stuffed shirts in Government offices in Lusaka. At the same time, I am sad today. I felt pretty tired last night and retired to my bed at 8.30 and asleep almost immediately. Rain fell most of the night though not heavy. This morning I feel gloriously rested and relaxed. The *ulendo* last night was in noisy, happy form as there was so much meat in the camp and they are still jubilant this morning, but Ian is going away and the old man is going out on a litter. I keep thinking of the old man and of the beautiful young buffalo cow. You have no idea how beautiful she was. Both of them are victims. She is dead, but the old man is not very well this morning, complaining of pains all round his middle. How much is internal bruising? We are going to stay here until midday and get across the river and pitch camp again this evening on the other side, so that we can make an early start tomorrow across the Chifungwe Plain, so much like Salisbury Plain, and beyond is an area which Eustace calls the South Downs.

⟨The injured native complained of severe abdominal pain. I hope he has no serious internal injuries which may result in peritonitis. We made a machila (canvas) stretcher and at 7 a.m. Ian pulled out with his carriers and a stretcher-bearer party of six men. I accompanied him to the ford which was a foot deeper than yesterday. The river was running very strongly and it was touch and go getting the injured boy across. I arranged three men at each end of the pole and four others supporting the canvas hammock. Ian assisted. The party made the crossing and then the carriers returned to their loads. On the next journey one carrier dropped his load but managed to recover it and get ashore; two others got into difficulty and were able only to keep their feet. Ian went in to help but was twice swept away by the stream. Sandiford, a man of fine physique, was magnificent, going to Ian's assistance and time and again helping a carrier and his load to safety. The river was hazardous and only picked men, of courage above the average, were capable of crossing – it would have been madness to attempt it with a big *ulendo* of which the majority would be faint-hearted and of poor physical development. Lack of heart is more important than inferior muscle in undertakings such as this.⟩

The *ulendo* is still busy drying buffalo meat over the fire, and if loads were heavy before they don't mind adding to them if it is meat for themselves. Indeed, the *ulendo* is now in great form. We have not got across the river today for it is too high and swift. The boys have worked like hell on this their holiday felling trees on the river bank to form a bridge, quite on their own initiative. Each boy carries an axe, which is a piece of car-spring steel burned through a shaft which is part root, part stem of some small tree. They have got through trees two feet across the trunk, and one much more than that. Eustace and I went down to the river this afternoon to see how things were going. We could get out so far but there were still 15 yards of swift current to the other side. A large tree on the other side had been started and we thought we had better go across to direct the felling. We had to leave our clothes on the bank, go out along the fallen trunk, make our way through the branches to the twigs and then let go and swim like hell, hoping you would strike the bank the other side before you were swept down. Eustace managed it, but I knew I could not swim with enough strength to cross 15 yards of current as swift as it was. I said so and asked for a pole to be extended from the bank a few yards farther down which I might catch if I didn't manage the bank. So I let go of the twigs and swam harder than I have ever done. It was rather a delightful sensation in the warm water being swept down on the current, but my concern was with the headway I was making across. As I thought, I could not make the bank where the boys were working, but I grabbed the pole all right and was hauled to where I could climb out up the roots of a tree. So Eustace and I worked the rest of the afternoon with the boys, and our clothes on the other side. The big tree came down but a heavy branch weighted it wrong and it fell into the river parallel to the current. The boys dived in and swam across beautifully. They fetched a rope – got it across. We hitched this on a branch and the boys pulled it tight on the other side, level with the surface of the water. Eustace and I came across hand over hand, the current carrying our bodies out level downstream. All in all it was great fun, but I was quite glad to be getting my clothes on again safe and sound. The river is gradually going down, 8 inches since last night, but we shan't get across because a nephew of the young Chief has died and he must be buried with the usual wake-like proceedings. The young Chief and seven newly recruited carriers must attend, so we are plain stuck. Eustace just hopes it won't rain, because in Africa you get across a river when you can, and you don't wait till tomorrow morning. Eustace is furious: the military man is baulked and he isn't the kind of guy who runs a campaign on those principles.

About half the *ulendo* presented themselves tonight for medical attention. Imagine me pouring a tablespoonful of ordinary lamp paraffin down two black men's throats as a cure for sore throats. The usual run of abrasions as well, and now a string of four with scabies, who have to be given ointment but Eustace lets them rub it on themselves. One has neck ache he says, and Eustace says, 'Get out'. There is a lot of lightning tonight. Even if it rains on the escarpment and not here, we shan't get across the river. The drums are going tonight far away at the wake.

9 March 1956

Another day of small adventure. We broke camp this morning and I was quite glad to leave the low lying jungly place in which we were. We marched two miles to the ford over the Mutinondo River and found the river down 4 inches from yesterday. Here I suffered a real humiliation, but I had to accept Eustace's discipline. He said his job was to deliver me in good order to my wife and family and to his Department. He sent 4 Game Guards into the river with a rope, myself in the middle of the string with my hands on the rope. Actually, it wasn't bad, the water came only to my armpits and the current was never so strong that I was in danger of losing my feet. We got every load across dry and were marching again by 8.30 a.m. We reached the young Chief's village, Kazembe, and there he was sitting in undress uniform among the huts. I was sorry for him because Eustace had been blasting that the young Chief had not met us at the river. Also, the Chief must have had a bit of a night of it, what with the wake, the beer drink and his three wives. Eustace tore strips of him, and the Chief said he wanted to bring some carriers along but they were grave-digging. So Eustace said we had better have our breakfast. We had a walk round the village meantime, and found the women keening round the hut where the corpse was, just like a Highland death. The arrival of the Game Guards and 40 carriers soon put an end to the professional grief and they were soon laughing and talking. Imagine what Eustace said when the young Chief came to say that our grave-digger carriers had decided that the grave they had been digging was not propitious. It would be necessary to dig another somewhere else. In other words, we wouldn't get them out today. So some stuff had to be dumped and we marched, the young Chief [Sandiford] in the lead, heading for a crossing of the Chifungwe Plain, which is like Salisbury Plain and about the same size. We moved through a mile or two of 10 foot grass and a little forest and gradually came up on to the Plain itself. It is magnificent open rolling scenery of grass about 7 feet high, but the leafiness of the grass is only shoulder high, so one can see through the flower heads. The ground is pitted everywhere with buffalo and elephant tracks and is therefore very rough. The grass itself has to be pushed through. We saw one large bull elephant ⟨tusks 45 lbs a side⟩ and two bull buffaloes. The going is heavier than anything we have done so far and we were marching through the heat of the day, which was terrific. The *ulendo* was cheerful and kept well together, being scared of buffalo and elephant charging them. I sweated my clothes wet through, and then at

3.30 when I expected another hour of going, Eustace said we would camp. We were on a brow where two old trees stood and below was a wet patch where water could be got. We soon had tea going and did I enjoy it! The tents were just up when a thunderstorm broke, of great intensity. The wind nearly blew our large tent away and the water poured right through the floor. No joke, I can tell you, and there are great storms floating round the Luangwa Valley. Any way, it passed off, but there are drifting curtains of heavy rain which will be filling the rivers. ⟨We had come to high ground where the grass was shorter and less dense than usual. There were three trees, the only ones in a 3 mile radius and there was water conveniently close. What was lacking was firewood and poles to support the ridge rope of the boys' tents and canvas storm-shelters . . . While camp was being prepared I cleaned my gun and rifle. The horizon to east, north and NW was obscured by heavy rain storms and presently, in fact when I was flying my hawk, the storms reached us. Soon the tent was being pounded by driving rain which was followed by a full gale of wind. In a few minutes the cracking reports of free, wind-whipped canvas indicated that the tent pegs had been drawn out and the outer fly of the tent was flailing free of its moorings. All hands were called to the rescue and the situation saved but it was touch and go. The floor of the tent was like a cattle kraal. The storm quickly passed leaving the plain sodden. A bright rainbow threw its brilliantly coloured arch across a sky full of menacing black clouds while to the westward the distant Muchinga Range was obliterated by a dark veil of falling rain.⟩ After the rain, several lovely kestrels came hovering and zooming near the camp, and there were three Montague's harriers sailing about, and one vulture. A covey of guinea fowls flew into the trees by camp just on dark. These rolling prairies or downs are such a delightful change after the jungly growth. I came for miles through these grasses without seeing any legume and then I found a tall-growing, small-flowering pink vetch. On this brow where we are camped the grasses are shorter and have been fairly heavily grazed by buffalo. Tonight as we sit in the darkness we are hearing lions and hyaenas. The kestrels are either Dickenson's or Newman's. ⟨As darkness fell a flock of guinea-fowl flew up into the trees; having no alternative roost they were obliged to use them, much as they resented sharing occupation with us.⟩

Eustace tells me I am one of the few white men who have crossed the Chifungwe Plain and one of the mere handful who have crossed it in the wet season.

10 March 1956

A wonderful calm night on this lovely plain. The lions and the hyaenas came near in the night but went their ways. ⟨They came so close to the tent that I was afraid one of them might take the hawk off her perch.⟩ We were marching by 6.30 o'clock and have done 10 miles to the south side of the Plain. It has been gruelling going, for the heavy thunder rain of yesterday evening has made this open-floored grassland a quagmire. We walk ankle deep in black loamy [cotton-] soil which is just wet enough and dry enough to ball up on the boots so that one is carrying several pounds on each foot. There are also holes in the soil that you can't see because of the long grass and now and again down you go into them. Insofar as one does see them, each man pats his buttock as a signal to the man behind to look out. This is the universal signal in black Africa when walking in single file. Yesterday, because the ground was hard and not showing our footmarks, the forward two or three of us – the young Chief, Eustace and myself – would bend the grass over with a hand as a sign of our passage, for the grass closes up and leaves little trace. Today, in our crossings of the plain, we saw three elephants, three rhinos, two roan antelopes and about a score of old buffalo bulls scattered about. We startled one in his wallow in our track, but he merely got up hurriedly and went off into the long grass. We also saw several Montague's harriers, Bataleur eagles and a pair of Martial eagles, very large and fine, soaring on the thermal currents. We came eventually to a great riverine flat of the Mupamadzi River, which looks for all the world like the Thames watermeadows at Hinton Waldrist. There are no villages and the south bank (we are on the north) is in the southern part of the Luangwa Reserve. I saw a herd of 15 elephants pass along in this riverine savannah, and when I went up the river this afternoon (with the young Chief deputed to take care of me) we saw buffalo bathing in the river (60 yards across) and a herd of puku grazing. These antelope are a bright chestnut colour about the size of fallow or rather smaller. I asked the young Chief where we could bathe and he said, 'No, too much crocodile', but eventually we found a tributary which he considered safe and we both had a glorious bathe. One little pool I tried was as hot as I would have a bath at home. The tributary itself was tepid. I saw two flocks of half a dozen Egyptian geese, some very upright white crowned wattle plover. Guinea fowl and the local type of partridge flew in front of us as we crossed the meadow land. This place is full of life and I could hang around indefinitely getting upsides with it; unfortunately late this afternoon the thunder rain has come again and stopped any more exploration. We get away in the morning to cross over swampy country to the lower end of the Munyamadzi. ⟨We are now about to enter the low-lying *mopani* country which is inundated by flood water and ghastly to walk through.⟩ Lions again in the night.

11 March 1956

We are marching by 6.45 a.m. after the usual parade of Game Guards and Messengers. Last night we had ten carriers waiting for medical attention. I gave three a tablespoonful of kerosene for coughs and two headache pills – they probably have endemic malaria. I dress the shoulders of another who is showing lumps, and Eustace covers them with a length of crepe Elastoplast. The rest were scraped toes and so on. The one with belly-ache gets a hell of a dose of magnesium sulphate.

The going today was the hardest we have had yet and we have done a good ten miles. We have come through *mopani* swamp which is a clay soil honeycombed with elephant footmarks and the whole thing under water. Knee deep most of the time and then the effort of maintaining a balance. The poor carriers have had a hell of a time. Occasionally we would get on to a sandstone ridge for a mile and it was heavenly to walk aright. We saw elephant some distance away and at one of our stages there was one among the *mopani* trees two hundred yards away. He was alone and quite

On the north side of the Mupamadzi River on 11 March 1956 – 'at one of our stages there was one [elephant] among the *mopani* trees 200 yards away . . . and quite unbothered.'

unbothered. Eustace asked me if I would like to see him close and of course I said I would. So he and I and the young Chief got to within 50 yards of him ⟨tusks 40 lbs a side⟩. The elephant now looked at us with interest and I photographed him in his habitat. After a time this gentleman decided he had been looked at long enough and in the plainest of words he told us to beetle off. He extended his great ears and ran at us. The young Chief and I ran, looking back over our shoulder. ⟨Frank was imperturbable, as usual, and removed himself without unseemly haste and, one felt, only because it was expected of him.⟩ Eustace stayed put with his double-barrelled .475 rifle, just in case of trouble. At 25 yards the bull elephant turned round and trotted off into the *mopani*. He must have felt pleased with himself. Soon after this we nearly fell over a small cow rhinoceros and her darling little calf. She ran away parallel with the *ulendo*, which scared the carriers stiff. They dropped their loads and climbed trees like monkeys. One trainee Game Guard [Diamon] also did this and Eustace nearly choked with fury – 'How dare you climb a tree and wear the uniform of a Game Guard! You –' and here he went into a string of Bisa which I didn't understand. ⟨I pulled him down; kicked his arse and told him what I thought of him . . . ⟩

For myself I was just fascinated watching this sweet mother and child, the first I have seen at close quarters. A herd of zebra came close to us on another sandstone ridge occasion and galloped to and fro across our path, obviously enjoying themselves. Eustace said they would clear all the game ahead of us. We also saw roan antelope (2) and impala, about 20. Finally, back in the swamp again we came on buffalo, and as it was time the *ulendo* had more meat, Eustace and the young Chief and I stalked into the heavily elephant-browsed *mopani* scrub. There were buffalo all round us it seemed and I could see one bull looking at us on our left. Ahead, Eustace chose his animal, which was moving fairly fast, and with a superb neck shot knocked it over. It was as good as dead but Eustace put another shot behind the occiput and made sure. This time, so different from the other buffalo occasion, the herd stampeded, about 500 of them, and the thunder of hoofs was awe-inspiring. Both of us were sorry they should have done this. We were near the 100 yard-wide Mupamadzi River, and we stopped here and had breakfast of fresh liver and fried brain. The *ulendo* cut up the carcass and gladly carried the extra weight. They love meat.

[This account by Frank leaves out details which he may have thought best forgotten, but which are narrated by Eustace in a respectful and sympathetic way. Frank was clearly unnerved by the buffalo stampede and saddened by its casualties. ⟨At the second stage we saw impala; a few waterbuck and finally buffalo. I decided to try and get one and thinking that Frank, an experienced deer stalker, would appreciate the chance of taking a shot, sent him on with Sandiford to try his luck. I followed a little distance in rear. As we went on it became evident that we were closing with a pretty big herd. In a little while I could see the animals and heard others on our left hand. I noticed Frank raise the rifle but he did not shoot. He went forward a little way and stood up close to a tree where he broke some branches that intruded on his line of sight. I expected him to shoot but he didn't. Frank and Sandiford were so obscured in the undergrowth that I couldn't see what was going on. Presently a shot was fired and at once I ran up with the spare cartridges. The buffalo on our left were already stampeding before the shot was fired and now those ahead, which Frank was stalking, were streaming across our front in a closely packed column. Sandiford 'broke' the rifle; I pushed another cartridge into its right barrel then took the rifle from him. The main column was too far for a certain shot and I was reluctant to take chances with the grass and undergrowth as thick as it is now. Next moment a buffalo broke out of the bushes at a reasonable distance and I dropped it with a shot through the forward part of the shoulder and hastened its passing with another through the head. It turned out that Frank found the buffalo were partly obscured by the grass and did not feel confident in making a certain kill: in addition he developed an attack of 'buck fever' — understandable enough considering that for the first time in his life he was hunting one of the most potentially dangerous animals in Africa and very recently had seen a man gored by one. Not feeling confident and thinking Sandiford, whom he knew to be an elephant hunter, must be a good shot, he handed him the rifle. It was Sandiford's shot that I heard. When I realised that it was he that had fired I was sure he had missed for, at normal sporting ranges, Sandiford is an execrable shot. Of course he was confident he had scored a hit and went off full of optimism to look for a blood spoor. Meanwhile I ordered a halt for breakfast. Very soon Sandiford returned to say he had found two dead buffalo. For a moment I though perhaps he had enjoyed a fluke and his bullet had passed through one beast to mortally wound another beyond. Following Sandiford to the dead buffaloes, a cow and a yearling calf, the truth was evident. A deep spruit, running between high, precipitous banks, lay in the path of the stampeding herd which, like the Gaderine swine, had poured into it in the course of their panic-stricken rush. No doubt the majority had gone over the bank into the water head first while probably many had somersaulted. This evidently was the fate of the cow and yearling. They had been trampled and drowned. While Frank and I were sadly contemplating the scene of the tragedy a third body floated to the surface; that of a very young calf.>]

Another dreadful round of knee-deep stuff and two or three waist-high *dongas*. Even when you are on a bit better ground it is heavy if your boots are full.

The elephants have started in recent years to bark and tusk out the trunks of baobab trees, which are enormously thick and pithy inside. We had a stage at one such great baobab this afternoon and in the pith we found the broken end of a tusk weighing about three pounds. This was given to me. It will be a pleasant souvenir and a good paper weight. We have camped beside the lower part of the Mupamadzi a few yards back from the river [near Pelembe's village]: I hope far enough back because the banks keep calving into the water with a great splash. There is a long sandbank on the other side of the river where several waders may be seen, including the common sandpiper. A big bull hippopotamus keeps raising his head and saying 'uh-uh-uh' in *basso profundo*. Each night there is a considerable thunderstorm and tonight the lightning across the river is most spectacular. Eustace hopes the tent won't blow over if we get the preliminary squall. We narrowly saved it two nights ago on the Chifungwe Plain and again last night. We have seen several crested cranes about the river and some knob-head ducks. You know it would have hurt me a lot if that elephant had got serious and properly charged us, because Eustace would have killed him. Yet it was we who chose to go and stare at him from close quarters. He wasn't bothering anybody. Personally I think his behaviour was of a highly responsible order.

12 March 1956

The camp is not moving today. The rain in the night was heavy and we heard several calvings of the bank on our side. We have only ten yards now. Eustace is an ardent amateur falconer and I don't think I have mentioned yet that he has an Ayre's eagle (about the size of a buzzard or possibly smaller) which was caught robbing a chicken pen and sent to Eustace. It is still a young bird. He has brought it on this trip and it sleeps on a rail in the vestibule of our tent. It is as senseless as most birds of the raptorial tribe and the training consists of getting a set of reactions going. Eustace has not flown it at wild prey yet, but it comes 100 yards to his fist. This morning he took it out with us at 7 o'clock in the hope of trying it with guinea fowl. I was rather bored because it is all a tremendous amount of trouble for so little. When the bird did fly at guinea fowl, the latter dropped to the grass quickly and the hawk went into a tree. Then it has to be coaxed down. Give me elephants. When the weather got too hot for hawking we were able to moon around a bit and see some natural history. The Mupamadzi River is in flood today and I saw a huge hippopotamus cow and her calf come out of the long grass, walk into the river and settle herself in the deep water.

Eustace Poles' view of the bird-watching pool on 12 March 1956 near the Mupamadzi River.

We also saw two very large crocodiles. How impressive they are! I also saw and heard a greenshank go from a sandbank. Once when going through the 10 foot-long grass trying to flush guinea fowl for Eustace's eagle I found a colony of small weavers like canaries that built their nests in the grass like harvest mice. The nests had an interesting opening in that it was fringed with projecting awns. The nests contained three eggs each, pale blue with dark blotchings and spottings of grey. The bird itself is canary-like. Another weaver, the buffalo weaver, builds a large communal nest at the top of a tree. The mass looks as big as an eagle's nest but it contains the large number of individual nests.

Another walk of 5 or 6 miles in the afternoon. Eustace and I stopped birdwatching at a lily pond of an acre or two and saw several pairs of pygmy geese and three or four jacanas. These latter are most gloriously elegant waders the size of an avocet and that shape. The head is black and white, the throat white and the whole body a brilliant chestnut colour with black wingtips and tail feathers. I have seen the fish eagle several times since we came into this Mupamadzi riverine area. It has a white head and neck, black body, and white tail. A spurwinged goose has rested on the sandbank over the river most of the day. We had to hurry home to escape a big thunderstorm. Lightning vivid as ever and two hours of torrential rain. The river is up and the bank continues calving in a big way. A string of sick parade again. One man has his shoulders covered with tropical ulcers, another a cut half across the sole of his foot, and the usual abrasions. Somehow these carriers are like patient black cattle, with about as much sense. ⟨Only Pelembe's village has brought meal and we have no reserves at all. We

The *ulendo* halted in the *mopani* woodland near the Mupamadzi River, 13 March 1956.

cannot afford to wait here and must set out tomorrow, hoping to replenish supplies at Nabwalia's group of villages. The chief is at Chalabila's, the next village but one – what he is doing God knows; probably instructing the headmen of these villages to withhold food. Nabwalia is the Game Department's confessed enemy who, unfortunately, has lacked a firm hand for far too long.⟩

13 March 1956

Some rain in the night. We are on the march by 6.15 a.m. and we finish our day's march by 1 p.m. and get breakfast before 2 o'clock. We have done about 12 miles through *mopani* woodland under very trying conditions. The clay of the *mopani* flats prevents water draining and we have been ankle to knee deep in water and mud for a good deal of the time. Even when you are on relatively dry land the mud is so slippery that you are exercising yourself all the time to remain upright. A few impala and zebra in the *mopani* and we pass through the village of Palembe which has no meal to sell to the *ulendo*. We are now heading for Nabwalia's village on the Munyamadzi [by way of Paison's and Bungolo's villages] and we get there around noon.

⟨How different is this woodland now to its appearance in the dry season. The trees are in full leaf and underfoot a short, lush growth of grass forms an unbroken carpet beneath the elephant-grey tree-trunks. Mostly the ground is waterlogged but here and there are islands of varying size, but nowhere extensive, that are firm and hard. Game tracks are everywhere but nowhere indicating large communities for at the present season the herds are split up and widely dispersed. The forest resounds to the constant symphony of bird voices; the sparrow weavers are most numerous and vociferous, but there is a steady undertone of the countless doves, while the obsequious long-tailed glossy starlings add their raucous cries to the medley of sound. Red-billed hoopoes and hornbills are other prominent members of the avian choir. We saw little game because Chief Nabwalia and his party had lately preceeded us, though the path, where it was not completely under water, showed the imprint of many feet; ranging in size from elephants' to those of field mice including those of hyena, civit cat and mongoose. Our third stage from Pelembe's was indeed a penance. On approaching the Mupamadzi we dropped to lower ground where the flood water stood deep in every hollow. The glutinous mud seized our feet at every step, frequently pulling off our worn-out shoes.⟩

The Chief, Nabwalia, emerges from a hut with two canvas chairs for Eustace and me. We greet him and sit down. Nabwalia takes little more notice of us: he doesn't like Eustace, who stops him and his people poaching in the Luangwa Reserve with muzzle loaders. Nabwalia is not the stamp of man that our young Chief Stamfordham [Sandiford] is. He sits there ⟨in rags⟩ scratching his thighs. Our young Chief was deposed by the Administration in quite arbitrary manner, because it was thought Nabwalia was enough. Nabwalia is a constant thorn in the Game Department's side, and such are the personal piques and aversions in the Colonial Service that the Provincial Commissioner will use Nabwalia to spike the Game Department's guns. I have learned something of the scandalous way in which the Labour Government pushed promising young Socialists into the Colonial Service and their conduct of affairs here is appalling. The African here should certainly be paternally protected for a long time, and even Lugard's doctrine of indirect rule is a farce here because the Chiefs are far too small beer to be left to it. Nabwalia grinned lazily when Eustace said he wanted a portrait and disappeared inside his hut to don another pair of shorts and a European jacket. Two of his wives appeared in coloury cottons and we both took just the photographs we didn't want. Nabwalia's sporting piece was lying across a box. Eustace picked it up and I noticed the barrel face and hammer face had about a twelfth of an inch play. How do people survive with guns like that? Incidentally the incidence of umbilical hernia in the children of these villages is appalling. The folk simply can't know how to tie an umbilicus. We left Nabwalia's eventually, passing through the village gardens of kaffir corn (which in these fertile river flats reaches to 12 or even 15 feet) then through stretches of knee-deep mud which was pretty foul and then up a kopje (one of three or four in this neighbourhood [and an old boma site east of Nabwalia's village]) to establish camp. The view of the valley from up here is tremendous and inspiring, forest as far as the eye can reach, the river, some lagoons, and the huts and patches of mealies and kaffir corn below.

We dine well on soups with chilli sauce and buffalo in various ways – its liver, its brains, its tail. We let the Africans have most of the red meat, because it would be so tough for us immediately, and we can't keep it until it softens. The carriers were busy all yesterday drying the meat over fires.

For myself, this steam heat of the valley floor seems to be doing me good. I have got fitter throughout the trek, of which this is the twelfth day. My feet are all right and I can walk on and on without feeling tired as I did those first days. My face has peeled and now I am browning my upper torso gently. I don't know how many pounds I have shed, but I imagine quite a few. Eustace and I have no mirror in camp, so I don't know whether my cheeks have hollowed to that pitch which you so greatly admire. I am doing my best to come back presentable. Even marching in the heat of the day does not upset me now: on the mornings of my first days of the trek I would wake up with my eyelids swollen up with fluid (first experience of this after a long walk in Sutherland last summer when I was far from fit) but it would disappear in an hour or two. Now there is nothing like that: I am out like a shot at 5.30 o'clock, fresh and presentable, and can do the day's march perfectly easily with the early morning tea and slice as the only food until 1 or 2 o'clock.

⟨Between Paison's village and that of Nabwalia the inundations and the tenacious mud were worse than anything so far experienced. Fortunately Frank and I are by now thoroughly hard and fit; even so, one feels the strain imposed on back and neck muscles. All the time one's feet slip back or slide to one side – each step is taken with tense sinews while all too frequently every muscle is called upon to save a fall.⟩

It seems also that some of the Department were opposed to my being handed over to the tender mercies of Eustace as they thought I might finish up being carried out. But there you are. I can keep up with the lean wiry Eustace and have not had to ask for any preferential lengths of marches. Eustace and I get on very well and I so much admire his general standard of conduct. It is that of the gentleman all the time. I know he approves of my capacity to travel and of my happy acceptance of the camp life and routine. Indeed, I am very happy in it all and am so glad to feel so well. I am learning so much every day. For example, there was a scorpion under my bed this morning. You learn always to tuck in your mosquito net.

14 March 1956

No rain last night. The camp on the kopje was soon struck and the carriers filed the half mile to the river for another dug-out canoe crossing. It would take half a day, so Eustace and I, the young Chief and Joseph (Eustace's head native Game Guard) went into the *mopani* forest beyond the river and below a string of kopjes. Eustace showed me a sulphur spring of clear boiling water emerging near the foot of a kopje. The effect on vegetation was scarcely noticeable. The spring is called *Chongo* which means noise. When we are on the march and the carriers start talking, Eustace or Joseph or the young Chief yells *Chongo* down the line and they shut up. We found very fresh elephant spoor and it was evident they were big ones, so Eustace said we had better have a look at them. We tracked them nearly three miles and found them in the *mopani* – eight big bulls. Eustace has his licence for two elephants and out of this lot he picks an old animal with big tusks. Apparently the art of elephant hunting is to get up among them and shoot at under 25 yards range. Eustace motioned to me to come along, so along I went. The elephants were slowly moving away quite unknowing, and there were Eustace and I just behind them. Their own slight sound of movement and gruntings hid any sound we might have made. Then they stopped: Eustace ran out to the left and fired his great .470 at the heart of his chosen elephant. He seemed so small below its bulk. The elephant was evidently hard hit and Eustace gave it another almost immediately. Now the great creatures seem all round us and so close, but they did all go in one direction at right angles to us and to the left. We heard a great crash and thought that must be the elephant dropping dead, but the forest was thick and we saw only elephants moving ahead. We ran for a quarter of a mile and lost them as they split in two groups. 'We shall have to go back,' said Eustace, 'and pick up the blood spoor.' This we did and found the Game Guards pointing in the opposite direction at two elephants 75 yards away. Obviously one must be the wounded one and Eustace went after them. They moved off to the left and we (the Game Guards and I) went to where they had stood; and there was the dead elephant, not 25 yards from where he had been shot. Eustace had got him in the heart. But so elusive are they and the bush so thick that we had erroneously gone after five elephants. The two had stayed by their dead friend to help him if they could. Eustace has seen them lift an injured one to his feet, using their tusks one each side. Well, this old bull was about 80 years old, born possibly the year Livingstone died or soon after.

[The same elephant shoot as described by Poles: ⟨We found a herd of impala under the eastern flank of the sandstone hills where Chongo has its source, and saw a single waterbuck which Sandiford wanted me to shoot. I declined to oblige him. We carried on, crossing the fresh tracks of some big bull elephants. The footprints of some showed their owners to be of great age. I was thinking how pleasant it would be were we not trekking to such a rigid schedule; in which case I could go after them and perhaps meet a heavy tusker. As we were back tracking the elephants I put these thoughts out of my mind.

Half an hour later we saw elephants a little distance to the eastward and I decided to get nearer and have a look at them. We found them to be bulls, about a dozen and all big animals. The tusks of those we could see were disappointing. One very large bull had one very nice tusk, but that on the other side was broken off half-way along its length. We followed the troop which was moving along slowly, feeding. It was difficult to know what to do with Frank. I offered him two alternatives; to stick close to me or fall right back with Joseph and Mofya. He chose to keep with me. Just then the troop, which hitherto had been rather scattered, closed up and turned in line into the wind. Taking advantage of the chance I hurried up behind them and was nearly mid-way along their line when they turned left-handed into single file. There was one elephant with a nice, evenly matched pair of tusks of fair weight and I decided to shoot it if I could. It was a little in advance of me so I hurried on to try and draw level for a broadside shot. I was now within 20 yards of the animals, but before I could overhaul the one I wanted, those behind and a little nearer to me, turned towards it.

I was obliged to make up my mind quickly. I had only a moment for a diagonal shot behind its shoulder, raking towards its opposite foreleg before it was masked by a much lighter elephant. It was not the shot I would choose, but if I declined to take it I should be obliged to withdraw and adopt fresh tactics. Meanwhile the wind could not be relied upon, and should it change to slant towards them the chance of a shot would be snatched away.

I decided to seize the chance, and fired a foot or so behind its left shoulder. It tucked itself up; and swung away, giving me no chance of firing my second barrel. I hoped it would turn round and so expose its right shoulder, but instead it went straight away for a little distance and, next moment, was covered by the great bodies of its companions, who all turned left handed, and came running down towards where Frank and I stood.

I was obliged to get out of the way, reloading as I ran. Seeing the elephants were running past and none were chasing me, I stopped and turned. The troop was running through some very thick cover where there was a narrow gap which each crossed in succession. Recognising the elephant I had fired at, I gave him a snap-shot as he rushed across the narrow opening.

Sandiford and I followed, and almost directly heard a hard thud and the bubbling gurgle that is characteristic of a dying elephant. But there was no sign of a dead or dying beast in the thicket from which the noise had come; instead all we could see were elephants far ahead, moving fast. We set out after them at a run as hard as we could go, over dead logs through mud, honeycombed by deep pot-holed foot prints. We ran until I was staggering.

The elephants had gone off right handed and two had diverged a little to the left. We fancied the right hand one to be ours and followed. Soon we found we were at fault; that which we were now following carried the same shape, but lighter tusks than the one I had fired at. We turned back and followed the elephants which had turned to the left. There being no blood on the spoor, it was decided to return to the starting point and there seek the tracks of the wounded one.

Frank was never far behind, and had since caught up with me. Turning back along the spoor we soon came up with Joseph and the others. At that moment I noticed an elephant disappearing into thick bush.

We went straight after it, imagining it to be the wounded one. It was soon evident there were two. Both were travelling very slowly, and we soon shortened the distance between us without having to run fast. A chain of deep pools and a swampy patch delayed us, but Sandiford found a way through and once more we were following.

I was unsure that either elephant was the one I had wounded, but by now we were pretty close and the next few minutes would be decisive. Just then Joseph shouted. Keyed up as I was, I felt furious with him because the elephants, hearing his voice, forged ahead at full stride. Sense then dawned on me, and I realised that no one so experienced as Joseph, would under present circumstances, raise his voice except for an urgent reason. It might be because Frank or one of the men was in trouble, but in that event I should have expected to hear desperate calls and yells; more likely he had discovered the elephant lying dead. At first Sandiford and I were sure that such was the case, but the shot had been a difficult one, and I did not feel so confident as would have been the case had the elephant been standing at a better angle when I fired.

We turned back soon to meet Joseph who, with Frank, was standing beside the body of a dead elephant which we had all passed closely in the long grass without seeing. My first bullet was sufficiently well placed to have killed it, but the second was equally good, both raking diagonally forward into the chest where they had cut into the heart or the great arteries which branch from it. A crimson carpet of arterial blood, blown out from the trunk stretched back along the spoor.

It was a big, old bull with a pair of tusks that justified my choice. I judged the tusks to be about six feet in length, and although they were rather slim, their points were well worn showing the ivory to be old and solid. After cutting off the tail and resting a little while, we moved on.

Frank was sad at the killing, though he had enjoyed the excitement of the hunt. I feel a similar regret at the death of a noble animal, particularly an elephant, but am not sufficiently a hypocrite to deny that I enjoy elephant hunting; nor can I afford to disregard a Game Ranger's perquisite.>]

I was so thankful he did not suffer, for I do so love elephants. So does Eustace, but he thinks there are rather too many from their recent attacks on the baobab trees. So he takes his two a year and chooses them as far as possible. The tusks of this one will bring Eustace about £140 which, as he says, helps him to go to England from time to time. If the white man were out of Africa completely, I suppose there would be no need for elephant control, but if the natives were left free with muzzle loaders, the elephants would have hell. Best leave it all at that, but my heart was with the poor old elephant. The Game Guards were delighted and excited: measuring the length and girth of his tusks and trying it round their own thighs. They are such children. On our way back to camp we went into the kopjes and saw a large black snake disappearing into the rocks. Eustace thought it was a cobra. The first snake I have seen in all these miles on foot. We also sat and watched a herd of 50 to 60 impala. How graceful and yet exaggerated they are! Back to breakfast by 12.30 but had to wait because the food hadn't come over the ferry yet; as Eustace says, 'The munt thinks arsey-tarsey.' The *munt* incidentally is the Biza name for a person, and old-timers like Eustace call the natives munts (the men or persons). The plural is Bantu, the name of the whole race. Munt seems to be a good descriptive word: it is truncated, stopped short, and that seems to me how these people are – cut off short in the making.

This afternoon Eustace held a session with Chief Nabwalia. The Counsellors and the *kupassu* (chief's messenger) arrived across the river first and sat on the ground near us. The Chief came later and was offered a box as seat on Eustace's right. Eustace's clerk and interpreter Dabster Makalulu stood before us. Eustace told the Chief that he looked forward to a period of friendship with him and his people, but that could be only if Nabwalia saw to it that the people obeyed the game laws and helped the Game Guards instead of hindering them. Nabwalia scratched his thighs and grunted. Eustace explained that if the Game Guard shot an elephant in the course of control, 800 lbs of meat were to be sent to the boma at Mpika by carriers, together with the tusks, and that the rest of the meat was for the villagers. More grunts. If a villager shot and killed an elephant raiding a garden, all the meat and the ivory were to be sent out of the valley by carriers and none was to remain. This measure is designed to stop flagrant shooting of elephant on the pretext that they are raiding gardens. Nabwalia scratched his thighs and asked how much meat the villagers would get. 'None, I'm telling you,' said Eustace, 'that is exactly what I am telling you. This measure is exactly what we are going to do in the Highlands if our standing committee gets going.' Nabwalia gave it up. A counsellor came forward and on his knees offered Nabwalia his little snuff gourd. He took a good pinch and the counsellor retired backwards. The interpreter explained to Nabwalia what the Bwana Doctor Darling had said; that the law which was being enforced here with elephants was going to be applied in Britain as well in relation to crop-raiding buck. Nabwalia scratched and showed no other sign of interest. Finally with grunts he rose and we said the proper goodbyes to him and to his counsellors.

Then Eustace and Joseph and I went off to the elephant as hard as we could go and found it to be all of four miles way. We passed some of our own carriers and villagers coming home with meat from the head and feet, all in great form. At the beast itself the skull had been cleared and our young Chief was in high fettle directing the axe-men working carefully to extract the tusks which are inset a good two feet in very dense bone. Villagers were there in a state of near ecstasy, covered with blood. One tusk came clear and was taken to a small fire to shrink the nerve pulp to facilitate its extraction. Elephant tusk nerve is valued as an aphrodisiac. More animism, I wonder? We came home at the same high speed of travel, by way of the kopjes and did it in an hour. We have done about 18 miles today, 8 of them at forced pace and I thoroughly enjoyed them. Have not been in such good form for years. Sincerely hope the *Daraprim* does keep the malaria away because we have been badly bitten by mosquitoes. Have been worse bitten by one damn thing or another today than during the rest of the trip. This morning at the hot spring I felt a most intense burning pain on the underside of my left forearm; a hornet had stung me. Eustace got his pipe out of his pocket and wondered whether it was dirty enough. He withdrew the mouth-piece and scraped out the tobacco juice, which he put on the sting. The hurt went out of it rapidly and I have no swelling. The tusks came in last thing tonight. I forgot to say that while we were at the elephant this afternoon one man wanted a peeled stick for carrying meat over his shoulder and to this end merely tore the bark off with his teeth. The young Chief, so thoroughly enjoying himself, was using a large knife and bent the point considerably; into his teeth it went and in due course came out straightened. The elephant is indeed a very old one; Eustace now says 80 to 90 years, so as a stripling it may well have been in the valley when Livingstone crossed.

15 March 1956

On the march by 6.30 and the day's journey finished by 11 o'clock [half a mile east of Nkonto's village]. Through villages and *mopani* forest and across streams. Rather easy going, thank goodness. The villagers greet us and the women ululate and the little boys come to see as much as possible of everything. Eustace's majordomo buys meal from the villages and engages two carriers for the ivory for tomorrow's march. Later in the afternoon Eustace and I, the young Chief and the chief Game Guard walk out over some little pebbly hills covered with poor forest growth of *Brachystegia* and *mopani* and lesser shrubs. There was remarkably little life, though we were close to elephants most of the time. Eustace flew his hawk a couple of times to the lure, but he is fed up with having no open country to fly her. I don't think I shall take up falconry. Was tired tonight and retired early, after a fine dinner of guinea fowl.

16 March 1956

A fine night and very heavy dew. Much of the march (away 6.30) was through dense long grass and we were wet through all the time. We also had several *dongas* and swamps to cross and were often up to the middle in muddy water. Occasionally we get glimpses of the Munyamadzi River and very fine it is. At one village the headman offers Eustace, the young Chief and me a pinch of snuff, which he dispenses out of a little bottle or phial. It is not bad snuff at all, made of tobacco and some of the brown interior pulp of the fruit of the sausage tree. These sausage fruits hang like giant salami. The Munyamadzi has banks 30 feet high here, and there is almost constant erosion and laying down afresh. Sometimes oxbows form sickle-shaped lagoons some way back in the bush. They contain lots of crocodiles and Eustace says the lagoons hold a lot of waterfowl in the season. We have camped today in dense long grass under trees. The carriers have hoes or mattocks and clear the ground for the tent and our tables and chairs. Some of the [Chilenge] village women came and helped today.

Forgot to say I saw a waterbuck doe yesterday at 60 yards and she kept still enough for me to have a good look at her. Grey in colour and a stripe on the cheek and round the tail patch. Also walked up to within 20 yards of a warthog sow before she saw me. When she did she was off, though she did not see me moving.

Out on an 8 mile walk through *mopani* forest and scrub with Eustace, the young Chief, Joseph and Makuka [Kukuka]. Saw only zebra all afternoon. But it was one of those very hot sunny afternoons and quite lovely to be out now that I have got fit. Eustace wanted to fly his Ayre's eagle but we saw no guinea fowl and the large partridges all got up wherever Eustace wasn't, and they are too sudden in mounting and away for this fairly slow bird at getting off the mark. Eustace stayed flying his hawk at the lure and the young Chief and I came back to camp. The village we passed through is in his territory and one or two of the younger women were winnowing corn. The young Chief bantered them in Bisa and you could tell their shy delight. It wasn't necessary to know the language to see what is humankind the world over. So very few Africans have any English, so it is not easy to communicate. Plenty of elephant signs today. They seem to come down to the river area at night and trek back into the *mopani* early in the morning. Saw the first evidence today of elephants having passed through a patch of kaffir corn. They had made surprisingly little effect walking through the 10 to 15 foot high plants which are a yard apart. Hell of a sick parade tonight: several cases of malaria; one cough; one badly swollen foot which looks to me as if a metatarsal was cracked or broken; another young boy with a swollen heel; one old man with cracked shoulders; one boy with belly ache with eating too much meat.

17 March 1956

On the march by 6.15 and stopped at 12 noon [at Kapambo's village]. Wet through as usual but we also passed through some pleasant *Brachystegia-mopani* scrub in addition to the miles of dense grass. Several deep, greasy *dongas* with knee-high water in the bottom which are very trying to the carriers. We stopped for ten minutes to look at a huge fir tree which must cover a quarter of an acre. It is one of the most magnificent trees I have ever seen. Each branch is as thick as a very considerable trunk. Probably hundreds of years old. Getting back to the resting *ulendo* I notice the oldish carrier with the badly swollen foot. He has been carrying part of the tent and a long pole and I can see he is in great pain. I can't boss Eustace's *ulendo*, but I do call his attention to this fellow, knowing well enough there is nothing we can do and he has got to get home somehow. The meal situation is already tricky. We can't hold up fifty men till this man gets better a fortnight hence. Eustace arranged that he should have a lighter load, not much more than 20 lbs and that is the best we can do for him. For myself, I would willingly carry his load for a stage but I could scarcely do this while there are Game Guards walking light. Also, we have to get the man home and if we show sympathy he will peter out right here and we are no better off. At least we are on the way home and the *ulendo* is still meat-cheerful. You see the revolting chunks of dried buffalo and elephant on top of each load. Tomorrow Eustace and I and the young Chief and Joseph or Makuka will cross the Munyamadzi and make a trek up the Luchenene stream into good elephant country. Dabster Makolulu will get the *ulendo* over the river and take them two stages up the other side, which will be a short march for the carriers and I shall feel a bit happier about the old man with the foot. We shall then climb the escarpment again, steep stony going, and I know his foot is not going to get better. One just has to harden the heart. So many people must have had similar harrowings in the course of the war.

We walked through a poor country of pebbly low hills this afternoon, with poorish *Brachystegia* shrubs and trees, and a lot of *Bromus* and fine *Aira*-like grass. Yet elephant had been through it recently. Today we passed through the dirtiest village [Bongo's] we have seen yet. People very wild. If it doesn't rain tonight Eustace thinks we may get across the Munyamadzi with a waist-high paddle for 60 yards.

18 March 1956

Three o'clock this afternoon and Eustace and I have just finished breakfast. He tells me it is Sunday: I didn't know. Up at 5.15 a.m. and all the *kutundu* down at the river bank by 6.15. The young Chief and I take the river first, he giving me a hand. The current is very strong and the river higher because it rained hard last night. Eustace follows between two Game Guards. It is exhilarating slowly getting across, though when one goes over the shoulders it is a little frightening for a poor swimmer. The great strength of the river is felt up the whole length of the body. The water is now only to the waist and you know the crossing is over and only another 15 yards to go. Eustace is slight and though he swims well I am glad he is between those two Game Guards. The carriers have been watching us and are nervous to start coming over. The loads on their heads are really a help in keeping their feet on the bottom. They go farther upstream and come across diagonally and have a rather better crossing than we have had. One of them dances over like a ballet dancer and I marvel at him. Eustace says if you can do it, it is the best way. Watching this fellow it looked so easy. Once the carriers are over they are so delighted they begin to play like little boys, going into the water again and swimming and fooling, playing ducks and drakes with pebbles from the shingle. They are happy, and apart from casualties in much better condition than when we left Mpika. Eustace told his majordomo this morning to recruit another carrier from the last village we passed, as the old boy with the swollen foot would be unable to carry. I was so relieved at this decision. When the ivory comes over the river the carrier recruited to carry one of the tusks for the next stage starts complaining that he doesn't want to go. The corvée and the pressgang is not allowed; the carrier is still complaining tonight and crying, but Eustace tells him through his majordomo interpreter that he will carry to Mpika or get no wages for the day he has already put in. I think this gentlemen will probably run away in the night. One of the carriers bought a young bitch from one of the villages on the Munyamadzi and had it tied to a stick with bark string. (Many of the loads are tied with this, the carriers making it themselves.) The young bitch is one of these tan-coloured, smooth-coated dogs of Africa, rather attractive and intelligent. Well, the more this young bitch looked at the river and saw the others crossing, the less she liked it and managed to escape, string and stick and all. Carriers on both sides had a good laugh over the 60 yards of river, and you could hear the fellow dashing through the long grass after his dog. He caught her because the stick snagged on something, and she was brought across, struggling at first, then quiet and observant, and finally of course she was in the water, and she swam hard alongside the carrier. She seemed as pleased with herself as everybody else once she was on dry land. This is our final crossing and much to Eustace's surprise and relief, we have lost no gear.

Eustace, the young Chief [Sandiford], a Game Guard trainee [Diamon] (the one who dived into a tree when we met the rhinoceros, but who is a good willing boy) and I now left the *ulendo* ⟨to go ahead to Chimbwe stream, near Kapatamoyo's village and make camp⟩ and went back into the hills over a stretch of flat forest. We got into the Luchenene drainage: this river is fast and noisy and comes down the Muchinga Escarpment in a fine white cataract, then through a fine wide glen and ultimately across a plain into the Munyamadzi. We walked for almost six hours over these pebbly hills and through the thickets of long grass. The pebbles are water-worn quartzite varying in size from an egg to one's head. Travel over them at fair speed is hard going. Today we had torrential rain for a couple of hours and the hills ran water. The complete wetting through with such thunder rain is a little chilling but when it stops most of the discomfort is at an end.

⟨We came to a high scarp overlooking the Luchenene whose swollen, turgid stream swept towards us round the broad bend. The view was lovely. Beyond the river the landscape is broken by a mass of tree-covered hills, which increased in height as the valley tilted towards the steep slopes of the Escarpment. The range, roughly lying north-south, stretched away through fragments of steamy cloud, often veiled by rain storms. Ahead of us, west by north, a mare's tail of foaming water cascaded over a bare rock face high up the face of the Escarpment. This is the Luchenene tumbling into the valley from the plateau. Curtains of rain were falling close to the face but as we watched, the rain, partly obscuring the cascade, stopped. At the same time a rent in the clouds allowed a shaft of sunlight to illuminate the tumbling torrent.⟩

We followed a new elephant spoor and eventually came up with it. Eustace and I went to within 20 yards and the old boy never knew we were there, the wind being right. Oh, how beautiful their movement is in this environment! We saw two more big tuskers down in a grass flat on the Luchenene, but on the wrong side of the river. In any case, the grass there would have been too long and dense to stalk elephants in safety. We got on to an elephant path for 200 yards or so – a much better going – but suddenly the young Chief stopped. A rhinoceros was walking along the path towards us. He kept on walking and we deferred by walking into the bush again. We came on another rhinoceros which had just finished its bath in a wallow of creamy mud. It was a big beast, now flapping his ears about and rubbing his great long face against a sapling. I had a nice long look at 30 yards range. He had a double wallow, side by side, each rectangular – one in use today and the other not. Eustace would like another buffalo for the *ulendo* and we spoored one for a mile till we reckoned we were only two hundred yards from him. Then the rain came plump and destroyed the spoor very quickly, so the rain saved that old solitary bull buffalo. We got back to camp at 1.30 p.m., the *ulendo* having come six miles and set up camp again while we were doing our long trek. No sun at all this afternoon and everything is dripping wet, so Eustace and I are calling it a day. The sick parade came early and did not take us very long. The old boy's foot is bigger than ever and it is obvious now there is an abscess. We have made a magnesium sulphate compress and hope it will draw it.

Forgot to say we saw a herd of 12 [11 Lichtenstein's] hartebeeste at close quarters today. They crossed an elephant path in front of us and we had managed to stop and crouch down before they saw us. They are rangey angular antelope with high withers, chestnut in colour like so many others here. ⟨It is unusual to see even as many as eleven of the animals together in the low-veld country where it is unlikely they have ever been plentiful . . . ⟩

19 March 1956

It rained torrentially all night and still going hard when we rose at 6 a.m. We took our early morning tea at some leisure and as it did not seem that it would let up, Eustace went to get the boys moving. Half an hour later they were still under their tarpaulin shelters. Dabster, the majordomo, said he couldn't get them out from under the tents. 'What?' roared Eustace, 'then pull the tents off them and then they will be out.' This had the desired effect and once they were on their feet the carriers were quite all right. The tents weighed terribly heavy being so wet, but we got under way in the heavy rain by 8.45 o'clock. We waded knee deep mile after mile, and had a few streams in spate to cross. These took us up to our middle and the strong current was rolling boulders over our feet. We got into long grass 10–12 foot high, which meant an alluvial flat and we were still waist deep. In fact, we found the Munyamadzi [at confluence with the Kachikwila] was running far above its banks, about 15 feet higher than when we waded it yesterday. We did a good mile waist deep through this high grass and had a few glimpses of the great river, 20 yards away, when we passed the places where the river was running into the country and had flattened the long grass. It was terrific, a roaring foaming flood with plumes of water being thrown into the air. One carrier fell with 45 lbs of meal. We reached higher ground a little later and continued knee deep and occasionally ankle deep. Eustace was so glad we crossed the river yesterday, for who knows when we would have got across, but he was still worried about a stream we still had to cross. The rain abated for a time and when we did reach the stream [Luavisi] it was 40 yards wide and running at a great rate. But it had gone down 4 feet as we could see by the grass, and we got across waist deep. The *ulendo* was behind us and the rain was torrential once more. The stream rose as we stood by it and as the *ulendo* came across, straggling badly, two or three of us got into the bed of the stream and helped the men a little or at least gave them confidence. The expressions on those carriers' faces were an interesting study. Those on the right side immediately became cheerful and called to the others. The man with the dog provided light relief for black and white. Eustace heaved a sigh of relief when all were over. Another half hour would have beaten us, I think. 'Nothing can hold us now,' said Eustace. But the rain continued torrential as we began our climb up the escarpment. The elephant path led us up the edge of a steep hill and now we followed the *ulendo* as the path was well defined and we could help any laggards. One man nearly went over the edge, but I saw it was going to happen and managed to grab the load and steady him. The *ulendo*, however, is elated at having crossed all the rivers, and as they are now in such good muscular condition, they climbed the first thousand feet of the escarpment [Musondeka Hill] in grand style. So did I, for my legs don't get tired any more. We had now reached a forested shelf and within half a mile we pitched camp soon after 2 o'clock. The rain abated but we were surrounded by Scotch mist. Once the tent was up we got out of our soaking things, and were having breakfast soon after 3 p.m. It has been a hard, adventurous day in a watery kind of way, but here we are safe and sound. Up here on the escarpment in the continuing rain, it is about as warm as it is in a south-west windy storm in the West Highlands in summer, and that isn't very warm. I shall enjoy the warmth of my sleeping bag tonight, the first time I shall have been able to sleep in it rather than on it. The rest of the bedding is rather wet. Am impressed by the immensity of natural phenomena today – the rain, the river, the constant roll of thunder and the dark brooding quality of the weather. Judging by the grass at that last river, the way this 12 foot stuff was flattened by the current, this is the highest flood of the rainy season.

⟨Frank is magnificent and really enjoying himself. As he says, 'This is old-time African travel such as few are privileged to experience in present times.' His cheerfulness and his ever ready helping hand wherever needed, particularly at river crossings, has done much to sustain the morale of the carriers who, even when exhausted and dispirited, find it impossible not to return his encouraging cheerful smiles or whole-hearted laughter.⟩

20 March 1956

Camp struck and on the march by 6.15 a.m. Started with a climb, then a flat bit and more steep climbing, more broken ground and up another steep hill, a climb of 1,500 feet in all and we are at the top of the pass through the escarpment. Eustace called a halt for rest and everyone is delighted. 'What a view it would make if some of these trees were down,' exclaimed Eustace, 'right over the valley to the scarp on the other side, bordering with Nyasaland. Wouldn't take long, you know, as we have nearly fifty men here.' So as we were not making a very long march today and it was only 10 o'clock, Eustace got the axe men going. He was busy with his Leica cameras and lenses – he takes inordinate trouble in getting his pictures – and I bossed the timber operations.

⟨The boys got to work with a will, and thoroughly enjoyed themselves, as natives always do when there is something to destroy. Sandiford plied his axe like Unslopa-gaas of old, and even the pampered native messengers lent a hand and set-to. After about two hours our clearing revealed a very fine view indeed. The slope below the saddle falls steeply into dense forest, which extends over the broken hills that range westward. In the middle distance Musondeka's crown is poised above the ridge which includes the final descent of the escarpment. Yesterday Musondeka seemed to retreat as we struggled upwards towards the drier ground at its foot, then, when we reached the foot of the spur, it towered above us, menacing the weary, muscle-straining carriers toiling upwards like over-burdened ants. Now he seemed to be peering up at us surprised to find he had failed to defeat us; that in spite of him, this string of heavily loaded men had won to the top of the pass.⟩

My biggest trouble was to get any order into the men. They were so happy and cheerful and would start felling anywhere, and I found myself yelling to get some of them clear of falling timber. The men would laugh like anything to see their fellows jumping clear and instead of a curse from the near-victim, he too would roar with laughter. One had to keep them just where they were needed or they would wander off and knock any odd tree down. Some of the trees fell across the path and I had to show the axe men just where I wanted them lopped for clearing out of the way, otherwise they would just hack (expertly enough) anywhere. When it came to shifting the logs I would take my place among them and each trunk went with a wild cry of laughter. You never saw such children. We felled nearly 100 trees from about two acres of ground and sure enough a simply magnificent view of the Luangwa Valley was opened out. Eustace took several photographs and the carriers sat around enjoying themselves. One of the two dogs in the *ulendo* began to yap and some wag among the carriers cried 'Chongo, chongo' like Eustace and the whole lot of them had a good laugh. Eustace says he has never had a *ulendo* so cheerful as this one. Good enough. Eustace says the carriers will be calling this place *Chitemene docatera*, the doctor's garden, but I say Poles' Peep is much more likely.

We camped at 1 o'clock at the twin palms, the place where we stopped on the first night of our tour [at the foot of the Bangwe Hill in the Valley of the Pregnant Twins]. These two palms go to 80 feet and are in a softish spot among these escarpment hills. A rocky hill goes up another 1,000 feet behind the camp and I set forth to explore it during the afternoon, along with my personal servant, Martin, who is a Chewa from Nyasaland and the nephew of a chief. How different from three weeks ago, when I doubt whether I could have done it! Now it was little trouble, only sweat and a bit of puffing.

⟨How different Frank's condition to that when he set out three weeks ago. Then his muscles lacked tone, his feet were soft and he had not fully recovered from the debilitating effect of a serious illness. Now he radiates good health and fitness, enjoying the exercise of co-ordinated muscle which respond to every demand. He tells me he is having a unique experience and a happy time, which makes me glad. Frank is a great man and a good companion.⟩

The trees at the top are stunted and strewn with Spanish moss lichen. There were also a few tree *Euphorbias*. Two klipspringers bounded away from the last 50 feet among the rocks, and a nightjar got up at the very summit. Martin says its Bemba name is *loombe*, a good name I think. The views from the summit are magnificent. The Luangwa Valley was full of mist and rain now, but to the south were miles and miles of *Brachystegia* woodland and then some considerable hills 50–60 miles away, and some more 100 miles away. Three ravens over camp at tea time and I called them down to about 40 feet overhead. Their call and silhouette just like ours. The top of the hill, Bangwe, is interesting in that fire does not reach it. The extra humidity of a mountain top also produced a variety of ferns which did not occur 150 feet lower. I would like to camp near here with a good cache of grub and explore these escarpment hills, in which there may not be a lot of game except klipspringer sparsely on the tops, some sable antelope, rhinoceros and occasional elephants and zebra. Almost all of Africa I have seen is a fire climax community. Variety in any one vegetational habitat is not great.

A long sick queue tonight but nothing serious. When we started there were many old septic wounds which have cleared up with sulphamilanide powder packed in. Same with all the tropical ulcers. Our old boy with the swollen foot is much better tonight. It burst last night under the magnesium sulphate pack and the swelling is down. One or two 'wraxed' backs, as the Highlander would call them, and all we can do is to paint them with methylated spirit. Anyway, we finish up with a healthier lot of fellows than we started with. The pity is that Eustace isn't going on another tour, because he could re-recruit over 40 fit men.

21 March 1956

Marching by 6.15 and the carriers keeping well together and a good pace. A straight march across the wooded plateau with its clear swamps (*dambos*) and the inhabited country which is devastated by *chitemene* cultivation. All this plateau country has been cleared of game by the natives and now the birds are going. We reached Mpika by

12 noon. Then a sense of anticlimax. This has been such a wonderful trek and it has given me back my health. All the rest of the trips I shall do this time will be short, car-borne, air-borne and featherbedded. Eustace stands alone in his tireless liking for foot treks deep in. I owe much to him and have learnt much from him. I am all for the early start and military style.

22 and 23 March 1956

Dined on evening of March 21 with the Fox's at the Boma, which is six miles away from Eustace's house. Fox is the District Commissioner. They have two sons at school (Downside!) and two small girls here being educated on the correspondence system by their mother. Seems to be working well. On the morning of the 22nd I motored up to the Boma to send a cable letter for Jamie's birthday, and send the Luangwa batch of diary to you, and to collect well over £300 from Fox for Eustace. I had to count it all, including £50 in silver. Then breakfast.

Just before 12 noon I went out alone for a long walk and got back soon after 5 o'clock. Eustace feels now he can let me out alone and I am thankful for the chance to see things in my own way. I headed for a distant long ridge of hill on the Congo-Zambesi watershed and reached it by going a little off course to get on to some lower hills. By following the ridges of these I kept my bearings better. You can't see the hills if you walk through the scrub *Brachystegia* woodland of the valleys and as the sun is bang overhead for a few hours, you don't get marks very easily. I think I could have done it all right on my generally accurate sense of direction, but even so the scrub is thick and awkward. The hills have trees to the summit but thinner, and except for the rocks it is easier travelling and there are magnificent views afar and over the nearer valley woodlands. I must have done about 15 miles and was struck by the emptiness of the country. Saw two duiker bucks, a family of baboons, three large dark grey owls, and I put up two nightjars from the highest point of the ridge which was my objective. I saw a few drongos (a blue-black long-tailed starling) and a kite hovered inquisitively above me for a mile or two. There was no sign of game on the ground and no game paths. The plateau or high veld is hunted out by the native with his muzzle loader. Some of these muzzle loaders are marked 'Tower' and are from the Tower Armoury, being the weapons left over from the Napoleonic wars. Eustace doubts if they will ever wear out. Since this last war cheap shotguns are being imported and sold to the natives. Guinea fowl and partridge have almost disappeared from the plateau country in consequence. Administrative policy here is weak and generally pretty poor, and the Socialist Government did untold harm to the Colonial Service and to the general situation here. My considered opinion is that those English people who have come here with capital since the war to farm, like the Widmores, are just dupes, and dupes of the Government. Some day not too far distant, there is going to be trouble and everything indicates a handing over to the African. It is a criminal policy, and people like the Widmores will be on the run, almost like refugees. A body on the lines of the Indian Congress, and called Congress, is on the work here and Socialists from Britain come out to the Copper Belt and preach Trade Unionism to people who have no

notion yet what it is all about. The present Conservative Government, as you and I
have often said, are just Socialists, and I think I shall call them the Doggo Socialists
in future. This is a long way from the *Brachystegia* scrub woodland I am walking
through, mile after mile, and which cries out as a fire-bedevilled beaten up country.
I ask myself, what was the original vegetation of all this plateau before fire was an
annual affair? I also note that this *Brachystegia* and other few species of trees are fire
resistant. The grass below grows tufty and does not cover the soil well. The soil itself
is red and deficient in humus, as it would be with all the detritus burned away and not
rotting down. The high ridge, about a mile long and which I walked, is obviously much
less burnt and the grasses and herbs form a closer mat, almost hiding some of the
stones, so that the going is rough. Unfortunately I don't know the species of trees and
shrubs up there, but they have that top of the hill quality – low spreading with plenty
of lichen. The rocks are also covered with crustose lichens (quartzite up here) but lower
down the rocks are more of a schist. I liked that ridge and found it hard to come down,
but I didn't want to cause any worry to Eustace by coming in after dark. The darkness
descends suddenly just after 6 o'clock and it gets light equally suddenly about a
quarter to six in the morning. The birds begin to sing at 5.30 a.m., bulbuls and shrikes.

Today, the 23rd, Eustace got me out at 5 a.m. and we set off before it was light
in the Land Rover, together with James the cook and the wherewithal for a comfortable breakfast, and Mukuka. We drove 30 miles and climbed to over 6,000 feet. It was
wettish plateau country where the fires do not rage quite so hot. In these areas there
are some relics of the earlier vegetation. We climbed a kopje and looked down on
perhaps 30 acres of bits and strips of what is called *mushitu*, the relict forest, clinging
to the wetter places where the fires did not reach it so readily. These dense, high bits
of woodland were oases in what is now little more than desert of wet moorland. It
is quite wonderful: here is tropical forest with all its variety of life forms, a true climax.
There is deep black humus, dense undergrowth and lianas that you have to cut your
way through, epiphytic growths on the trees, a wealth of ferns and mosses, some giant
sedge, and a very characteristic bramble, *Rubus*, which was only in flower. Some of
the trees were magnificent, and the birds were numerous and vocal. I had that thrill
once more of entering a vegetational complex which was pristine and right. This forest
may be called *Syzygium–Xylopia* swamp forest ⟨dominated by *Syzygium* sp. in conspi-
cuous stands up to 60 feet in height, with which *Mitragyme stipulosa*; *Fagare* sp.,
probably *F. macrophylla* and *Canarium* sp.; *Pygeum Africanus* and other tall trees,
together with a shorter marginal growth of *Syzygium cordatum* (C.G. Trapnell) covered
much of the adjacent plateau⟩, but one may be quite sure that in possibly modified
form it must have spread over much of this area at some remote time. Lightning fires
would destroy bits every year no doubt and provide that variety of environmental
conditions which enriches a whole country. These patches of *mushitu* should quite
definitely be preserved but it is evident that even now the fires are nibbling at their
edges and occasionally running in. After the feeling of elation I felt the opposite way,
in that I was probably looking at beauty which may soon be gone. There are little
patches here and there elsewhere on the plateau, but no considerable expanse. Eustace

and I took a lot of trouble taking time photogaphs in the *mushitu*. When we emerged at 10.30, James had breakfast ready for us and I was thankful.

On returning from the Luangwa *ulendo* two days ago I beheld myself in a mirror. The nose is a bright brown-red, terracotta perhaps – the rest of the face and neck are brown and rather less bright; the physiognomy is somewhat attenuated so that the width of the head is greater than across the chops. My Mexican belt sits easily round my middle when tightened to the uttermost hole. I shall strive by exercise to maintain these lines, and when I return on April 14 you must not press me to too many second helpings of the pastry I like so much. I have never felt better in my life, and I am sure one of the reasons is the warmth. I do so enjoy it. Some of the muscular effort in the Luangwa trek was pretty gruelling, yet I never had an attack of cramp in the night, which I so often do have at home. Warmth quite definitely suits me, and I am quite sure it suits you. It leaves one time to think and be relaxed.

24 March 1956

Up at 5 o'clock and away in the Land Rover to Luitikila swamp by 5.45. Less than 10 miles away and then a lot of walking through long grass and then through giant sedge 7 feet high and calf deep in water. We did several miles of this and the midges were dreadful. Occasionally one could get a view of a great saucer-shaped hollow with reeds, papyrus, rushes and occasional pools of water. Pallid harriers flying over the reeds, and here and there saw pied kingfishers. Bird life not abundant. Ant hills rose from the swamp edges and provided places where small trees could grow. Saw three hippo in one pool but they are shy. The hippo keep these places from just filling up and they make roads into the river and the swamp. If the hippo go, the water is eventually lost under the terrific build-up of vegetational detritus. Some bright spark has started African peasant farms at one point near the swamp and as the hippos come out to graze and have damaged some of the crops the cry is for extermination of the hippo. They are doing such a good job. As Eustace says, 'Every hippo killed here means a bulldozer less.' There may be 10 to 20 hippos altogether in the swamp. ⟨We flushed an outlying situtunga without getting even a glimpse of it – a number of these unusually interesting swamp antelope are still about. We found several places where they had been lying up.⟩

25 March 1956

Left Mpika at 6.30 a.m. for Kasama in Con Benson's Land Rover, which had come down for me 137 miles the day before. The road was bad and one gets a good deal of cervical jolting. About 60 miles out from Mpika (in the interminable plateau *Brachystegia* woodland all the way) we found a car well ditched and down to the axles in mud. I changed my clothes, and my driver William and I tried to help the man out. He was a dark guttural-spoken fellow whom I took at first to be a German, but apparently he was a Boer. We couldn't get him out despite the added help of an African who came down the road with his axe and his muzzle loader, evidently out for his Sunday morning's shooting. William then remembered there was a Roads

Department caterpillar 3 miles back, so we took the unhappy guy back there. The driver was in camp with his African wife and children. He was a very big-made halfbreed and I liked him. Even he had but little English, though I could make contact with him easily. His three children were quite delightful and so delighted when I gave them a banana each, and for which they thanked me most politely. The big man looked at me and smiled in a wondering sort of way. He couldn't place me. They had a very nice well-grown bitch pup of the yellow breed; English didn't matter with her, for her doggy language is a universal one I happen to know, and she and I parted on the best of terms.

About 90 miles from Mpika we came to the Chambezi River, a broad beautiful waterway with a fine pontoon. An island in the middle of the river bears an inscribed stone, and a field gun. It was here von Lettow surrendered the German forces in 1918. I looked at it with a little more than ordinary interest. An elderly African woman on the pontoon spoke to me in Bemba, asking me for a lift I suppose. I thought and hesitated, not knowing quite what to do. One of the men on the pontoon who could speak a little English said, 'Woman Chief', and then I noticed her several ivory bangles on her left wrist and knew she must be someone considerable. I slipped round to William and asked if it was in order for a Government vehicle to give a lift. 'It is not in order, but the bwana can decide.' So the bwana said she should have a lift the 16 miles to her village. William let down the backboard of the Land Rover, untied his cycle and let her in. I took my cue from William. If he, a Bemba, put her in the back, I had better let her stay there (apparently I was right) but I took up one of the Dunlopillo seats from the front and handed it back to her. She nodded, unsmiling. We stopped later at a path leading to her village, where a bunch of women were waiting for her. William undid the back canvas and tail board for her to alight: she rapped a little peremptorily on a rope which William had tied between the body and his cycle, and William hurriedly untied it for her easier passage to the ground. A boy of 14 or so came forward and bent his knees; she pointed to her bundle and the boy deferentially took it up and preceded her with it on his head. The women curtsied. The Chieftainess turned to me, curtsied and cupped her hands in thanks, but no smile. I bowed and raised my cap, and I saw the procession go slowly along the path. She is, says William, the mother of three Chiefs as well.

Reach Benson's house by 12.15. He is a scholarly type who moved from being a District Commissioner in Nyasaland to being Provincial Game Officer in the Northern Province, Northern Rhodesia. Extremely keen on birds and has just produced a new check list of species in the Territory. Has done much collecting for British Museum and really knows his stuff. Wife very intelligent, a botanist, but a continuous prattler. Two girls, $7\frac{1}{2}$ and $4\frac{1}{2}$, very spindly – trouble getting them to eat; little one a confirmed thumb-sucker.

Out in the afternoon to see another *mushitu* but it is not comparable with that at Danger Hill; the fire has been in and there is much less epiphytic growth on the trees. Benson says the *mushitus* may also be looked upon as relics of the last pluvial period which corresponded with our last glacial. On again to some extremely rocky, low kopjes. The boulders were tremendous and some of them were climbable; cactuses, ferns, epiphytic orchids and stunted trees, and *Vellosia*, a branched, fibrous-stemmed plant with bunches of lily-like leaves coming from the brown stems. A fascinating, haunted kind of a place. We halted under an overhanging rock and there on the face which varied from sandy to heliotrope in colour, were two little antelope-like animals painted, very faint and dull heliotrope in colour, and over the whole was an ochre lattice. Was this lattice a net? If so, are those curious lattices in the Lascaux paintings also nets or stylized symbols of them? Benson says a people like Bushmen lived here until the Bemba came out of the Congo region in the late seventeenth century. They were called the Akifoola and it is still not quite certain if they are extinct in a region of rocky hills farther south near the Portuguese border. No one sees them but evidence of them is found from time to time. Rather like the Indians of the Californian High Sierra which were thought to be extinct, but one boy was left in 1926.

26 March 1956

Out this morning to visit the Chishimba Falls, of the Luombe River, about 20 miles west of Kasama. We drove in the Land Rover 16 miles to Chilabula Mission, run by the White Fathers. This Roman Catholic body has been here longer than any other denomination and this mission is built of home-fired bricks and tiles in the Spanish style, so very satisfying. The Bishop himself lives here. Two fathers greeted us cheerfully, provided a guide in the shape of a cheerful little elderly African and invited us to take morning tea with them on our return. Off in single file at a good pace, passing the girls' school where the Mother Superior came out to greet us. She wears a habit of grey and white surmounted by a white solar topee and wimple. At the pace the little man walked through the bush we came near the Falls in under an hour, to the lip on the left bank. Here the little man bent his knees and clapped his cupped hands in obeisance and salutation to the spirit of the falls. They were quite magnificent, a river 20 to 25 yards wide of crystal-clear water plunging 100 feet and then much lower in rapids. A beautiful rainbow formed below. Falls like this have a tonic effect. The adjacent vegetation on the right bank was of *mushitu* type and very lovely. I took several photographs and waded out into this glorious river.

To see the Provincial Commissioner and his Deputy in the afternoon, Heathcote and Thomson. They drew me out about my visit to the Luangwa in the wet season and about conservation policy in general. I talked for an hour and a half, and later that evening when the Bensons threw a party for me, Heathcote talked further, and Thomson (a St Andrew's man) told me he was thankful I had talked as I had and that my talk had made a great impression on Heathcote. This doesn't mean much in action because Heathcote has ambitions and is on the point of possible considerable promotion, and is most anxious to play safe.

27 March 1956

Not much of a morning. Went into Kasama to get a box of stores to give to Eustace, who has been burdened with me for most of a week. Forgot to say that yesterday

Desmond Vesey-Fitzgerald came from Abercorn and was staying the night also at the Bensons. D.V-F. is a most cheerful fellow and a very good naturalist. He is with the Locust Control outfit and is very much interested in game-carrying capacity. We walked and talked a good deal and I hope next trip to go and see him. One of the guests at the party was a young Forest Officer called Kerfoot, very keen and on top of his job; an ardent conservationist and was at my lecture to Forest Officers at Oxford last year. Also met an agricultural officer who seemed to be rather a complete ass. Another agricultural officer was more or less a refugee from the Sudan. He was bitter about the hand-over which he said the great majority of the Sudanese did not want, and the lower Nilotic tribes had implored us not to let them down. We promised not to and then we did. Eustace says we did just the same in Burma, throwing away the devoted loyalty of the Karens. Ian Grimwood also told me we are throwing away the real loyalty of Pakistan and playing ball with India in Nehru's way simply because India is more powerful than Pakistan. Who are the clots who create the reality of 'perfidious Albion' in this way?

Away in the Land Rover in the afternoon back to Mpika. Rained nearly all the way — $5\frac{1}{2}$ hours, 137 miles — and the road was in much worse condition than on Sunday. We passed one lorry only in the whole distance.

Forgot to say yesterday that when we returned from the Falls to the Mission, the Bishop himself, a Monsignior, welcomed us. He is a handsome elderly Frenchman. There were Dutch, French, Belgian and one Scots father present. The Bishop, I feel, has no illusions about the African, but that is as it should be. Our tea and biscuits were a pleasant and polished interlude.

I saw an interesting thing on the way down from Kasama — the aggregation of butterflies of various species at one spot in the road. Benson says the attraction may be leopard's dung, a great favourite, or excrement of some kind, or if it were the dry season, some moist spot. I photographed one aggregation with the 13.5 cm. lens.

28 March 1956

A day of waiting for the aeroplane. I went a 10–11 mile walk in the morning and spent the afternoon fiddling about. The airplane came at 5 p.m. instead of 1 o'clock, a little single-engined 4-cylinder Auster, not nearly powerful enough for this territory, where heat, convection and thermal currents may lessen the density of the air, and you start at 4,000 feet anyway. The pilot, an able man of 35 is chucking in because the Government is so piddling, objecting to the expense of 2-engined aircraft, hedging him round with regulations and so on. I would much like to fly the Luangwa River just now, but the Administrator says the Luangwa Valley is forbidden flying. And this pilot, Lenton, says it is dangerous with a little plane like his, because he can't climb back over the Muchinga Escarpment with any ease or certainty. I had already suggested back at H.Q. that the Game Department should have one or two planes and pilot-biologists on the Alaskan model. In the red lechwe country of the Kafue Flats, the native *chilas* (illegal drives and mass slaughters) are bringing the animals near to extermination (250,000 to 25,000 in 25 years). I suggested that an aerial patrol would keep that situation in order, and it is the obvious thing, but His Excellency the Governor has expressly forbidden aerial patrol as it would possibly upset the tribes and would not be sporting anyway. I see no reason, however, why my ultimate report should not strongly recommend aerial patrol of the game areas. Once more there is a nervousness which is not justified in view of the African's inability to organize.

29 March 1956

The morning was not propitious flying weather, but we took off in the little Auster at 11.45, i.e., Ted Lenton the pilot, Con Benson and myself. The going was bumpy and my first round of sickness came after a quarter of an hour and I was sick every quarter of an hour after that for the three hours of the flight. Bile makes one's throat so sore. We flew to the marshes of Lake Bangweulu, the great indeterminate water which Livingstone discovered. The marshes are now under water and many of the marsh villages are inundated. They lie abandoned until the dry weather comes. The present villages are on small island sites. We could see the people coming out of the little round huts to see us. These marsh people get around in well-fashioned dug-out canoes, poling from bow and stern, and they are very fast. We were going to see the black lechwe, which are found nowhere else but here. I suppose they are a race of the red lechwe, but they have some black in the face and their horns do not grow so large as in the red lechwe. There were once an estimated 150,000, now only 15,000, determined from aerial count. We came upon them in the marshes, feeding belly deep on the grasses that grew through the water. How lovely they looked in their chestnut coats, great herds of them. Throughout the marsh termites have created small round islands of nests, and these raised places may grow a shrub in the middle and give room for two or three lechwe to lie down in the dry, or even half a dozen. I suppose we saw several thousand lechwe, and we also saw two poaching canoes, one with a dead lechwe in it. (The species is now totally protected but not effectively.) Ted Lenton the pilot is a very able intelligent young man who lives his life in a rather modern way. He is dead keen in taking care of animals and has a real dislike of poaching *munts*. So he set to to dive-bomb each canoe as we came upon it, once one way and then the other. The two men in one canoe lay down, but the other two went overboard as we stooped at them. On the return run, just their heads showed above the water. The marsh people fish in Lake Bangweulu and send their catch to the copper belt. Lechwe meat is eagerly bought there, so the poached meat goes out under the fish in the baskets. We flew over as far as the Luapula River before heading for home. No other animal can use this marsh environment as effectively as the lechwe. The tsessebe inhabits the slightly more dry land part of the marsh.

⟨At 2.45 p.m. the plane returned and I never saw anyone look so ill as Frank Darling. He confesses to be habitually air-sick. In Lenton's opinion he becomes so ill that he should give up flying altogether. He retched until he became exhausted.⟩

This evening, the Provincial Commissioner Heathcote came in (from Kasama) with Fox the District Commissioner (Mpika). We talked a lot more about Lugardism and colonial administration. I feel that the Africans of Northern Rhodesia are not yet sufficiently differentiated in class for indirect rule to work.

⟨Heathcote (the Provincial Commissioner) and Fox (the District Commissioner) came for sundowners. The conversation turned to Colonial policy and Native administration. Frank was constructively critical, particularly in regard to education, – believing there were too many free benefits particularly education, thinking it preferable to subsidise rather than provide the service free. He deprecated a system which failed to encourage pride in handicraft but merely led to an overall low standard, little in advance of the 'Three Rs', which engendered contempt for manual work, while suggesting false notions of clerical skill, leading to a dangerous inducement to abandon rural tribal life for the temptation to adopt an urban environment. He felt there was a great need for a strong paternalism in Government's policy to ensure a stable future. I referred to the Matebele and Mashona Rebellions, reminding how Southern Rhodesia was caught unawares and unprepared, and suggested the recent Mau Mau outbreak in Kenya had taken the local administration by surprise. Heathcote I think was impressed, though guarded in his replies and opinions. I think he would prefer to see a stronger rule. 'Brer' Fox lay low and said nothing.⟩

30 March 1956

Eustace went flying with Lenton over the Lavusha Manda, so I went for my last long walk into the Congo-Zambesi hills which I have come to know fairly well. I left at 7 a.m. and got back for 10.30 breakfast. Then packing and refuelling the Auster, then a long heavy shower, and finally it was 1.45 p.m. when we started.

⟨. . . the plane left very heavily laden, so that I was relieved to see it safely airborne. At 4 o'clock I flew the hawk again, and this time she took to the soar and, while at a great height and already off the airfield, another eagle mobbed her. The two drifted away, further and further, and when the wild hawk broke away, Rankin who was the higher of the two, continued to drift down wind to disappear from sight. I waited an hour in case she should return – a very improbable event in the circumstances. The remainder of the evening was spent seeking her; luring and blowing the whistle, and I was obliged to return unsuccessful and empty handed. I was more than sorry to have said goodbye to Frank, and now the additional loss of my hawk left me desolate. I read myself to sleep but woke up at 1 o'clock. The image of the two hawks soaring and drifting away remained etched on my retina, while the sound of ghostly hawk-bells rang in my ears.⟩

The flight took three hours, mostly over the plateau and occasional ranges of hills, the ground being this curious mosaic of *Brachystegia* woodland and bog (*dambo*). Then the plateau broke away eastwards into pretty rough wooded country and I could see some good falls. Finally into alienated land, i.e., land sold to European farmers, and on to Lusaka, and I was not sick though I felt devilishly uncomfortable all the way. Ian Grimwood met me and brought me out to his house to stay.

31 March 1956

Ian and I set forth to the Kafue Gorge this morning. The Kafue River bridge is less than twenty miles away and shows a deep strong current of clear water. It has come off an immense flood plain of grass, and because the flood plain acts as a reservoir the river runs strong all the year round. Some miles below the bridge (a war-time London Thames Bridge) the river begins to fall rapidly between steep hills, and fifty miles again downstream it debouches into flatter country and then into the Zambesi. Strangely enough, the course of the river in the gorges was not known until 1949, when two Wesleyan missionaries walked through in five days as a holiday jaunt with a purpose. We came back several miles from the river and then bashed along a track for a dozen miles or more in the Land Rover and came to a small river in fair flood that we did not care to cross in the car. So we ate our lunch, waded across and walked four miles to the river, crossing several streams on the way. We then followed the course of the river for three or four miles through the wooded hills. It is a fine river by any standards, tumbling along and eddying among rocks, always with a good body of water. I was interested to find that hippos evidently came out of the river and grazed paths as far as 400 yards from the water. We had pleasant sunshine all our walk but coming back we found one of the streams was in flood and we had to wade up to the shoulders. I had some difficulty in getting across with my photographic gear, the current being so strong. None of the other streams were up much, so it must have been a very local and heavy plump.

1 April 1956

Easter Day [in Lusaka], but no feel of it. I rested most of the day and felt the better of it.

2 April 1956

Away by 8 o'clock this morning to the Blue Lagoon Ranch. My companion is Gerry Taylor, *quondam* Regular Army, well bred but very worldly. Has had a lot to suffer physically though originally a strong man, but he refuses to forego the pleasures of life and lives for the moment. Is now one of the Game Rangers, and is very fond of animals both individually and as stocks of game. I can take him as I find him. The Land Rover was well packed and two African Game Guards disposed themselves as well as they might in the back. We went westwards out of Lusaka, at first through thinly farmed country and then into bush with occasional African villages. When we came on to black soil instead of red, acacia trees became much commoner and ultimately the scrub was almost entirely acacia with grass below, like a heavy savannah or prairie forest. Further still we came to the great flood plain of the Kafue and the trees disappeared. It was a sea of long grass. Our journey was a rough one of 130 miles, and now a few miles away was a low island with fig and cassia trees where the house of the Blue Lagoon Ranch (75,000 acres) has been built. We got bogged a mile and a half from the island, but I was glad of the walk in the sun. The Ranch is owned by a partnership of four, two of whom are Sir Alfred Beit and Lt Col. Ronald Critchley,

D.S.O., M.C.. Critchley is the one who manages day to day affairs. He is a queer egg; well bred, good cavalry regiment, 6'5" tall, lots of courage, a playboy who uses a lot of money. Whatever he had he has now spent, most recently in getting clear of his second wife (first one also a divorce). The case made one's hair stand on end, but there it is. He married Erica last week. She is another colourful personality, the only child of Mopani Clarke, one of the pioneers in Northern Rhodesia. He carved out vast estates for himself and became very rich. Erica is 45, petite and slim, reared as a princess in the bush and as an ornament in London and Paris. She has always said and done exactly as she has wished, and she still says and does exactly that. Her manners and address are perfect, but if she doesn't like the company and thinks it bogus or hypocritical, she comes out with a stream of language and sentiment that would leave a sergeant-major chittering and incapable. A Colonial Office type, a Sir William Fitzgerald, thought he was on to a good thing and married her in the early 1930's. He seems to have been a mean-souled one and very much concerned about his career. Erica despised him and embarrassed him so much by some of the things she said at her own and other people's dinner tables that he was ultimately glad to divorce her (and believe it or not, take alimony from *her*). Erica is an expensive type and she began to get through her own money. There were plenty to help her do that. She married a card-sharper and general trickster called Michael Lafone, who would have been deported had she not beaten the authorities by giving him one of her farms and thereby establishing a landownership qualification. She has just got rid of Lafone at much expense and is left now with little more than old Mopani Clarke's largest ranch. Now, having married Ronnie, she is having to keep him meanwhile but nothing upsets her except meanness and bourgeois hypocrisy. She is one of those women who have to exist and should be accepted as a corrective and tonic in society.

I suppose I am a bit of an odd fish in this foursome of Ronnie and Erica, Gerry and me, but I fall into it for four days quite easily, much enjoy myself, and find my companions extremely pleasant ones. They were most kind and hospitable to me; talk was good, food was good, the days out in the sun were delightful. That evening before sundown Ronnie and I walked out into the immense flatness and looked at some steers which were being corralled for the night. There were a dozen or twenty bulls among them, some Afrikander and the rest Barotse. Apparently the Barotse cattle survive better than the more improved Afrikander. Two days later I looked through a big herd of cows and calves of what Ronnie calls his native cattle. They were in excellent condition and though showing no uniformity there were some very nice cows among them. I should choose them every time, especially as Ronnie tells me the herd produces 80 per cent of calves against the Afrikanders' 60 per cent.

3 April 1956

A glorious morning and 180° of an horizon all ways. We set off in Land Rovers across the flats towards the Luwito which is a lagoon or tributary of the Kafue. We did perhaps three miles before we reached the first sign of the flood at the bottom of the long grass. So we set off on foot complete with Erica's and Ronnie's dogs: one a dachshund bitch, one a pointer bitch and the other a Great Dane bitch. The level of the water was soon such that the dachshund was swimming, so Erica went back with it. Then the pointer was swimming and had to have a rest now and again by letting down its hind legs and resting its front ones over the Great Dane's back. Finally only the Dane could keep its head above. The floor of the swamp was quite hard and of course the water was warm as we wandered about hip deep. We saw a herd of red lechwe half a mile away but we could get no nearer. The birds were magnificent – the huge wattled cranes, crested cranes, marabou storks, spur-winged geese, sacred ibis, egrets and many ducks. We got back to lunch, a most wonderful curry. These two believe in living well. Erica herself cooks well and she can manage her Africans. She is strict and occasionally lays into them, but they stay with her and I noticed particularly that when she commanded them she said 'please' and 'thank you.' Ronnie never seems to raise his voice to the Africans, treating them shortly but kindly. Gerry Taylor is also strict but kind and I could see his Game Guards liked him. We went off again later in the afternoon into the flats and the flood and did not get back till after dark. This time we took another direction and got farther on dry land. Then walking through the flood water to a lagoon, on the over-side of which was a herd of about 200 lechwe. We reached to within 200 yards and got a good view through our binoculars. The animals seem attracted to the flood edge and the first two feet of depth of the flood. The swallows took advantage of our passage through the grass in following us and hawking the insects we put up. Erica's coffee after dinner was very good and Ronnie's very old brandy was quite the best I have ever tasted.

4 April 1956

Away this morning with the Land Rovers and the dogs; first back into the dry ant hill country and the drier edge of the bush, then out again to where the flats start quite abruptly. We stopped there for lunch as there were trees for shade. The cold duck and ham, and the lime drinks went down very well. The coffee then revived me from drowsiness – if some of my companions' humour did not shock me out of it – and Ronnie and I Land-Rovered into the flats for a mile or more. I then took to the water on my own and continued another mile till I was 200–250 yards from some thousands of lechwe. Wind and sun were right for me, but the sound of one's passage through the water carries as if in a whispering gallery and the animals heard me and began to move along the lagoon. There was a forest of horns ahead of me, moving across my vision, and the loud susurrus of their legs in the water. It was a quite wonderful moment, standing there among the grass and water lilies, part and parcel of the lechwes' world. I turned back reluctantly. Erica is quite wildly pro-game and anti-human. Ronnie is pro-game also but is fond of his shooting. Now, Erica curbs his shooting quite heavily, and Ronnie puts up with it with amused tolerance.

5 April 1956

Gerry Taylor and I left the Blue Lagoon Ranch at 9 o'clock this morning, Erica having packed enough food for our lunch to feed an army. Actually, it fed our two Africans as well when we stopped at a Tsetse Fly Post 70 miles on. Back then to Chilanga and tea with Gerry's wife and three children: 7, 4 and 2. They are quite the nicest trio I have come across and the two-year old Jonathan is terrific, frightened of nothing, devoted to his father and has the wickedest of smiles. The worldly Gerry's history is neither here nor there, but his admiration of his wife and hers for him is beautiful to see, and his delight in the children and theirs in him is equally beautiful.

6 April 1956

Gerry and I away at 6 o'clock for the airport, to fly the Kafue Flats with Ted Lenton in the Auster. Ronnie and Erica had asked us yesterday morning if we would drop some meat for them as we flew over the ranch and if we would ring her lawyer and ask if there was any urgent message about the sale of Erica's big ranch. The lawyer last night had said he had a message, so we called at his house in Governor's Lane for it. He said this opportunity was heaven-sent because a Jo'burg buyer wanted yes or no to his offer by midday. His letter said that acceptance was to be shown by raising of both hands, 'no' by raising one hand. We were in the air by 7.30 and as the morning was overcast and cool we had a steady flight for the first hour and a half. I was in good form till the last hour, when I lost my breakfast as usual. Ronnie was on the lawn as we flew over and dropped the letter. A second run and we dropped the meat. Back again in ten minutes to find Erica on the lawn as well and both she and Ronnie had both their hands high above their heads. As soon as we returned to the airport, Gerry 'phoned the solicitor (10.30 o'clock) and so Chikupi has been sold. We passed over it on our way back to Lusaka, a lovely house and garden, a white-painted stable square and clock tower, and all the appearance of gracious living.

We probably saw most of the 25,000 lechwe which are left in the Kafue Flats, and they do not occur elsewhere. It was estimated twenty-five years ago that there were 250,000. The drop may not be wholly due to persecution by illegal *chilas* or hunts, but also due to parasitism. Most of the lechwe on the north bank of the Kafue were strung along the lagoon in close masses. They ran when we flew over at about 350 feet, but not for very far. The lechwe on the south bank are more or less concentrated now on the Lochinvar Ranch and are further hindered by a fence which the owners (a firm of butchers) have erected. When we flew over these animals I got the impression that they were not in such good condition, and what was particularly interesting, they did not run so readily. Back to Lusaka shortly after 10.30 and am impressed with the badness of the site of Lusaka – in an area of flat wet ground, sewerage almost impossible – even the swell new shops and banks on the Cairo Road have buckets at the back. Only Government House and the best residential area on the slight hill can enjoy what I consider to be the greatest benefits of civilization, indeed almost the only ones. Back at the Grimwoods I felt so sleepy I had an afternoon on the bed. Took them to dinner at the Blue Boar afterwards. This pub-restaurant is out in the country and fills a definite need for dining out. Quite well done.

7 April 1956

All social today. Am invited with the Grimwoods to the Parade of the Northern Rhodesian Regiment which is going away to make room for the King's African Rifles home from Malaya. The weather chose this day to rain with a fine misty rain all the time. The parade was cancelled (I was so sorry for the poor Africans who had rehearsed and drilled so much and who dearly like the Parade) but the reception party afterwards was not cancelled and I had also been invited to that. Jolly nice too: I drank gin and tonic-water and talked hard to Vaughan-Jones who was there, and to Price, the Deputy Secretary for Native Affairs, so much nicer a man than Stubbs, who is the Secretary. We lunched at the Ridgeway Hotel which is on the hill – new, and the very last word in modernity and ton. Anyway, it was very well done if expensive, and the African service was good. They must have put in a lot of time training them to this pitch.

8 April 1956

Sunday and nothing of importance. The Dinwiddies were supposed to come to tea, but only Mrs and two children came. She is a daughter of John Bartholomew the geographer, and he is a son of Dinwiddie of BBC Scotland.

9 April 1956

Away with Bill Steel, the tsetse entomologist, to the Zambesi Valley, nearly 90 miles away. The Zambesi is crossed at Chirundu by a fine suspension bridge. The river was 420 good paces across at this point and is a quite magnificent river, tree-lined as far as I could see, silent and strong. I walked over into Southern Rhodesia just for the sake of it and to look down the river; two and a half miles on the Northern Rhodesian side we struck off westward in the Land Rover along a poor track and in 9 miles came to a Tsetse Fly Supervisor's post. He is the man on the ground who keeps check on the tsetse position with the help of 34 fly boys. His station looks out to the Zambesi, which is about 1,300 feet above sea level at this point. We had dropped through the escarpment hills for 3,000 feet to reach the river. We then went out for another 7 miles farther into the bush to the Lusito River, which we waded and then walked about 5 miles to another river now dry. We passed through a good deal of *mopani* scrub cropped by elephants, and occasionally small patches of closer, more riverine-like vegetation which were apt to be tsetse fly loci. We walked down the dry watercourse and observed some spectacular water erosion of 20-foot red-earth cliffs on the western side. Elephants, cows and calves, had been playing in the sand the night before. We saw no game at all. Was so glad of this 10 mile sharp walk in the sun. No hat or shirt now, as I am quite acclimatized to the sun. Home 7.15 p.m. and Vaughan-Jones and his daughter came to dinner.

10 April 1956

Down to Mazabuka today with Parnell and a veterinary surgeon to attend a meeting about the influence of fish traffic away from the Kafue River on the spread of tsetse fly. It is evidently a considerable factor, and as the fish traffic from camps on the Kafue River has grown up mushroom fashion in the last five or six years the danger grows. The camps have sprung up in the National Park because fishing rights were not extinguished and Game and Tsetse Departments are keen to get rid of the traffic out of the heart of the Park. The Veterinary Department is equally keen, as outbreaks of trypanosomiasis are getting much commoner in what would be fly-free areas. Once more the Provincial Administration is slow to act lest they upset the Africans.

The Director of the Veterinary Department, James Swann, took me to see the Station's herds of cattle. The Department has 90,000 acres, 4,000 being on the Mazabuka level and the rest going gradually to the Kafue Flats. I saw a nice herd of the big rangey Barotse cattle and some nice Kenya zebus. There was also a herd of cream zebu type, Borans from Kenya. All these East African zebu-type cattle are originally from India. The Indian influence on East Africa is profound. They were there long before us. Even the general language of Swahili has a great deal of Indian in it. These Boran cattle were really beautiful and they thrive well here. What on earth people want to fiddle about with European breeds for, I don't know. Mazabuka and district has the best farming land I have seen as yet in Northern Rhodesia.

11 April 1956

Long talk with Vaughan-Jones at the Government Offices on policy. Then to lunch with Magnus Halcrow, the Deputy Director of Agriculture. He was a contemporary of mine at Edinburgh in the 1920s. He had also invited a District Commissioner called Button, who is one of the few conservation-minded people in the Provincial Administration. So more policy talk. To Mount Makalu in the afternoon with the grasses I collected in the Luangwa Valley. The Director and the ecologist seemed quite pleased with them, but the whole press had suffered from inundation.

*   *   *

Thus ended the account of Fraser Darling's first tour, and it leaves a curious impression of having been on the set of an epic film re-living the experience of Livingstone. Within reach of four-wheel drive vehicles and light aircraft, Frank had what appeared to be a twentieth-century trek in the African bush in a surviving enclave of nineteenth-century Africa in the Luangwa Valley. To have covered the same route in the dry season, however, would have been very different. There would have been none of the excitement and danger of fording swollen rivers or trekking endlessly knee-deep through inundated flood plains. Where there were only foot prints in the mud there would also have been tyre marks in the dust. The illusion of Livingstone's Africa might have been shattered. Yet this was not a rehearsed piece of theatre, and there were no stunt men or 'doubles' for the main characters in the action. Frank provides word portraits of his travelling companions, giving an extraordinary light and shade of character. He took a distinct pleasure in watching the antics of the African porters. He identified with them and laughed with them in a way that was not open to his expatriate ex-military companions, who did not quite know why. Frank kept his illegitimacy and underprivileged upbringing strictly to himself, but he unconsciously revealed it in the respect he had for poor people who wore their culture with dignity, and took delight in sharing a joke with them. He loved animals. Around him at home he had dogs, a cat, peafowl, doves, and bantams; the porter who acquired a dog in the crux of the trek was a man after his own heart.

The shooting of the buffalo and elephant were awful for him. Eustace Poles could be forgiven for thinking that Fraser Darling – the man whose name was more than any other, except Landseer, linked with red deer – was a practised deerstalker and a good shot. Stalker he was with telescope, camera and notebook, but not with a rifle. Though he always approved of the shooting of game for the pot and liked venison with a good claret, he was deeply opposed to shooting for pleasure (in which was included trophy shooting). In all my talks with Frank or my reading of his books and papers, I have no knowledge of his having shot a deer. His son Alasdair similarly has no recollection of his father ever having shot a deer during his years of deer study in the forests of Wester Ross, nor does Louis Stewart of the Red Deer Commission who was one of his stalkers in the Red Deer Survey of the 1950s. He was in favour of culling old and ailing stags and hinds, insisting on the rifles having telescopic sights, and was always realistic about the need to shoot humanely for conservation purposes.

Franks's failure to pull the trigger on the buffalo, and his passing the rifle to the African gun bearer to do it for him on 11th March (Poles' account, not mentioned by Darling), chimes with his innate gentleness towards wild animals, and his tender thoughts of the slain buffalos and elephant, as well as for the old African injured in the charge of the dying buffalo cow on 7th March. The latter episode was chaotic, cruel, and violent in comparison with the shooting of the second buffalo on 11th March, which was carried out personally and safely by Poles. However, the buffalo stampede on that occasion caused by the shooting resulted in the drowning of a cow and two calves and possibly severe injury to others (Poles' account, not mentioned by Darling).

The two accounts of these events convey the great difficulties and dangers of hunting big and ferocious animals in thick bush or long grass, and of how easy it is to make serious mistakes. Poles, the Game Ranger and seasoned hunter, gives straight factual accounts of what he saw and did; Darling the non-hunter and animal lover, leaves a patchy and emotional record of experiences which, given a choice, he might rather not have had. He consoled himself with the thought that the elephant was old and that, despite very awkward angles of shot, Poles' aim was true to the heart of the great beast. He reasons with himself rather unconvincingly about the justification of killing the elephant – damage to baobab trees, and the perquisite of the white Game Ranger of two tusker elephants annually 'to pay the fare home to UK on leave' in ivory! As if that was not travesty enough, these hunts were enacted in tribal lands, in the sight of tribesmen who were forbidden to kill elephant and other game species. The warning delivered by Poles to the mischievous and slovenly chief Nabwalia and his tribe for poaching game in their own country took place while the elephant which Poles had just shot was being publicly cut to pieces nearby. The longer term adverse consequences of the arrogance and self-interest of the white hunter in the education of tribal Africa should have been obvious to Fraser Darling. His comment about the big game getting hell from the African muzzle loaders to supply the copper belt butchery and the Asian ivory trade was justified. But what possible reason could be given for a white man's perquisite in ivory, when demonstrable self-denial was being demanded of the African in the conservation of the elephant? The decimation of the African game is caused largely by habitat change exacted in many ways by man, and not necessarily by direct predation to quarry species *per se*. However, in human terms, the science of the matter is underlain by a moral and ethical code of conduct of which Frank was acutely aware. That he did not articulate his thoughts fully in his Journal is made clear in his concluding remark as "Best leave it all at that" (page 40).

Frank was not his great self in these sporting interludes. He was a stranger in a shooting party, but did not declare it. He was a good natured guest of expatriate ex-soldiers who had been attracted to game management because of its implicit connection with hunting, the control of large wild animals, and law enforcement. Theirs was one way into wildlife conservation. Many ex-soldiers, like Ian Grimwood, did great work and were highly honoured by the conservation movement world-wide. However, theirs was not the way of Fraser Darling, who came to the scene through a compassion for nature, a natural affinity for animals, scholarship, and research. It was for these very qualities that he was respected by the game men whose thrill in the hunt he tried hard to share, but failed in his heart to appreciate.

It is one of the paradoxes of Fraser Darling's life that, having so little military aptitude himself, he admired it in others. In his African travels he was hosted by a succession of distinguished ex-soldiers whose personal qualities he openly applauded. He felt flattered by the admiration of these men, and this led him to make mistakes as in the buffalo and elephant hunts, and, more seriously, into the compromising of his judgement when coming to perceive the application of his science to the administration of the African people.

Among his colourful array of hosts in this first tour were Ronnie and Erica

Critchley at the Blue Lagoon Ranch. He clearly thought that Averil would greatly enjoy meeting the Critchleys, and provided her with an especially lengthy and vivid description of them with information gathered by hearsay. Ronnie is now eighty-five and lives in Australia with his fourth wife. Erica died some years ago. His nephew, Robin Pellew, who knew the Blue Lagoon Ranch well, writes: 'Fraser Darling does seem to have regarded Ronnie as something of a playboy, which was probably true, but at the same time he pioneered wildlife conservation in Zambia both before and after independence, including the establishment of the Zambian Wildlife Society. The Blue Lagoon was run with military precision as a wildlife sanctuary with a major anti-poaching campaign to protect the red lechwe until it was handed over to the government in the late 1970s as a national park, since when it has steadily gone downhill, partly as a result of the Kafue River hydro-electric scheme which limits the area of lechwe habitat.'

Frank enjoyed the company of such worldly people and sharing for a short time the largesse of their life in the African wilderness; the same qualities and taste for good living in a different setting which he created for himself at home in the Berkshire countryside. Erica later put an effort into an educational unit in the Kafue National Park, and sought Frank's advice and help in fund-raising. He said in a letter to her dated 1st December 1960, from his private address in New York, that she should prepare an impeccable memorandum and use it as fishing fly: 'Having become aware of several random casts made by several organisations in several African territories over the fishing pool of the United States, I realise that the fishermen have not taken the trouble to learn the behaviour of the fish. You can't do much in this pool with a piece of string and a bent pin. You are fishing a fast running stream for salmon and trout, not a carp pond.'

CHAPTER FOUR

The Kafue Basin and Barotseland

August to October 1956

O N HIS RETURN to Britain, Fraser Darling was faced with a gathering crisis at home. He was at the centre of a web of his own making. A six-week disconnection from his normal life resulted in a backlog of work to be dealt with against his clamant duties at Edinburgh University in the busy summer term of examinations, and those of the Nature Conservancy's Red Deer Survey in times of census and reporting. Above all, Averil was seriously ill. She and the three children (Francesca was only one year old) required attention which he hardly had time to give. His priorities in making further foreign ecological reconnaissances, though applauded abroad, were deplored at home. They were wrong for his work in Britain, and for his wife and children.

Through all his years of shuttling to America and Africa in the 1950s, Averil did not accompany him, nor did Christina in the 1960s. His plans of having Averil to live in a grass hut in the Luangwa Valley must be seen for what it would have been − her first time abroad with him after years of lost opportunities. Then it was too late. The reasons for the double life seem based on his obsession with his own work, and with his sense of being a messianic figure among his peers, who had ascended by his own gifts and endeavours from the pit to the sunlight. Friends close to the family at the time predicted a break between Frank and Averil who, on her death bed, gave the custody of their children's estate into the hands of a trustee and not to Frank.

Frank was justly proud of the way his teaching was being received by those whom he most respected. His students in Edinburgh flocked to his lectures in the first year, but found the second year very much a re-run of the first. The game managers of Africa on the other hand valued his visits for the lucidity of his ecological interpretation of their problems; he opened their eyes to what was already staring them in the face, but they had failed to see. His charisma with the first year students was generated at the same intellectual level as that of the game men. He was teaching in broad principle, rather than in detailed analysis. His ecology was holistic, lit by personal experience on a world scale, and appreciated for its sense of application to human affairs. Whatever may have been Frank's motivations at this period, they resulted in his leaving his family at a time of dire need and of his failing in his two surviving jobs in Britain.

In four months he was back in Africa, in Kenya, for ten days *en route* to his second tour in Northern Rhodesia. The interlude in Kenya relates to affairs in that country, and laying the foundations of his tour in the Mara in 1958. Accordingly, the record of these ten days from 18th to 27th August 1956 is presented in Journal II (pages 149–67). He flew from Nairobi to Lusaka, seeing Lake Bangweulu so much smaller than when he had been there in April, and many grass fires. He was met at the airport by Ian Grimwood who was his host for the night at Chilanga.

28 August 1956

Busy at the Secretariat. The Colonial Office and the Nature Conservancy had cabled for my release at an earlier date than arranged and we had to go through programmes and telescope and cut, and then arrange an air passage back. So I am to come back in 23 hours by Viscount, reaching London October 13 at 5.45 o'clock. Then to see Dr Fosbrook, Director of the Rhodesia – Livingstone Institute. Ian and I had lunch there. Very fine chap. He is the greatest authority on the Masai and he did his best to give me a statement of the position, which should help me to make better judgement when the time comes. Back to Chilanga and Ian asked the Parnells and the Maclarens in for 7 o'clock drinks. I was tired out and had the devil of a headache.

29 August 1956

Away this morning by 7.30 with Ian and another Land Rover and trailer bringing the *kutundu*. We drove to Mumbwa, 132 miles from Chilanga, first through miles and miles of *Brachystegia* woodland and finally down to lower land nearer the Kafue where the soil was richer, growing *Chipea* and other trees. There are even European farms round Mumbwa, more or less a pocket in native lands and game country, and bounded on one side by the Kafue flats. The season now in Northern Rhodesia is dry, but it is also the spring, and the effect is to produce an impression of a delightful mixture of spring and autumn. The atmosphere is that of dry warm September at home, and is most invigorating. The trees are coming out in leaf, some a very delicate green, and some red which ultimately turns to green. Many flowers are coming through the dry brown floor of the woodlands and bush, a beautiful convolvulus that flowers straight out of the ground, a pink and white lupin-like flower, a brilliant geranium-coloured flower and many others. Also, I think, many birds are getting ready to breed, as they are looking very brilliant and there is quite a bit of song. We had lunch at the house of one of the Game Rangers, Peter Whitehead, whom I had met last time. Very typical Australian. Then out into the bush about 20 miles to camp. The party is Bill Steele, the tsetse chief, Andrew Crosbie, a Scottish vet who is gaining experience of what the Game Department is doing in tsetse work, and an elderly Boer called Abram Bothmer and myself. Bothmer is a big quiet man who looks like a Red Indian with grey eyes. He was very anti-British during the war, joining that ox-waggon organization. Game and Tsetse Department took him on for building the 57 miles of game fence between the Kafue Flats and the hinterland, and he was to do buffalo control. Nobody took any notice of his politics, and as he is a scrupulously honest and straight man, the time came when in seniority and ability Abram should have received the position of head of the unqualified tsetse staff. Abram did not expect to get the position because of his politics – no Nazi would have given it him anyway – but he was given it, and he was so impressed with such civilized behaviour that he decided his political notions were mistaken. So he laid them aside utterly and completely, and Abram continues to be a good servant, much respected. Lying under the stars tonight, as I shall do for the next six weeks, until the night before I return home. There is no need for tents now.

30 August 1956

A long day out to get a notion of tsetse control work and policy in this territory. Northern Rhodesia does not subscribe to the game extermination policy, but to a much more subtle and ecological way. The research on tsetse control was mostly done in Tanganyika by Swynnerton and Jackson, and their methods are being applied here. The tsetse fly cannot live through a very wide range of environmental conditions: it is susceptible to variation from an optimum temperature – moisture situation. In nature *Glossina morsitans* favours bush with dense thickets such as form on and round ant hills, and the Lusaka thorn thickets. These thickets are the breeding ground and the more open ground in between is the feeding ground. The situation in this area heretofore was that game migrated to the Kafue River Flats from the tsetse-infested back country. Villages were scattered here and there and the game took tsetse on them from the back country to the Flats where no tsetse bred at all. The technique here was to put 57 miles of game fence across the migration route at a point where it was felt that riverwards the country was either free of tsetse or could reasonably be cleared by discriminative clearing of bush. It is found in practice that disruption of the fly's habitat is the best way to reduce or even eliminate the tsetse concentration. Bill scarcely bothers the game, because the fence has indeed succeeded in breaking the migration, and he finds the discriminative clearing so disruptive of the fly's sensitive optimum that fly elimination has been successful. A certain amount of game within the fence was shot, but the flats buffalo were left alone except for some reduction, a herd of 50 eland were left intact, and such small buck as reedbuck and oribi were left alone. Furthermore, the villages outside the fence in the fly zone which had been riddled with sleeping sickness have now been moved inside the fence, where there is plenty of good land and water bores have been made. The villages themselves are a factor in fly clearance, and now they are able to keep cattle. A further sector of the hinterland is being cleared of fly as a result of discriminative clearing of lusaka thickets and the constant running of transect fly counts to find out the centres of infestation. Bill Steele himself is a Durham Geordie, indefatigable and with plenty of brains but now, as he says himself, no longer an entomologist but a tsetseologist only. We have been bush-bashing all day in the Land Rover, which has revolutionized getting around in Africa.

31 August 1956

More bushwhacking, and a call at Abram's house which he has built himself in the African fashion, of home-made unfired bricks, done over with adobe and whitewashed.

So very delightful with its large verandah, the whole thing deep-thatched. Mrs Bothmer is the big, stout Dutch vrouw, with all the household abilities of such women. When the solemn Abram came into the veranda he went and kissed his wife so nicely and then she brought tea and home-made biscuits. She can only just speak English and apparently this embarrasses her somewhat. Then we took aboard more stores and came further into the bush, spending the night by the Nausenga River. What is almost unheard of, a thunderstorm built up in the evening and there was a shower of rain. It has helped slake the dust. Before darkness I went a walk alone along the Nausenga and watched the birds — jacanas on the water foliage, a small cormorant, a brilliant green and chestnut finch and several kinds of warblers I don't know at all, but their behaviour is so well known to me. One was large with stripes on the back and a long tail. Another was small and very chestnut, with a much used typically shaped warbler tail. The fish here in this muddy stream were barbel. On the way back to camp I kicked over a few termite heaps and saw what a variety of other fauna are to be found — a slug the size of our grey garden slug, grey ladybirds, sometimes other sorts of ants, and other beetles. I also found spongelike structures in the nests of a dirty pink tinge. I know so little about termites.

1 September 1956

Thinking so much about you today. Away early into pleasant sunlit open bush, seeing a variety of small buck and a pair of jackals. Then down to the Kafue Flats. Miles and miles into them by Land Rover, flat as a pancake and the horizon all round you. It was like being on the low tundra of the Yukon — Kuskokwin. The birds are breeding now — some black and white plovers, many coursers, the great bustard and the little bustard, and what was most interesting, a short-eared owl more chestnutty than our own. At the shallow lagoons still left by the floods were egrets, sacred ibises, open-billed storks, the lovely cranes I saw last time at the Blue Lagoon Ranch, and spur-winged geese. You see, the tundra and the flats are the same habitat, only the species are different. Instead of caribou there were the lechwe, but in addition there were several herds of zebra, and once back on the anthill zone before the bush proper, were the gazelle-like oribi. We passed through several Ila villages of cattle keepers, livers on the open flat expanse. Also visited another great reed bed of some thousands of acres called Eeasha, where the traditional buffalo *chila* [hunt] is held.

2 September 1956

Again out in a great sweep, and camping at a delightful spot on the Lukinene River on the game side of the tsetse fence. The river is fringed with *Raphia* palms and I have done a bit of exploring off on my own. This large area of country is now empty of people and abuts on its western edge the Kafue National Park. In effect this is a buffer zone, fairly lightly stocked with hartebeeste (Lichtenstein's), kudu, roan and sable antelope, waterbuck, impala, reedbuck, oribi, and elephant. There is also a stock of 4,000 buffalo. Abram was engaged to reduce the buffalo and he shot 2,000. He tells me that when he began there were very few calves, but now the stock is back again at 4,000 and there are many calves, showing high productivity. Here is the same old story told to me by one who had merely observed the phenomenon and had no notion whatever of how important it is. It seems to me that a population can reach a ceiling and what might be called a post-climax phase, and then if it suffers some disaster, the stock becomes much more productive. Bill Steele shot four guinea fowl with one shot yesterday, a welcome help to the pot, and today he has shot an impala which should carry us through till Tuesday when I am handed on to Peter Whitehead. We have done a big traverse through the bush today looking for the 4,000 buffalo, but no sign. Abram went home yesterday and returned this evening, bringing me a bag of eland biltong from his wife. This must be because I asked him to thank her for what I had had already, and saying how much I had enjoyed it.

3 September 1956

Away soon after 6.30 a.m. and back at nearly 3 p.m. on a rough bush-bash searching for the buffalo. No trace, and it is an interesting problem where they have gone. We saw a dozen or more elephants of the 30 or 40 that must be in this sector, and had some delightful glimpses of hartebeeste, kudu, waterbuck, zebra and warthogs. It is tiring work bumping over the *dambos*, pock-marked with elephant footmarks and buffalo passage in the wet season and now cement hard. Then there are continual termite nests, also rock-hard, and antbear holes and so on. Sometimes you wonder how many discs one has slipped, but the best way to take this travel is to relax and to do little more than balance. The bush is so delightful in the fresh though hot sunny air; sometimes open parkland or savannah, sometimes woodland in which it is easy to get about, sometimes closer thickets like an oak – hazel copse. And you know it goes on for miles and miles and miles. I love it all. I am very fond of biltong: the meat is dry and hard and stiff as a stick; one sits around paring off pieces with a very sharp knife and reflectively chewing away. Abram's biltong is eland and is beautifully made. For lunch when out in the bush it is wonderfully satisfying. Only lean meat is used and must be cut in the direction of the muscle: it is put in salt to dehydrate it and to give it the salty flavour, and the strips are then hung up to dry. The drying takes several weeks to complete. Nothing seems to attack biltong in the way of decay organisms or flies or worms. When the wet season comes it, the biltong, should be put into airtight tins to keep the damp air away. Abram says buffalo makes the best biltong, with eland next and then hartebeeste, but of course other buck can also be made into biltong. I shall bring some home.

A lovely fire in camp tonight and so thoroughly enjoying sitting around in the velvet night. When I wake up in the night I look upwards into the branches of the trees and an optical illusion makes the clear patches solid and the dark patches appear to be the dark sky. It is as if some fire-glow is reflecting up there, and the apparent solid branches and patches have highly contorted forms as if they were the background of a ballet of *Comus*. The other fellows tell me that they also get this optical illusion.

4 September 1956

A very long day today and quite tired at the end of it. First ten miles to the banks of the Kafue River where I had seen on Saturday a canoe being made from a felled acacia tree. I wanted photographs of this, and now two men of the Lozi tribe of Barotseland were working on the dug-out canoe with axes. These men used to raid on the Ila tribe, and now under the Pax Britannica they have fishing rights on the Kafue. These two men were now making their canoe in Ila territory. They were a most polite couple and expressed their pleasure that someone from England should take an interest in their boat. The tree is felled and the trunk sawn clear. Then as it lies, the rough outside shape is fashioned with axes. Then the trunk is turned on to the future keel and further shaping of the gunwale is done. The hollowing out is done in small sections, at first leaving several bridges across from gunwale to gunwale. Later these are cut away.

Then back to camp which had now been struck and we set forth northwards to Mumbwa, nearly 50 miles along a bush track. We saw scarcely any game, though the range is excellent; but as the whole area is uninhabited it will doubtless build up a good game stock. Abram and Bill want to know very much where the 4,000 buffalo have gone, and for that matter so do I. Our journey was through lusaka thicket and *Chipea* woodland and then into *Brachystegia*, with occasional kopjes of red granite, the rock of this area, very coarse. Reached Peter Whitehead's for lunch and had cold roast wild pig. Absolutely delicious. Had a delightful bath and then Peter and I set forth for Myukweukwe in the Kafue Hook. There is a ranger camp there just inside the Kafue Park, but in a part from which the public is excluded. The Kafue Park is 8,000 square miles and fairly remote. The public is confined to a few defined roads and the game enjoys the rest. Myukweukwe is 90 miles from Mumbwa westwards, and along a fairly straight road. We saw some sable and some roan antelope near the road after the sun had fallen. The sable is really a magnificent antelope, the bull after five years becoming quite black, with some white on the face. On Sunday evening at the same time we saw three very good kudu bulls. They are tall, and their horns very showy. Too dark for photography. We reached camp at 7 o'clock and had a beautiful meal of fish, bream fresh from the Kafue, caught by the Game Guards. Lying in bed tonight I could hear the hippos bassooning in the Kafue River immediately below the camp. The country is woodland of mixed kind with sparse underbrush, and back along the road were some large *dambos*, i.e., the damper hollows grassed over and devoid of trees. Just now the *dambos* are dry and their silty grey gluey soil is set hard after the passage of elephant and buffalo and the attention of ants. They make very rough going on foot and worse in a Land Rover. I have done miles and miles of this *dambo* bumping this week and if I have any vertebral discs unslipped I am more than lucky.

5 September 1956

Out with Peter Whitehead by 7 a.m., walking in the Hook of the Kafue. We saw several small herds of puku and impala, and saw spoor of a few buffalo and a few elephant. There were plenty of hippos to be seen in the river, which is very fine here, nearly 300 yards wide, the second river of Northern Rhodesia, the Zambesi of course being the first. I thought the herbage floor of the Hook very poor indeed, practically bare just now, and I was not surprised the game was so sparse. The trees are not magnificent except along the great river. We heard lions roaring on the other side of the river. Tsetse flies were numerous. Back at 1.30 p.m. after twelve miles or perhaps a little more. We spent a delightfully lazy afternoon sunbathing and bathing in small lagoons of the Kafue. We can't bathe in the main body of the river because of the extreme plentifulness of crocodiles. I found otter sign on the rocks and saw several fish eagles, cormorants and black ibis. Also, across the river, we saw a leopard on an island. He saw us as well and coughed in his antagonism.

The sable antelope is a creature of the upper drier woodland; the allied roan is more catholic in habitat. It may be found almost anywhere. The hartebeeste or kongoni is also ubiquitous but prefers the woodlands nearer the *dambos* and the edges of the *dambos*. The puku likes river edges, whereas the impala likes the woods proper. The water buck is never frightfully far from water but it grazes well out from the rivers. The same applies to the reedbuck. The oribi likes the open places and drier. It was common in the anthill country behind the Kafue Flats.

6 September 1956

A heavy and gruelling day from 7 a.m. to 6.30 p.m. We struck eastwards and then northwards and did nearly 100 miles of bush-bashing, most of it through woodland of mixed kind and patches of lusaka bush. The tsetse flies were terrific all day. The range was better here but game was very scarce, a few sable and roan, a few impala, kongoni, waterbuck and reedbuck, and one or two zebra. Peter cannot understand where the game has gone. I too should like to know. We came to the Kafue River where last year Peter was trying to fix a pontoon. The whole area here is uninhabited and unvisited. After a lunch of tea and biltong, we set forth through 10 miles of country which is supposed not to have been crossed before. We relied largely on the sense of direction of Peter's Africans, but they confessed themselves lost ultimately and were very depressed. The going was the worst I have ever had, and Peter admitted to the same. We had one long dig-out of the Land Rover, and saw a hippo walk down to the river. Then we passed through a lot of thick bush and came to a *dambo* two miles and a half wide. We set off across it, dunting through dried elephant tracks and then the usual *dambo* surface of dried out vegetation clumps covered with cemented anthill earth. Here the whole thing was exaggerated and our passage was painfully slow. Once over we were into such thick bush that we had to use the axe. On again till we reached a stagnant stream and wove our way along the bank between the trees. Finally we came out where we had intended, at a bit of the river called Kafwala, where there is white sand on the bank, some rocks and rapids and the great beautiful braids to two miles. Peter thinks the current a little too swift here for crocodiles, but where we had lunch you occasionally saw these unpleasant creatures raise eyes and snout above the water and look at you. In all this passage we saw no game at all. Once we came out where we had hoped to come, the Africans began to giggle and be extraordinarily cheerful, just like children. From here it was a long and tiring journey back to camp, about 40 miles. This morning we nearly ran over a green mamba, 8 or 9 feet long. It rose up to about 4 feet and came for the car. Peter backed and I tried to get a photo, but unsuccessfully. It went off into the bush much faster than I could run, still reared up. It is certainly a most impressive snake. Think of the thousands of miles in Africa I have had to travel before seeing this creature. Yesterday I saw a 2–foot iguana drop off a tree into the Kafue. Peter is about done tonight. He is an Australian of 33 years, formerly a horsebreaker and rodeo rider. Had three bad falls necessitating long periods in hospital: broke his neck and shoulders, and three vertebrae are now fused as one. Flew in the South Pacific during the war. Now very happy as a Game Ranger here. Once had a wife but the thing came unstuck. Very likable boy, good clear eye in a handsome face. His colourful speech very like Jock Marshall's. He is kind to his Africans. Have enjoyed his company.

7 September 1956

Southwestwards this morning to a cracking good diesel-engined pontoon over the 600 yards of the Kafue. What a magnificent river it is! The pilot was a nice halfbreed. After a total motoring of 25 miles or so we walked for three and a half hours in lovely park-like bush with incipient *dambo* formation. Again practically no game; just a few waterbuck, impala and oribi. The range was of moderate quality and the large trees had not suffered unduly from the elephant. Had a wonderful bathe and general clean-up in the rushing Kafue below the camp this afternoon. The food situation wasn't too good this morning but Peter has rectified this by catching a beautiful large bream from the river and shooting a young hartebeeste bull tonight. So we dine in style. Hippos doing their basso profundo tonight.

8 September 1956

Spent a solid five hours doing correspondence about red deer, Serengeti and what not. Heartily tired of it and the only saving grace that I could sit in the sun with my shirt off. Had a cup of tea, some paw-paw and an inch or two of biltong and went out for a walk by myself down river for two and a half hours. Thoroughly enjoyed pottering about, especially when I saw some hippos across the river 500 yards away. I made a few lion roars at them without effect, but a little farther up river from where they were lying with their ears, eyes and nose out, I certainly affected about 20 hippos who had been ashore beyond the fringe of metate reeds. They splashed into the water and felt safe. Then I started making various grunts and nonsensical sounds from my tree on the bank. The interest of 30 or 40 hippos was immediately obvious and they came gently over the great river to investigate. I continued the solo and got the lot in a great semicircle from 15 to 50 yards out from where I was in the shade. Sometimes I varied the sounds to squeaks and my well-known signature squeak, and you could see the immediate recaptured interest. Occasionally they yawned and showed the great cavern of their throats. I left them to it. The banks of the river seem to be a series of hippo docks where they come from the water and this seems to help to widen the bank zone of metate reed and a tree that has roots in the mud but the upper parts of the roots are bare because of the action of the flood when the river is up. Where the river is swift between the barrier rocks of granite there are no crocs or hippos, but as soon as a calm stretch comes both are found. I then went away from the river into the bush where there seem to be *mopani* among the *Acacias*, *Chipeas*, and *Brachystegias*. I saw several puku and bushbuck. One of the latter, a buck with a nice head, let me up to 30 yards simply because I kept on squeaking at him and the experience was new enough to hold his interest. The hippos come half a mile inland to graze, and in some places more. Plenty of evidence of elephant, but not within the last week or two.

This has been a good camp and I am not very anxious to move on. I have a feeling I may not like the company so well as up to now. Peter has been telling me stuff about administrative affairs in his District, and once more I cannot understand what the Colonial Office and its Governor here are doing. Since I left in April he has personally ordered the shutting down of rural police posts. The result is proving disastrous and the fact is not being lost on the African. There are one or two Chiefs indoctrinated with communism and sedition.

9 September 1956

Away this morning by 7.30, first to Kafwala which, although a name, does not signify anything else. It is just a point on the Kafue River where the river tumbles through a rock barrier. Then a further 20 miles of bush-bashing of trying kind to reach the point on the Kafue opposite the Lufupa confluence. We depended on our two African Game Guards to steer us, but I had that curious feeling they were going wrong and taking us too far up the river. We stopped on the bank and Peter and I had lunch of tea and biltong; the Game Guards walked still farther up the river. It was quite delightful sitting around in these parklike surroundings, with herds of puku grazing in the distance like fallow deer. The trees are very good here and add so much to the scene. Peter and I then moved down river two miles, looking for a camp on the other side wherever the riverene vegetation gave us chance. At one point a hippo had just come out of the river and I took some photographs of him very close. One side of him was badly slashed and I imagine he wanted to get out of the river for a while so that the fishes would not be biting his open wounds. The range here is pretty good and not over-used. Fair sign of elephant. Farther back on this up-river move from Kafwala I had noticed what is so common in a dry-season country, the over-use of the half mile strip beside the river. We heard a shot from across the river and replied to it. Then we saw the camp, and soon a youth called Barry Shenton, a Game Ranger, and a little dark South African called Ernest, an assistant Game Ranger, came over in an aluminium dinghy for us. The view up the wide river from the other side was superb, a fine calm stretch tree-lined. It is intended to site a small one-party tourist camp here when a moderate track has been made to the place. So far, as on the other side, the going is only fit for Land Rovers and it breaks them up. The tsetse flies have been quite dreadful today. Shenton has nets out in the river and we have plenty of fresh bream for supper. Peter also fished at dusk with a spoon and spinning reel and caught several silver barbel. The labour squad here, for the rondavel huts and so on, are being fed on mealie meal and fish. There is a fish eagle's nest at the very top of a 50-foot tree on the other side of the Lufupa. It contains one young one. Awake a good deal in this night, having a bad round of sciatica and thinking about you all at home. A lion roared far away, a lioness purred very near at hand, and I think she must have had cubs along because there were smaller squeaky noises. The frogs also piped incessantly. Was dreadfully homesick.

10 September 1956

Yesterday when Peter and I were going along the river bank, a bee decided it was angry with me. I tried to get rid of it, running and what not, but no good. Peter also clouted it, but it came back and stung me on the bump behind the ear. Peter removed the sting but it has hurt ever since and ached a good deal today. We struck camp by 8 o'clock and moved north up the Lufupa. It is a good river for water birds – fish eagles, cormorants, darters and several kinds of herons including the Goliath heron, which certainly is huge, with an impressive wing span. The river banks get low in places, almost horizontal, and at such spots we saw hippos out of the water and crocodiles sunning on the banks. Crocodiles and hippos do not seem to be antipathetic to each other. Egrets, ibises and storks walked delicately among them. While we were looking at a mineral spring there was the sound of a Land Rover and along came Ansell, the Provincial Game Officer, who is the superior officer of these two chaps I was with. Peter, who left us at 6.30 this morning, had warned me that I might not like Ansell, and immediately I did not like him – a Cornishman who struck me as being a Welshman, with eyes too close. A voice like a rasp and never stops talking, even when he eats. His actions are quick and not as economical as they might be. When he puts a hat on he looks like a German about the eyes. When he drives and you are sitting beside him, he never stops talking and looks at you almost all the time. The result is that he wanders about and gets his wheels into elephant tracks. There is something almost crazy about him. I am supposed to be with him for eight days and he tells me he is going to tack on for part of the tour I am doing with Ian Grimwood. We came up to Moshi, a small rest camp for one party and where this Ernest Tulyard is living in a grass hut with his wife and child, a girl aged $2\frac{1}{2}$ years. The wife is real Dutch who has studied in Italy and has a doctorate. She is very brave and keeping her form well. Poor Ernest was a staff sergeant in the South African army, born in South Africa and Anglo-French. He was offered a commission if he would change from English Church to Dutch Reformed. Standing to attention before the selection board, he made no answer of any kind, and they got the idea. Thereafter he got out of the Union as soon as possible, both from army and country, and has got this job as a junior game ranger in Northern Rhodesia. The poor man is most anxious to learn more about wild-life management. His little wife provided lunch for us, including an apple strubel and I did my best to show her deference. Her husband was away with us last night and will be away for several nights, and last night the lions were around. Not that the lions are any trouble, but a woman from a different set of conditions altogether dropped into such a position with no man of any kind around. We came on after lunch to Mtemwa, another one party camp. All the way from the Lufupa – Kafue confluence we have passed herds of puku up to 50–80 strong. They are whole-coloured bright chestnut in colour, rather larger than roe deer and a little smaller than fallow. There have also been a few kudu, impala, reedbuck and roan antelope. Here and there we saw a few wildebeeste. You may have noticed I have not mentioned this species until now, the reason being that the Kafue is a barrier to them. They are not found within the great bending of the river which flows from east to west, then north to south from the Hook, and then west to east until it joins the Zambesi. The discontinuities of distribution of African game are striking and puzzling. There are no giraffes in Northern Rhodesia except in the south-east of the Luangwa Valley. At Mtemwa the Lufupa is tree-lined and full of hippo, and the country round is green and parklike, some of the trees being enormous figs, under one of which I spent the night. This country is the beginning of a flood plain 25 miles by 5 miles, including a relatively small swamp which is the beginning of the Lufupa. The area is called the Busanga swamp on map, but the Busanga is really this lovely plain on which many game species graze and many birds breed. This year is anomalous. There was a record rains in the spring and they continued a month later than usual. Therefore the country has not dried up and the game has not concentrated as usual on such chosen pastures as these. It is still in the bush in small parties.

11 September 1956

We left Mtemwa by 7 o'clock without breakfast to go up to the Lushimba camp in the middle of the Busanga Plain. There was trouble in getting up there because certain patches of the plain had not dried out. Ansell was almost unbearable, rasping away as he drove me along and almost spoiling my enjoyment of a magnificent group of roan bulls and a dozen or twenty buffalo. Indeed, by looking at me he landed three times in elephant footprints and at that point I got out and said I would walk the few miles to camp and follow the wheel tracks. We have had to dream up some obscure trouble with my back in order to get away from Ansell, and Barry Shenton plays it up beautifully. So I had a glorious five mile walk up to the point where the Land Rover could go no farther. Ansell was away prospecting a route to the camp. We got a poor breakfast by 11.30 a.m. Ansell has not brought a cook with him, only a skinner. Ansell has by hard work made himself the authority on small mammals, especially rodents in this territory. He has done a splendid job of work and is continuing. I wish the work took more of his demonic energy. I then saw a heliograph from the camp made by Barry and Ernest with the driving mirror. They had set off much later than us and had got right to the camp by another route. I walked over and had some more tea and arranged that we should come over here to sleep. Ansell, for some reason best known to himself, had engaged 16 carriers and these had made their way up to Lushimba. So I took over a dozen of these to Ansell's Land Rover and loaded each man with stores and *kulundu*. I rather enjoyed this handling of a bunch of Africans on my own and I had them all smiling. The day, in fact, was pretty well wasted. But I have taken command of the situation and am going to run the next week or so my way. To start, I have decided to spend an extra day on the Busanga Plains.

12 September 1956

Out by 7.15 with Ansell to count the red lechwe population on the Busanga flood plain. This place is an outlier of the species. Ernest came too, and a Game Guard, and as we moved in single file I managed to keep Ansell quiet for nearly two hours. The plains are rather like the Kafue Flats in miniature. We walked out over the types of ground which make up this habitat. Our camp is on two or three anthills on which grow fig trees, a few acacias and scrub. Then down to burnt upper part of the flood plain, rather akin to mild *dambo* conditions. Down almost imperceptibly to a heavy laid grass stratum, very wide. As one treads over it, it goes down to ground level and is tiring to walk over. Then to the slightly inundated, but you can't see that it is until you tread into it. Then into thigh deep inundation with the grass layer still floating as a dense mass. It does down under you; you don't go through it. Then the few patches of open water with worn banks where the birds congregate. These patches are but a few yards across. Twice we saw large crocodiles lying on the banks and they slid down below the floating vegetation. Somehow, this observation does not add to one's enjoyment of wading thigh deep in this stuff. However, I am still here after a good deal of it. The birds are a delight, large flocks of black open-bill storks, flocks of large white egrets and flocks of small white egrets; flocks of black and white ibises, some spectacular saddle-bill storks, pairs of crested cranes, of wattled cranes, and some big blue herons. A black and white kingfisher hovers and fishes like a tern over the small channels of water. There are two or three kinds of plovers, one of which makes a noise like a stone curlew, and there are pratincoles that are tern-like in shape. Overhead fly a few kites and Battaleur eagles and an occasional fish eagle. There is also a lark-like bird on this flood plain but I have never got a good look at it. Sometimes a flock of whistler ducks wings by making their quite charming little sound, and at the edges of the pools are little flocks of pink-eye ducks. Both species are darkish brown in colour and not spectacular. There is a small stock of lechwe on the Busanga Plain, now well protected. We set out to count these and found 155 in all, but as there were less than twenty males it is probable that there are a few more males in a flock somewhere. Apparently there is a true papyrus swamp twenty miles up river from the Busanga and a few lechwe there. The lechwe are pretty antelope and the size of fallow deer, orange red in colour with white below; the feet are adapted to the soft and often flooded habitat. When they run, it is a short-stepped ambling trot, the head being held forward and downward. We also saw a herd of twenty buffalo bulls, those lovely black cattle of Africa, that could play such a useful part if they were truly valued. As it is, they will go with the rest of the game unless a true conservation policy is accepted. We got back to camp at 12 noon, having walked hard. I put on the pace in order to kill Ansell, who is in soft condition. The result was what I had hoped. He rested during the afternoon while I got out northwards with Barry Shenton and Ernest Tulyard. We got into a phase of the Plain where there were numerous large anthills making islands on which grew *Raphia* palms and a few trees and scrubby tangle. In between the ground was soft and at one place where the Lushimba stream passes through we had to walk almost waist deep for 300 yards through *sud*, the vegetation floating as a closed mat on the surface of the water. The area was quite empty of game.

* * *

The only person to whom Fraser Darling took a dislike in his African travels was Frank Ansell who now lives in retirement in Cornwall and has read this account. As is apparent from the Journal, Fraser Darling was 'warned' about Ansell by Peter Whitehead before the two met. A check has since been made with others who were present at the time, and it seems certain that in the two Franks there was a direct clash of personal styles. The clash was obviously kept within bounds. Barry Shenton, who still farms in Zambia, writes of that trudge which the two Franks and he had across the Busanga flood plain:

> FD never slackened his pace, and it was all I could do to keep up with the great man despite my youth. Ansell on the other hand dropped back, and fell behind and returned to camp. Though I had become aware of the building tension between my two companions, it was not until we were relaxing in the woodland and FD remarked on the sudden peace and tranquillity of the moment, that I realised what was behind the blistering pace which he had set across the flood plain.

As is also apparent from the Journal, Fraser Darling held a high respect for Frank Ansell's scientific contribution in Central Africa. Rightly so, for Ansell rose to become Deputy Director of Wildlife and National Parks Service of Zambia, and was the author of four key books and twelve major scientific papers on African mammals. In his Journal in the midst of much personal invective Fraser Darling states that Ansell 'has done a splendid job of work [in studies of small mammals] and is continuing.' In *Wildlife in an African Territory* he stated that 'one Officer [Frank Ansell] has already done remarkable original pioneer work on small mammals, and as he is the acknowledged authority in the Territory on a little known but ecologically important group, he should be allowed to follow this work for more of his time than at present.' It is clear, therefore, that this personality clash was not allowed to damage the true value of Ansell's contribution to natural science in Central Africa, and the high regard in which he held him as a scientist.

Frank Ansell's 'basic' commissariat was probably at the root of the matter as he now writes:

> I plead guilty about the catering – no matter how rough I lived in the bush myself I should have made proper provision for such a distinguished visitor. But if he had anything else against me why

didn't he have the guts to say so to my face? I have never worried much about what anyone thinks of me and do not object to honest criticism. But I do appreciate fairness and integrity. Such scurrilous writing reveals more of Fraser Darling's character than about mine – the biased invective, for which I alone appear to have been singled out, betrays a meanness of spirit quite unworthy of anyone in such a position. Darling's professional stature remains but clearly there is need for reassessment of him as a man.

Frank Ansell had the misfortune of being at the receiving end of Fraser Darling's wrath. Had he known him better Ansell would probably have acted differently. For, despite his great enjoyment of pioneering in wild places, Fraser Darling relished his creature comforts: for example, in his Scottish days life in remote places was greatly enhanced for him by – and perhaps might never have been tolerable without – 'Bobbie's' good housekeeping and cooking, followed on occasions by a good cigar. He had a short temper which he regarded with resignation. He felt that such was inevitable in one who, though illegitimate, had Spanish roots in a forebear who had fought in the Peninsular War. He was a man of mature taste, as any visitor to Shefford Woodlands House would know, and was clearly appreciative of the special touches of hospitality and gracious living in the bush extended by most of his hosts.

*   *   *

13 September 1956

Ansell has got a nice burst blister on his heel, so he has had to stay around camp all day – a great relief to us. Typical that he should come unprepared even for blisters, but I was able to patch his heel with Elastoplast bandage and get him comfortable. The two other chaps were delighted because they want to get out with me and ask questions, which is quite impossible if Ansell is there because he never stops talking in that awful rasp of a voice. We walked westwards across the Plain to the bordering *Brachystegia – Isoberlinnia* woodland. The reason I wanted to come here was that the wind blows from the east throughout the dry season, which means that when the Plain catches fire the fire goes smack into the adjoining woodland and produces the typical fire climax *Brachystegia* forest on which I commented so often during my first tour. Shrub growth was almost non-existent as such, herbs were few. We saw a score of wildebeeste with 5 zebra and 4 roan antelope bulls; nothing else. A Martial eagle flew after a lesser bustard for a mile or more and nearly caught it but the bustard lifted each time the eagle stooped. Back by noon and after lunch away eastwards over the plain to the woodland bordering it on that side. We had to walk through water calf depth, but the surface looked most gloriously green. Still another similarity with the tundra on the Arctic Ocean in Alaska, where the lovely green sward is ankle to calf deep in water which cannot drain away because of the permafrost in the ground beneath. The mile and threequarters took 45 minutes to cross to dry land again. Here as usual were a few anthills and finer grasses, a 50–100 yard zone of *Erythrina* and *Parinari* acting as a skirt to the forest proper. The object lesson I had planned for my two young friends worked according to plan and I felt rather pleased with myself. I should have looked rather an ass if it hadn't. Anyway, as I had hoped to show, here was a woodland of the same type as that across the plain five miles away, but being to windward of the plain it did not get the fire rushing into it. This woodland burns sometimes but not often. As I have said, there was the protective skirt of *Syziacuum*, and the forest itself was quite beautiful, the trees good and not malformed, and in greater variety than on the other side. The shrub layer was 5–6 feet high and dense enough to hide game at 50 yards. The herbaceous growth was also richer than across the plain and there were woodland grasses. I saw two species of butterflies I had not seen elsewhere. This woodland is within the area of the Kafue National Park and I asked these chaps particularly to keep this place free of fire for as long as possible, and should they have botanists coming over, to show them this lovely contrast. I asked Barry to explain the situation to the Game Guard. He began so to do by asking the African if he knew why the two kinds of wood we had seen that day should be so different. God made it that way, he supposed. Barry continued to explain and you could see the idea getting across. This has really been a splendid day.

14 September 1956

Leave Lushimba this morning and down through Mtemwa to Moshi where we had a bite of lunch with Mrs Tulyard. Ernest is staying behind with her. The poor woman is cracking; a hyena came into her grass kitchen one night this week, and yesterday she found a small snake in the middle of the floor. Personally I think the Game Department has slipped up badly just dropping them out here without a house. On then another 90 miles to Kasempa through uninhabited woodland. Kasempa is a boma, set among wooded hills and very pleasant. It is in the Kaonde country. They are not much of a people and this whole area holds less than two people per square mile. The Kafue Park itself is 8,000 square miles and uninhabited. We spend tonight in the Government Rest House, a little whitewashed room, a bed and chair and table. Adequate; seven shillings and sixpence the night. Cicadas going hard and the nightjars.

15 September 1956

Busy writing till midday. Away after lunch to Chizera, 101 miles, through continuous woodland with only occasional villages. Ansell went on ahead with his Land Rover and trailer, and Barry and I followed in Barry's Land Rover. Twenty miles from Chizera ballrace in the nearside front wheel ground to bits owing to grease not having got through to it. We spent an hour and a half taking the hub down, discarding the wheel bearing that was chewed up and putting everything back so that the wheel ran on one bearing and a flange. We then drove at 10–15 miles an hour and had no further trouble. This mishap put us into the dark and I was much interested to watch the pennant birds – nightjars – which were on the road and flew up and away only just in time. As it was, one female just lightly hit the wind screen and was dead when I picked her up. The male only has the black and white pennant feathers streaming from the secondaries of the wings. I saw one displaying on the ground to a female, pirouetting it seemed, and wings spread. The wings are long. The stone coloured plumage with chestnut spots is lovely. We spent the night at the rest house. Ansell was comfortably installed and had just finished his whisky when we got there. In another couple of hours he may well have sent a driver back in his Land Rover to see where we were.

16 September 1956 (Sunday)

Away after breakfast to Kabompo, 75 miles. The road quite awful; everything aboard Ansell's Land Rover and trailer. Ansell drove with his usual bad judgement and after 27 miles he broke the drawbar of his trailer completely and utterly. So we left his skinner along with it to look after the gear and came on with a little gear and Barry to cook, over the pontoon at Loloma (Kabompo River), and to Kabompo boma. Met by Game Ranger, Paddy Dunn, a most likable type, immediately dubbed in my mind as 'master at a public school'. Born and reared at Darjeeling, Royal Naval Volunteer Reserve in the war, on Atlantic run and torpedoed four times; distinguished service; then seconded to Indian Navy; went alone aboard a mutinous cruiser at Karachi which was firing on the town. Talked them out of it. Then giving general education to young naval cadets — so I wasn't far wrong in my thumbnail assessment. Entertained to dinner by the District Commissioner, Herbert Stewart. We had managed to get rid of Ansell by sending him back with his Land Rover the 58 miles to the trailer to get food and gear. Left at 3 p.m. and was back by 7.15, so he must have got airborne or something. Naturally he was back in time to dominate the conversation. We had all groaned when we saw the lights of his car come on to the boma. I had gone out in the late afternoon with Paddy Dunn to the Chigarta Rapids of the Kabompo River. Fine rocks and pool. I caught a tiger fish with Paddy's rod and spinning reel. These fish have formidable teeth, and are very fast, and fight hard when on the hook. The country is still woodland: villages rather more numerous, of immigrant tribes from Angola. The people file their teeth to points and are pretty primitive. Witchcraft is supposed to have a firmer hold here than anywhere else in the Territory. Only so-called Christian Mission is Plymouth Brethren. These types come and set up as traders, make a packet and get out. The one here has had three convictions for flagrant breach of the game laws. Have met the Education Officer of the province, J.H. Mackenzie, fairly newly out from Stornaway. Was a classmate of Colin F. Macdonald, Murdo Macdonald and Callum R. Morrison, a very nice Hebridean, son of the minister at Gress, Isle of Lewis. Looks 50 and is about 40. We talked together and he told me how proud Colin was to have been our best man and the godfather of Richard.

17 September 1956

Land Rover for 25 miles northeastwards and then a long walk up a *dambo* narrowed and almost closed and we struck left-handed into the woodland which had changed character from the open fire climax *Brachystegia* to *mavunda* forest. The main tree species is *Chrystoserphyllium* and there is a dense undergrowth of shrubs. This forest is really an outer of the Congo equatorial forest and is on the dry Kalahari sand formation, with a rainfall around 50 inches. The *mavunda* forest is really the dry counterpart of the *mushitu* which I described in my last tour. Fortunately there are many square miles of this forest left, stretching northwards to the Congo. It is the chosen retreat of the elephants and the buffalo of this whole area, and a pretty safe retreat it is. Paddy Dunn tells me there is a dry season migration of elephants eastward into the Lunga game reserve, a place very remote chosen specially for its habitat value of *mavunda* forest, with some open plain country as well. There is also a local south – north migration about October–November by elephants to the Maiyaw area for the *masuka* fruit. As regards the main eastward trend, the elephants come from as far west as Portuguese Angola. By the way, the *dambo*-cum-river bed we had come up was called Makalandende. On the way back I was so pleased to see a pack of wild dogs, 9 of them. They were at a waterhole in the *dambo* and on seeing us they rose and half barked half growled at us from 100 yards but were otherwise not much troubled. They are about 26–27 inches at the shoulder, black and chestnut in colour with vague whitish markings along the back. The ears are rounded and held up and are black with chestnut inside. They moved to our side of the *dambo* and lay down unconcernedly, but as we did not go away – they lying in our path – they thought they had better go into the bush, where we saw their heads towards us through the grass. These dogs are diurnal and must have a much greater influence on moving game about than the lion. Unfortuntely there is still a £2 bounty on wild dogs. The weather is hot here and when you are on the red sand the heat comes up at you as if you had opened an oven door. Paddy and I had had enough by the time we got back to the car. The evening at the Dunns, who have a boy just two and a girl baby of four months. Rather far out in the bush for Mrs Dunn's peace of mind, I think.

18 September 1956

Ian Grimwood arrived this morning after three weeks out on a bird-collecting trip. He and Birdy Benson are doing the new check list for Northern Rhodesia. He came with the drawbar and the springs of his trailer broken. As luck would have it, there was the wreck of a trailer on the boma at Kabompo so Ian's driver cannibalized it thoroughly and we shall be set up completely. What a blessing! otherwise our tour would go to bits. At 2 p.m. we said goodbye to Barry Shenton and Paddy Dunn, and by cutting out going to any more *mavunda* forest we are going west to Balovale on the upper Zambesi and then westwards to the Portuguese border almost. Ansell is coming with us, but the presence of authority in the shape of Ian Grimwood seems to quieten him a little. We covered 77 miles of atrocious road through woodland which became more dry-forest in type as we got further into the Kalahari sand. We came to a *dambo* in which there was a small lake of permanent water in the middle. We had passed water 15 miles back but the presence of a large village near had put me off it and we had continued. This was much better: the village was lower down, and on the lily-studded lake was a pair of pygmy geese. We had a delightful camp with the moon almost full. The drums in the village half a mile away kept going till after midnight and the sing-song monotonous voices of the chanters. Evidently a beer drink and a dance going on.

19 September 1956

Breakfast early and away, doing the 25 miles to Balovale by 9 o'clock. First to the boma to say how d'ye do to the District Commissioner, Rowland Hill, then to Sorenso's Stores for one or two things. Antonio Sorenso is a Portuguese who runs a most up-to-date store here – showcases and all the rest. I bought a bottle of Pedro Domecq Amontillado sherry for eight shillings and sixpence. Portuguese wine was in 5 litre demijohns at thirty-five shillings but I let this go. Balovale is quite a place, though I can't quite understand why. Maize and kaffir corn do not grow over this vast area of Kalahari sand and the food crop is the miserable cassava – manioc or tapioca – or whatever you like to call it. The boma is placed on an escarpment 100 feet above the Zambesi and looking over the vast plains of Barotseland and beyond. The great river is 200 yards wide even this far up and quite magnificent. Curiously enough there is no pontoon and we had to arrange with another trader called Rudge to put planks across a barge for our cavalcade of two Land Rovers and one trailer. All this took a long time and much talking on the part of the dozen Africans who undertook the task. Everything looked crazy, the barge once loaded took a horrible list to starboard, but the Africans and their paddles got the outfit across and they sang with happiness of success. We ourselves were paddled across in a large dug-out canoe. Then up a ramp of sand on to the plain and striking westwards; two miles to a Chief's village and we had to go in to pay our respects. The elderly Chief was living in a modern style of house with some Government furniture helped out with some well-made African tables and some large African wood carvings of storks and elephants. He explained that he was the deputy of the Chieftainess who was now too old to rule, but had he known we were coming he would have introduced us to her. Would we please call on our way back? All this through an interpreter and well done. Photographs of the Royal Family adorned the wall. Then away for 60 miles, first through very thin scrub and then into wide open plains of grass. We crossed one almost dry river and the line of riverine vegetation. On and on through the deep sand which held us back into second gear. There is no road here, only a track remarkably straight. Coming to villages we knew we must be near water, and to our surprise came a to a beautiful large lake from which rose a greenshank in full cry and a kingfisher perched on the mutate reed. This lake is really a widening of the Luitapi River, well stocked with fish and of great importance to the inhabitants of the villages here. I notice that these Barotse women meet you with a smile and a greeting, quite unlike the tribes east of the Zambesi where the women look so sullen and low in type. The women wear their hair in plaits about 4 inches long or less, all over the head. We pushed on through two or three more villages till we came to a more important one from which there seemed no further track. Many of the houses had windows, there was a native style courthouse

Crossing the Zambezi at Balovale (Zambezi) on 19 September 1956. 'The great river is 200 yards wide even this far up and quite magnificent.'

and there was a Chief's stockade. We were seeking the Lungwe-Vungu River, which is about 20 miles from the Portuguese border. This was Chief Chinyama's village and this gentleman came to ask us who we were and where we were going – again through his interpreter. We explained. Whereupon he said it was a difficult path and we must have a *kupassu* as a guide. This was just as well because it was bush-bashing of a most intricate kind for another 15 miles. I had been driving all afternoon and I was pretty tired when we reached an escarpment 20 feet above a lovely sheet of water with marsh beyond, just before dusk. Ian and I immediately said this was where we stopped for the night, though our guide said this was not the river we sought. In fact it is an ox-bow lagoon. The low sand across was striated with ridge and slack in the best style of the ecological textbooks. As we arrived the fish were rising and Ian got out his rod and spinning reel. At his third cast he caught a 4 lb bream (*Tilapia*) which we had for supper. What a lovely place this was – bush and parkland, water and marsh. The only game we had seen was a herd of 9 tsessebe with 5 young calves. This antelope is exactly like the hartebeeste or kongoni but carries a better head and is chocolate-brown in colour with almost a purplish bloom. Both Ian and I took to this place in instant affection and we felt inclined to accept the Chief's suggestion. He had said to us that what his little kingdom wanted was a few good Europeans to live amongst them. As it is Barotseland which has recently been bought by Northern Rhodesia it is being treated as strictly native territory where no Briton is allowed to live. Personally I think the Chief is right and entirely humane. The weather has been very hot, and I was sweaty and uncomfortable. Crocodiles or no, I ventured to think they might be full of fish and I ventured into the edge of the ox-bow lake for a bath. It was absolutely glorious and I saw no sign of crocs. Thereafter lots and lots of tea, and the moon rose full and clear into the warm still night. To bed by 10 o'clock but I suffered badly from sciatica in both legs till after 1 o'clock, despite the fact that I had taken a walk for half an hour in the moonlight and thought I had killed the wretched jumps. The frogs are singing splendidly and the nightjars churring.

20 September 1956

Up at 6 o'clock and after early tea we went our own ways. Ian fishing but I wanted to find the Lungwe-Vungu River. This I did find, inside a couple of miles of wandering through the bush. The river was a fine water 75 yards wide. Where I came upon it was upon the steep 10 foot bank of a meander, and across was a lovely white sand beach. Across were great stretches of marsh and up stream were stretches of high reeds and lush grass. I just sat down to soak it all in. Farther along the steep bank was a colony of carmine bee-eaters nesting in holes like sand martins. They were very talkative and would occasionally sally forth as a flock and hover over the water. I also saw the white-fronted bee-eater and the least bee-eater. I also saw a greenish large wren-like bird which Ian says is called *Eremomela*, very handsome. A fish eagle was nesting on the outermost branches of a tree over the water and extraordinarily well concealed. I returned to camp and found Ian just finished fishing – two barbels for our three boys. I suggested he should come back with me to the river he had not yet seen and this he could not resist. He brought his rod and with it I caught a 9 lb tiger fish. A fisherman in his slight canoe, standing and with a long paddle, skimmed down the river with polished ease, beautiful on that lovely sunny water. It was hard to leave it and come back to camp for 11.30 breakfast. I drove back the 15 miles to Chief Chinyama's village. There we were led into the court house, chairs were brought for us, and a *kupassu* brought water in a jug with three glasses on a tray and knelt before us. Honour obviously demanded that we should drink it, though it is highly unlikely that it had been boiled. The court house was of adobe brick, plastered. The roof was of beautiful Barotse thatch supported on eight poles. The walls supported the thatch as it came down to them and then there was a verandah outside. The walls did not reach the roof all round. Large spaces provided light and air. Within, the floor was sanded and a dais, being a slightly raised adobe floor, was honoured by an armchair, cushioned and covered by a material which carried a garish portrait of the Queen, Elizabeth II R., and carrying the Garter emblem and motto. The interpreter talked to us for a while and then the Chief came himself and made himself very agreeable. We were ready and rather anxious to go and I said to the interpreter that I much appreciated the building of the court house and that it would give me much pleasure to have a photograph of the Chief and his house of indaba. The Chief answered that he would be pleased to be photographed but he would wish to prepare himself. So we had to wait while he disappeared within his stockade. After only 20 minutes he reappeared in beaded headdress, lounge jacket and lilac-coloured foot-length skirt. He

Chief Chinyama's native-style court house between Balovale and the Lungwe-Vungu River, 20 September 1956.

walked slowly and with immense dignity to the door of the court house and turned. The interpreter had also put on his best Montagu Burton and stood a little way from the Chief, not near enough for the convenience of the photograph but I could scarcely ask him to move nearer to the Presence. We then made our farewells, promised to send the Chief a copy of the photograph in due course and gave him one of self and family in the garden at the Old Rectory. He was delighted.

I drove back most of the 45 miles to a lily pond where we had decided to camp as we passed it yesterday, at the edge of some thick bush and of a wide grass plain. A high-branching grass-like legume grew by the lily pool and was adorned by weaver birds' nests. We had seen two tsessebe bulls at stretch gallop crossing our road during the afternoon. They are probably the fastest of all game – superb in action. We had also seen a lone wildebeeste bull who twisted and turned and whisked his tail this way and that in characteristic fashion. And we also saw a steenbuck, light chestnut like an oribi but rounder and fuller in the ear.

21 September 1956

Up at 6 o'clock after the best night I have had since I came to Africa this time. The Angola mourning doves were calling, the shrikes and half a dozen more saying it was day. Two cocals, bronze-tailed ones, came to perch on a sapling before coming down to the lily pool for their morning drink. Then on the road the 20 miles odd to the Chief's village where we called on the way out yesterday. This time he had prepared for us. We were asked into the room where we sat yesterday and a servant laid cloths on all the tables and brought several teapots and plates of biscuits. His prime minister and counsellors were also present and we had quite an indaba. He asked, showing graphically by the pattern on his tablecloth, why he could not, as now, have direct access to specialist departments, instead of having to go through the District Commissioner. He called the specialist departments the legs and arms of government. He also told us that Europeans had been over the river and had illegally hunted his game in the Reserves which he and his people had set up to conserve the game. Ian and Ansell promised retributive measures if they caught the offenders and said that these new reserves, set up at the request of the Chief and his people, would be placed in the Game Ordinance by the end of year. It is really very promising that a Chief and his people should take this step, but there you are. Barotseland is not at all like the rest of Northern Rhodesia. These people are far ahead. The entry of the Game Department into this area is new and promises well, especially as the Chief pointed out through his interpreter that he was particularly pleased that in passing through his country we had had the courtesy to call on him and explain things. Now after our refined drinking of tea, it was time to pay our respects to the very old Chieftainess for whom the Chief was Regent. A *kupassu* with staff led the way, then the Chief, then Ian and I, Ansell, the prime minister and the grandson who was heir. We walked slowly across to a new large house, stockaded with beautiful bamboo fencing. We entered a large room with two or three wood and cushioned settees against the wall. In the middle of the room was a carved wooden heraldic lion about a quarter life size. In the right-hand far corner was the lion throne, a large wooden armchair, the front legs of which bore carved lions rampant. On the raised chair reposed the old Chieftainess, immobile, in a blue robe and with beaded headdress. At her feet sat two handmaidens and behind her head was inscribed on the wall '*Yesu Ikiye Mwata*', otherwise the room was bare except for a curtain hanging over the doorway immediately opposite the one by which we had entered and to the right of the Chieftainess and her handmaidens. We were shown to the settees and then I, Grimwood and Ansell were introduced separately and our status explained. The old lady remained still except for giving her hand to each of us. Each of us bowed. Then the Chieftainess spoke in a soft and musical voice and was quiet again. The interpreter explained that our compliments were appreciated and she regretted being so old as to be unable to take part in our talks. 'And I am glad of the care you will henceforth be giving to my game.' There was no doubt whatever of her sureness in the position in life to which she had been called. The Chief then took us round the house, which was built round the large central room which was so restrained and yet dominated by the old lady's small figure on her lion throne. The Chief and his counsellors and ourselves taking tea were people of a modern world. Here was something immensely older. The rest of the house was of almost ascetic bareness. Tables and chairs of the highest quality were there, all made of local wood by local men. The polish on a large table top made me envious, it was so good. The two bedrooms were bare except for the locally made wooden single beds, the one room for the two handmaidens of the night, and that of the Chieftainess. Her bed was simple but yet with more wood in it, and on it was a patchwork quilt of great beauty and intricacy. Dressing rooms with presses of local wood were adjacent. On our return to the large room where the Chieftainess was still on her throne, the interpreter had brought forth a visitors' book from the gold-coloured cardboard box in which it had been sold. The box was now rather threadbare but it was obviously an important part of the visitors' book. We signed our names, went up to shake hands once more and bowed ourselves out. We progressed back through the village to the Land Rovers and bade goodbye to the Chief. Along then to the Zambesi and another African crossing on the barge. The event was West Highland in character: first the man in the dug-out canoe did not think we should be back today, though we had most carefully made the appointment. It was an hour's job getting the barge crew together and another hour getting planks across and our Land Rover aboard. Just as last time, however inefficient and apparently haphazard the whole affair was, we did get ashore safely. Then there was trouble getting the car up the steep ramp of deep sand. We had to unhook the trailer and ten chaps helped Ian up. He went off immediately to the boma to see the District Commissioner, so I got all the crew to man a rope after putting planks down to the trailer and we hauled it up. The shouts of delight and the songs showed their spontaneous childlike joy. I was about finished, for it was extremely hot and I had pulled pretty hard. It was now 1 o'clock and we don't eat lunch. Ian returned and we hooked up the trailer again and went for petrol – another long job. Incidentally, while waiting about I had tried paddling myself on the Zambesi in the big dug-out canoe. Lots of fun. We got away from Balovale by 2.30 to go 65 miles to the Dongwe River. Ian drove the 50 miles to Mombezi and then I took over for the 15 miles down a track which is going to be a road one day. We reached the considerable Dongwe River after a 10 m.p.h. crawl and found a pontoon in operation, African fashion. The pontoon was pulled along by a bark rope by chaps going ahead in a dug-out canoe who also pulled on the wire hawser which prevented the pontoon going down river. Again all the fun of the midday Zambesi crossing. Two local damsels had paddled up in a large dug-out canoe and our Game Guard, Enoch, put the come hither on them and borrowed the canoe for the night. We camped a little way down river on the bank and in the moonlight we paddled a mile or two on the river. Perfect; and I felt so much better after having had a bath in the river.

22 September 1956

The place where we are camped is called Chiseya and we set off at 8 o'clock for Lukula on the side of the Zambesi where there is a Catholic Mission and also a branch of W.L.N.A. which is pronounced 'Winella', which is the labour recruiting organization for the Rand Mines. We travelled 44 miles along a track. We were travelling along the low scarp of a considerable flood plain, so we passed through many villages. Always the people were cheerful and waved, men and women alike; the children ran out and called to us. I like Barotseland. At Lukula we went to the Catholic Mission and were met by Father Edwin, a most helpful young man. We had been advised against going down to Mongu this way as there is no road, just last year's wheel tracks over the Zambesi flood plain for 80 miles and very hard to follow. Furthermore, the late floods put stretches of the road under water even now. Father Edwin insisted that we must have a guide, and what a good thing ultimately that we have had, for the route would have been impossible to find or to follow. We also went up to see a big man called Bailey, who ran Winella's show. He was the original, I am sure, of Barney's tobacco advertisement. He has been there but a short time, and he and his wife and a nine-month old baby are living in a mud hut and grass camp. Here was a woman quite untroubled and thoroughly enjoying herself, even to doing a piece of tapestry. Bailey had a small launch with a 30 h.p. outboard engine on it. He insisted on taking us on the Zambesi in it and we hurtled about at 12–15 knots. He built the launch himself – able fellow altogether. Back to the Fathers, who insisted on our staying to lunch. The guide came afterwards and we set forth on this most chancy leg of our journey. The first ten miles were a nightmare of rough weaving and then out on to the plain or along the extreme raised edge of it. Every village came out to wave and cheer. Some of the men had only loincloths and most of the children were so clad. The women wore blouses or blankets or not, presumably as they felt like at the moment. We got down to the flood plains of the Zambesi after 30 miles, only a matter of 3–5 feet and yet what a difference it made! We struck our first swamp after 37 miles and had to unhitch the trailer; help man-haul the Land Rover going its hardest in reduction gears, and then put a tow-rope back to the trailer and haul that through. Then a longer 2 foot-deep draw, but after testing the bottom everywhere we thought there was a spot where it could be rushed, and those of us in the water could jump to and help the outfit along. We did this with success, and rushed one or two smaller soft places. We camped at 40 miles at 6 p.m. on an anthill where grew a little scrub, an acacia tree 20 feet high, a clump of *Raphia* palm, and two tall *Hyphaenae* palms. These last rustle their great leaves in the breeze so that it sounds like heavy rain falling on a sycamore tree at home. These anthill communities are very interesting; the scrub holds a pair of shrikes. A dove or two comes to roost in the acacia, and in the tops of the palms the palm swift nests; also Dickinson's kestrel, and in our case here a pair of little hobbies. Sometimes there is the white-necked crow, and weavers may be in the *Raphia*. The night is cold down here and we are keeping a cheerful little fire of fallen palm fronds.

23 September 1956 (Sunday)

Away by 8.30 o'clock, shaved and breakfasted. The Barotse cattle are coming out of their bomas on to the plain, lowing, and the sound is good. Sunrise this morning was magnificent. Life altogether is good; even our progress in the second 40 miles of the road to Mongu. We did eight miles in three hours, for we had several stretches of water to go through and we had to test the bottom of each one and decide whether we should unhook the trailer or not. We got stuck in one long stretch with the exhaust under water, so the engine had to be kept going at all costs, though at the same time the fan was throwing water all over the engine and battery under the bonnet. At the next deep water we unhooked the fan belt, so that the fan was still. Furthermore, the petrol we got in the drum at Balovale had a lot of water in it and we had to drain our tank and petrol system to get rid of over half a gallon of water. The villages we passed through gave us enthusiastic welcomes and would embarrass us with still another guide to show us the best water crossings beyond. On and on and then the track became definable but atrocious in the sand. In fact, for 10 miles to Limulunga, which is 10 miles from Mongu, we were passing through a heavily sheet-eroded landscape, eroded by over-grazing. There were no villages in this stretch, but we saw one or two mobs of poor cattle. I had been struck, back in the vast Zambesi flats, by the absence of signs of erosion. Game, of course, is now non-existent. We saw only a pair of little oribi the whole way. We had noticed a ridge ahead of us perhaps 200 feet high, running for several miles, and coming nearer we could see it was sandy, naturally covered with woodland but now much cut into by cultivation. We made our last water crossing and straight up the ridge to the considerable village of Limulunga on the top. The track was just loose sand and we had to manhandle the Land Rover and trailer to get up at all. We found the village deserted, and with the erosion it was not a bonny place. But we noticed a considerable house within a stockade. Guide Edmund told us this was the palace of the Paramount Chief of the Barotse and that the Chief had gone to Lealui, 11 miles away, and therefore most of the village had gone with him. Having got up the ridge with so much trouble, we went down the other side through more eroded cassava gardens and then at the bottom we turned along a narrow strip 10–50 yards wide between ridge and plain. There was dense and continuous habitation along this 10 mile way to Mongu. I imagined there must be constant water seepage into this strip, making it potentially fertile, but in fact this narrow band is of extremely fertile loam. We saw banana trees badly husbanded, mango trees, kapok trees and food crops along here. The people simply cannot be got to move their huts a little uphill to the useless sand so that the whole of the valuable strip could be cultivated. We noticed also that we were out of the simple delightful communities of the Zambesi flood plain. This was a dense rural strip of sophistication. African women swing their behinds anyway, but when you bustle the garish-coloured skirt and spring it with a wired petticoat – well, the African woman's progress is something to see. Top it off with hat and parasol to match, and you nearly have to step off the track. It was Sunday afternoon and quite a spectacle. At one point a crowd was gathering under some trees and as we passed we heard the drums beating a quickening rhythm; a woman coming along the road smiled at us and swayed this way and that in an ecstasy of rhythm. You could see her saying as it were, 'My! what a beer drink and a dance I'm going to have'. Some of the men here wear brightly-coloured cotton kilts and swing them in good Highland fashion. In general, we were not greeted any more, nor did we receive many smiles. Only here and there would an old countryman raise his hat. Mongu was a typical boma of the larger sort and not attractive. It is on a little hill and overlooks the great plains. We drove out six miles to find a camp and struck luckier than we thought. We came down a hill to a flat plain over much of which was shallow water gleaming blue, striated with golden grass. We drove along the sand at the edge of the water and below the very low wooded sand ridge, and set up our camp under a great *mopani* tree. This *mopani* is characteristic of the deep sands: the one I met in the Luangwa Valley is the *mopani* of heavy valley clays. This camp was really perfect. I bathed in the lovely clear water and got into decent clothes to go back to Mongu to meet the District Commissioner. It was now 5.30 p.m. and our exertions of earlier in the day quite laid aside. In fact, we told the company at the D.C.'s what a trouble-free journey it was.

24 September 1956

We left our beautiful camp by the great pan of water regretfully and came back through scrub to Mongu. I sat in the District Commissioner's garden writing this diary and one or two more important letters. Then to meet Glennie, the Resident Commissioner of Barotseland, a stuffed shirt as far as I was concerned. Back to lunch with the D.C. and his wife, John and Irina Haley. I liked him – Oxford, incisive, intelligent, courageous, and he likes red wine. His place is Nigeria, but he got shot full of buck shot in some troubles there and he was sent here for recuperation. A very civilized and pleasant luncheon.

Then on the road southwards for Senanga. The track led along the edge of the plain for 25 miles as yesterday and conditions were so similar that had it not been for the sun's position I would have thought we were going back along the way – just continuous habitation on the narrow strip of black soil. Then we struck across the sand plain for almost 25 miles, and are camped in a patch of scrub near some perilous-looking water which nevertheless must make our tea and moisten the cooking of the evening meal. Coming over the plain today with its patches of marsh, even to a bit of papyrus, I saw the following birds:- black heron, squacco heron, greenshank, wood sandpiper, common sandpiper, wattled plover, blacksmith plover, a wagtail *Montacilla simplicimus*, coppery-tailed cucal, cattle egret and a hawk-like owl, white below and blue on the back that was eating a nestling bird on an acacia bush. Ian had never seen it before. Large bats are squeaking in the trees and a ginger-coloured flying drone termite, two inches long, is bumbling about.

25 September 1956

The 27 miles to Senanga took us along the edge of the flood plain again and through numerous villages much of a pattern. We called on the acting District Commissioner, Lawrence, who said he thought we might get through on the route we hoped to take. One man had gone this year and as he had not come back it was assumed he got through. We came down the east bank of the Zambesi for 15 miles and crossed on an up-to-date Diesel engine pontoon. Thereafter we struck west through the bush on a track left by the few cars which have gone through in the course of a few years. Where the track went through thick bush the trees had been cut. There were but few patches of deep sand. We did 35 miles from the pontoon, about 87 in the whole day. This part of the journey was across the Siwilana Plains, lovely parkland and woods and grassland laid out as if in the nineteenth-century English covert-shooting style. We saw no game for 30 miles and then began to see tsessebe in twos and fours, sometimes with a wildebeeste bull along with them. We also saw a herd of about 20 wildebeeste. At 35 miles we came to an area where in addition to the parkland, grass and wood, there were pans of water, some covered, some surrounded by reeds. This was where we decided to camp, though we were sorry to find depositions of salt round these meres, and in fact the water was pretty brackish. We endured this because everything else was perfect. Ian and I set forth after tea to explore the pans. The bird life was good – black-winged stilts, ruffs, little stints, common sandpipers, three-barred sandpipers, wattled plovers, blacksmith plovers, white herons and egrets. There was also the African reed warbler, which sings a little like a nightingale and very well. I also saw four Hottentot teal. We were excited to find red lechwe on these plains round the pans – 37 this evening, mostly does and calves. Our Game Guard, Enoch, caught some barbel in one pan and we had these for supper. Quite good. Mosquitoes bad and I got a good many bites. An acacia bush near the camp has a colony of spectacled (masked) weaver birds. Their nests were new and lichen-green in colour. The nests contained two eggs each – pale blue ground, reddish-brown mottling.

26 September 1956

Away by 7.30 this morning, doing a further 37 miles westward. The road deteriorated to deep sand except where it passed over anthills, where it was deep bumps and hollows caused by elephants' feet. The parklands stopped and we got into considerable forest where the elephants had done a good deal of pulling down of trees over the track. Our rate of travel was less than 10 miles an hour. The evidence of elephants is really considerable, though we never saw one. The shrikes here in Africa are an important family. I saw the crimson-breasted shrike, a really lovely bird, and the magpie shrike. The least bee-eater is rather like a large humming bird in action. We saw several groups of tsessebe and wildebeeste today before 11 o'clock, but you would say nevertheless that game seemed scarce. At that hour we came to a track called the Barotse cattle cordon. This line is along the Mashi or Kwando River, which is actually the boundary between Portuguese Angola and Barotseland. The Portuguese claim this territory and occasionally make labour raids. The British characteristically do not try to administer it. The natives live in primitive state. The cordon is maintained by the Northern Rhodesian Veterinary Department as a safeguard against the introduction of contagious pleuro-pneumonia in cattle. The line is patrolled from camps 10 miles apart by African guards. We turned south along the cordon which is deep sand running through the forest along the swamps of the Mashi River. We did a further 60 miles, getting at last into good game country. We saw a herd of 24 kudu, and within a few yards 8 more. A mile later 6 fine kudu bulls, some roan antelope, a few zebra, and lechwe by the pools. We had many halts in the deep sand when we had to get out and push to keep going. Finally the road improved a little as we came on to acacia scrub and forest, which needs a rather stronger base than sand. It is striking that the good high forest we have passed through should be on this deep Kalahari sand. We are camped by a lake less than 100 yards across, on which are pygmy geese. The frogs are new ones to me: they make sounds like the smart tapping of little wooden bells. Lechwe are grazing 100 yards away. I forgot an important record of this morning: we saw a giraffe in the forest. These animals occur only in the south-west of Barotseland, the westernmost part of Northern Rhodesia and in the south-east of the Luangwa Valley, the easternmost bit of Northern Rhodesia. This animal was not so reticulate on the neck, nor was its face so finely pencilled as in the Kenya forms. Villagers have come into our camp this evening, men clad in loincloths only, and they have sold us chickens at a shilling each and some eggs equally cheap. They have also helped haul the firewood together for our fires. Our men will probably keep the fires going tonight because on the site of our camp we have found the new footmarks of a lion.

27 September 1956

Away by 8 o'clock, having heard nothing more in the night than a hyaena clear, and to my ears, a beautiful sound. The villagers came to say good morning and goodbye, very friendly altogether. We continued down the cordon track, sometimes getting magnificent views over the reed flats and wooded islands which constitute the Mashi River. What a paradise to explore biologically. We came to the last cordon post after 33 miles, i.e. 93 miles of the river we have done in all. The African was away but his two wives were in the forest edge and we saw them coming back to the huts, each carrying two children under two years. In the fashion of this area, each woman was wearing only a knee-length skirt and when Enoch asked them where the track turned eastward into the forest they obviously told him to wait. Each woman and her children disappeared into her own hut and in less than five minutes reappeared clad in bright prints. Now they knelt down and cupped their hands in greeting. They remained kneeling while Enoch asked about the track. One of the women answered without haste and the information she gave us turned out to be good. The roads here are chancy things. The maps show tracks which aren't there and you find tracks which aren't on the map. The woman told us we should find no water till we came to the Zambesi, over 70 miles, though she could not give us the distance. She said it was all a man could walk in two days. Ian and Enoch could scarcely believe this, but she was right. She served us well. Anyway, we set forth on a track which took us through the uninhabited forest known as the Yando, going north-east. I was driving now and navigating by the sun. I thought we were going too much north and my companions thought so too. Enoch is such a clever African and is just like a schoolmaster. We thought perhaps the woman had put us on the wrong track. We climbed and had some trouble with the deep sand. Then on to firmer ground through a curious strip of what seemed devastated forest. The trees were young and had burnt a few years ago, but as there was no old timber dead or alive in this quarter- to half-mile strip, the relative bareness was natural. The path kept in this strip between the dense forest for over 20 miles and then we were in forest proper. Our pace for 40 miles was 12 m.p.h., very satisfactory in such conditions. Then we struck another and better track just as the woman said we should, and I turned right along it being quite sure we were now all right. Ian was still not sure but Enoch agreed with me. We continued for 10 miles and saw a village on our left. We turned in the quarter mile, but having got there found the village deserted. Hens, a dog and a calf wandered about, and belongings were around. Then we heard children's voices in the forest and Enoch went after them. He returned in a quarter of an hour spreading his hands wide. 'These people are quite wild. I go after them and they all run away, men, women and children.' He was quite disgusted. However, they must have run parallel to the road and we got rolling quickly; a quarter of a mile on we saw a girl and two little boys in the bush. Enoch was after them and called to them to stop, telling them we shouldn't hurt them. The girl stopped and the little boys slipped behind her skirt ready to fly. Enoch spoke Lozi to her and was understood. She said the road we were on would take us to Katima, the place we wanted. Enoch clapped his hands and said, 'Hurray!' and jumped about in his pleasure. Even the girl smiled and all was well. I drove on another 30 miles and more through acacia forest and lusaka scrub, when we were conscious of going downhill. This was promising and in another mile sure enough we saw the lovely broad blue ribbon of the Zambesi. And on its shores we are camped tonight. I have bathed at its edge and washed my hair, and am feeling good. When we started through the Yando this morning we passed through old gardens, and in this place alone we found the sand thrown up and dug out by jerbils, creatures like kangaroo rats. It struck me immediately that where you had the forest climax broken for cultivation, there you get the rodents dense. We saw no game on our passage but plenty of elephant sign.

28 September 1956

In effect, Ian's and my trip together is over, for we have turned east and reached the Zambesi just below the Mogambwe Falls. This morning we came along a road running parallel with the great river and reached Katima, which is actually out of Northern Rhodesia and in the Caprivi Strip. This is a strip of land awarded to Germany in the 1880s to give German South West Africa access to the Zambesi. There was no sense in the award but there it remains as a bit of the Union of South Africa which is no use to that country and little use to anybody. The administrator appointed to its care looks upon it as a political plum – very little work, a good game area, interesting archaeologically, and he is left very much alone. The man who has just retired is called Trollope, a prodigiously fat man who is a bachelor and a very cultivated type. Ian and I went to see him. He knew of me because of my interest in wild life conservation. He is himself extremely keen to take care of the animals. When, a year or two back, the Supreme Judge of the High Court of the Federation was in the Caprivi Strip shooting himself an elephant, he also shot a lechwe (protected) for the pot. Trollope is known to maintain an excellent intelligence service, so he soon heard of it. He was down on the Judge immediately and said as it would look rather bad for the Judge to be had up by him, he was prepared to let it go if the Judge would donate £200 to charity. The Judge accepted the ultimatum and paid up. Trollope has a lovely house and garden overlooking the Katima Rapids of the Zambesi. He is a born gardener and obviously loves doing it. The floor of his sitting room was of African parquet, rich and well waxed. He had shelves of good books, some nice pictures and one or two nice bits of china. He says how happy he is in Barotseland and I can understand it. And coffee beautifully served to us with just a wave of the hand. We are crackers to remain living in England. We then crossed the Zambesi once more on a pontoon and came north-east for 50 miles through a forest reserve to the Luanja River. The forest species conserved are *Baikea plurijuga*, a *Pterocarpus* and another I don't know. The forest grows on deep sand in a waterless area and is now divided up by 45-foot firebreaks, and fire towers are placed here and there. There is a dense and almost impenetrable thornbush undergrowth to the trees. As one might expect, an earlier Forestry Department tried clear felling and completely devastated the area. It has never come back. The present head, Colin Duff, whom I met last time, is a first-class man who is doing a fine job of conservation with selective felling. He has a rich sense of humour, which alone saves him from the frustrations he suffers from the Provincial Administration, generally known as the P.A. We are camped tonight by a swamp of the Luanja River, backed by secondary scrub of old gardens. There are villages across the swamp and only half a mile away to the side of us, this being the only water for quite a long way.

29 September 1956

On today at a gentle pace through large areas of forest reserve and some of which was cut over by the concessionaires. Eventually we got into *mopani* country which meant we were off the Kalahari sand and on the clay. We reached a place called Mulinga where a Tsetse Control post is being set up because the tsetse is spreading in here from the south. The man was building his house. We also met a Game Ranger called Mingard, with whom I am to travel for two days. We plodded out over elephant tracks for 10 miles to a long half-mile lake set in grass in *mopani* woodlands. Absolutely perfect eighteenth-century English parkland planning.

30 September 1956 (Sunday)

Away at 8 o'clock going over a lot of *dambo* ground much cut about by elephants in the wet season. Our rate of progress was four miles an hour, and after four miles the trailer attachment on Ian's Land Rover severed. By this time another Ranger, Bill Mitchell, had caught us up, he having a rendezvous with us at dawn today at Mulinga. Between us all we coopered a wire rope attachment and got along. The thing kept breaking, of course, but Ian wanted to get his trailer 30 miles on to where there is a road. Then he can send a National Park tractor for it from Ngoma. We had been coming nearer to a fire all the time and while we were crossing a large plain of unburnt grass we saw the fire 300 yards away. It came towards us quickly to 100 yards, and I was a trifle bothered whether we were going to get past. I know Ian was also, but of course did not give the slightest sign. I badly wanted to get up on the bonnet of the Land Rover and take a photograph, but we daren't waste the 30 seconds it would have taken to do. We have come into a rich game country of sable and roan, zebra, kudu, waterbuck and puku. Elephants are about. We have come 32 miles since yesterday to a pool called Tamva. I walked the last six miles to get a rest from the incessant jolting, and in order to soak up the country in my own way. Too much of this jolting makes me hurt at the back of the head between the ears. Thinking over this problem of burning in Africa, I see how far I am from an answer. The present general policy is to burn early in order to reduce damage. This helps the trees, but it also helps to establish acacia scrub which for a long time must limit grazing game. There is also the point that early burning makes the grass grow fresh in the dry season and there may be consequent impoverishment of the grass plant. If man were out of it, the fires would be lightning fires and they would not occur until just before the rains. Late fires keep stretches of country open for grazing animals because they are fiercer on young trees. My own feeling is that too much of Africa burns too often. The perfect fire incidence, surely, would be that of unaided lightning fires. But both African and European seem to act exactly like the Highlander at home. He can't keep his matches in his pocket if there is anything to burn. That great plain of barley-yellow grass that we came through yesterday was brilliant, but is now black and gone. What would have happened for a year or two had it not burned at all? I think that within the Kafue National Park of 8,000 square miles there should be some careful experimentation on the influence of fire. Those forests either side of the Busanga Plains that I described on September 13 are a good natural start, though I want to know more about grass fires, or lack of them.

1 October 1956

Away in a leisurely way from our camp among the *Hyphenae* palms. Then six miles of hard bashing till we reached a track on which we drove for nearly 40 miles to a camp at Kasha. Six miles before Kasha I got out of Bill Mitchell's Land Rover and walked by myself to Kasha. What a delight it was to be both alone and without noise. I saw some wildebeeste and a buffalo who were not bothered by my walking by. The cicadas sang in the tree tops, little lizards rustled in the dead leaves and the sun poured down vertically on my receptive back. All the way through *Brachystegia* woodland. When I reached Kasha I found it was an African camp, with two rondavels left vacant for European game staff. The African Game Guard was suffering badly from blepharitis and looked thin and ill. His children had very bad coughs and eyes running. I suspected bilharzia, as I did 15 miles back where there had been another post and the man sick. This was really a bad place to camp and I said so. And I did not want to sleep in a dark rondavel. So the Land Rovers were repacked and we went a mile down the *dambo* and camped on a slight rise under a particularly fine *Brachystegia* tree and near a lily pool which would provide water. Went a walk for 7 miles or so in the *dambo* system with Bill Mitchell. We saw only one kongoni and an iguana lizard, and some wattled cranes in the pools. I also saw a black rail, rather like a moorhen with long red legs. Bill Mitchell is a biologist, 46–47 years old, London graduate. Ian said I should find him rather like a stage curate. So he is, but a good man with whom I can travel easily. As Provincial Game Officer he does precious little biology. After dinner he turned on his searchlight lamp over the *dambo* to pick up eyes in the darkness. We found one small pair 150 yards away belonging to a spring hare, a rodent about the size of a large rabbit that hops on its hind legs and has a bushy tail. We walked over towards it and I had a good view in the lamp beam at 25 yards.

2 October 1956

We did a 35 mile bush-bash in the Land Rover today through the *dambo* systems. I think I have said before that the *dambos* are the grasslands among the woodlands and are very slightly lower, being the drainage system of the country. The *dambos* look easy going but if elephant have trampled about in the wet season or the white ants have built numerous little hummocks, the *dambos* can be frightfully rough going. We saw a few wildebeeste, kongoni and zebra, and then, about half way round along some *mopani* flats we came on a real concentration of game, quite the best I have seen in Northern Rhodesia this time. At one time we saw 6 eland, about 250 wildebeeste, 20 kongoni or hartebeeste, 50 zebra, 3 oribi, 4 reedbuck and 4 waterbuck. A little farther on were 200 buffalo. Rather earlier in the morning we had seen a pride of six lions at 100 yards. All were lionesses. They had a good look at us and made off. We also had a good look at them. With the main concentration there were also 30–40 roan antelope, and as we came back we saw a fine herd of 20 sable antelope. The bull was jet black and had splendid horns. We walked out for a few miles in the afternoon but saw nothing. The range here is in fine order. Water in pools is to be found in every *dambo*. It is a wonderful game area.

3 October 1956

Bill Mitchell and I left Kasha this morning and came up the old cattle cordon road to Ngoma, which is to be the great game-viewing tourist camp of the Kafue National Park. A fool Administration has already spent £48,000 in erecting what the Public Works Department considers are suitable buildings. It is hard to see where or how this amount has been spent except that a large number of Africans are working (sic) without any European supervision. Naturally, the game has gone with all the clatter and it remains to be seen whether dudes will be able to see it any more from the camp. The place is always alluded to as the gin palace. In 10 years' time it may be an African school. The future of the National Park is by no means assured, and the Administration (the P.A., as it is always called, i.e. the Colonial Office civil servants) pack the Advisory Committee and prevent any consolidating action. The Game Warden in charge at Ngoma is Len Vaughan, a gentleman aged 55 and shortly retiring. He is a charming man but unable to take the responsibility he has been given. He knows next to nothing about wild life management though he has been a hunter all his life and has known this country a long time. He is the kind of dead wood this Department will have to cut out, but make no mistake, he loves the animals and does his best, and he is such a nice man. He and his wife asked us to lunch. We were to occupy one of the chalets for a few nights but the thought appalled me, and I said, 'To the bush, please.' We came 12 miles down the way to where the Mushi joins the Kafue and camped by a lagoon by which were crowds of wood ibises, knob-nose duck, pink-billed and whistler duck, pelicans (pink-backed), spoonbills, black-winged stilts, several sand-

pipers, open-billed storks and a few spur-winged geese. Really a marvellous place and
the birds so tame. Back then to lunch most charmingly dispensed, and then out in Len
Vaughan's Land Rover to a place near the edge of the Kafue Flats, outside the Park,
where there are rather too many zebra. Vaughan shot two which we gralloched, cut
in half and were about to put in the car. Two natives (Ila) appeared from nowhere like
a couple of vultures and asked for meat. Vaughan gave them one zebra. You should
have seen their smiling faces! We came back for a cup of tea and then out in the Land
Rover again with the meat. Vaughan during the past five years has made a habit in
the dry season of feeding zebra meat to a pride of lions. By this means he has got them
tame. Sure enough at a point two large lions appeared and came towards the car. The
Game Guard heaved out two quarters of zebra 50 yards apart and the two lions settled
down to feed. We were then able to circle near them in the Land Rover and take
photographs. The lionesses did not appear. Of course all this is absolute damn silliness,
making a show piece of lions which I imagine may have to be shot when Vaughan
retires. All the same, through the years it has given Vaughan a chance to make some
interesting observations on leonine life though in an unco-ordinated sort of way. We
then went off into the bush and I was really interested to drive near to 200 buffalo.
What lovely gentle cattle they are, yet if they are wounded and made cross no animal
is more fearless and dangerous. We drove back to our camp where we had puku,
reedbuck and waterbuck around us unconcerned, and in the night a large herd of
buffalo, perhaps the 500 we had driven past as near as cattle would come as we came
back to camp, came by only 50 yards away. It took them a quarter of an hour to graze
by our beds, going down to the lagoon for a drink. The reedbuck whistled in the night
and were answered.

4 October 1956

We moved camp half a mile this morning to the bank of the Kafue to be nearer potable
water and in order not to upset the wonderful array of birds on the lagoon. We walked
lazily up the river for a couple of hours, watching the crocodiles in the river. I also
learned some of the riverside trees from Bill Mitchell. The sausage tree is *Kigelia*; there
are *Syzigium*, *Albizia* and *Adina* by the river, and in the true forest there was Northern
Rhodesian teak, not teak at all but a leguminous tree *Baikiaea*. The other dominant
species in the teak forest on the Karoo sandstone is *Entandrophragma*. Out in the early
evening with the old Game Guard, Mahoma, who has been lent to us by Vaughan.
The old boy is an Ila, tall, spare, deferent, and a good bush African. His top front teeth
are absent, as is the habit among the Ila, knocked out when they are children. I take
a guess that being without upper incisors makes them feel more like cattle, the Ila being
a cattle tribe. We were out for two hours, walking through poor *mopani* flat, *combretum*
scrub and bare hardpan. We saw little but a few waterbuck and warthogs.

5 October 1956

Drove up to Ngoma for breakfast with the Vaughans, then into the teak forest which
is within and part of the Park. The forest trees here rarely grow dead straight, but come

out of the ground at an angle. The bole, however, is quite reasonably straight, and the forest is really better than it looks. There is a fairly dense shrub layer, some of it terribly thorny, and some that looked and grew like hazel at home. Indeed, the forest is not unlike an oak – hazel copse on a larger scale and not quite so tidy. This forest is out of leaf at the moment and you get the impression of one of our oak – hazel woods in a sunny warm March before any leaves appear. The floor is of crisp dry leaves. We had come armed with an African device for calling up small buck like duiker. It consists of a length of duiker horn three inches long and hollow, open both ends. The rather wider end has spread over it and gummed to the outside a diaphragm composed of a spider-web fabric. The spiders make these dense white lozenges, about an inch and a half across, under the thatch of dwellings. The operator then holds the horn upright against the lips and blows down the open end. The resulting sound is a penetrating wail. We tried this tool at a dozen different places in our three and a half hour walk, attempting to call up a very small buck called the blue duiker, the presence of which has not been confirmed in this forest. We formed the opinion that the forest was pretty empty anyway, but we did call up a Sharp's stenbuck, which is a sweet little creature 15 inches high, bright chestnut with a few white spots. Its ears are veined like a leaf. We also called up a common duiker, and finally a tiny creature came running like a fairy through the undergrowth where we were crouched. The Game Guard and I just saw it and no more, and Bill Mitchell blowing down his horn missed what was obviously the very shy blue duiker. These three species came running as if urgently, but probably only inquisitively, and are gone again as soon. Birds were not numerous but I enjoyed seeing the cinnamon roller. Very little evidence of buffalo in the forest and none of elephant. Incidentally an elephant passed almost silently by our camp last night. On the immediate mud bank of the Kafue River below our camp I see the tracks of the water mongoose and of an otter. There are also slightly raised runs which are the tops of the runs of the mole crickets. The water mongoose lives on these, scratching for them here and there along the runs. There is a tree here, *Longocarpus capassa*, usually called the violet tree. The flower is very approximately lilac-like and the colour red-purple. The scent is violet-like. The roots smell like wintergreen and if the natives feel stiff or sore they rub themselves with an extract of this root. There are elephant shrews in the *Baikiaea* forest. As they leap along they bare places on the sand and you can tell their tracks from the succession of bare sand ovals about the size of tea plates.

6 October 1956

We drove about 10 miles this morning up the line of the Musa River, a tributary of the Kafue, into anthill and *Combretum* elephant-coppice country. The ground herbage was rather stamped out by the buffalo, and some of the anthill thickets showed typical cow-country appearance. We found rhino tracks and droppings at a midden by the Nakabusi *dambo*. Now, with Mahoma guiding and walking before us at a great pace we walked for three and a half hours, some of that time in the dry sand bed of the Musa. There were pools here and there, and in one we found a man, a boy of about

14 and another about 9, fishing. How happy they were! Naked in the pool, the man and the youth each had four spears and the little boy three, and these they threw together — I mean each group of spears as a salvo under the reed-lined bank or here and there in the pool. The method is entirely hit or miss, but the barbel are so plentiful that this method is as good as any other. I took a photograph of the party. They had spent the night at Nakabusi camp (a game guard's hut) and they said seven lions passed through at 7 o'clock last night. We went back via the camp, where I found the guard was a Lozi (by his filed teeth) and from Balovale. When I managed to tell him through Bill Mitchell and Mahoma that I was at the Lungwe-Bungu a fortnight ago, he was so surprised and almost incredulous, for it is a long way from here. The country round here was light *Brachystegia* and *Combretum*, with anthills and little bare grassy places that were too small to call *dambos* — good game country though we saw only a few waterbuck and hartebeeste. Fish eagles seemed very common.

7 October 1956 (Sunday)

Up to Ngoma after breakfast and away with Len Vaughan to the Nansila Flats where we saw the concentration of game the other day. But today we motored over a great deal more of the flats, which remind me of the water meadows below Hinton Waldrist. The flats are flooded in winter and quite empty of game, but at this time of year the water has gone and only a few pools are left. The hollows where the pools are often carry bushes on the edges, just like the clumps of hawthorn in the water meadows where the cattle go in to escape the flies. Here the buffalo must often do that. The presence of anthills allows the growth of some tall trees like the elms in the water meadows, and you get hobby falcons there also. We came into a herd of 600–800 buffalo and drove very quietly by these gentle creatures, who stood and looked at us with what I think are perhaps the loveliest eyes of any game. So gentle when unmolested as these are, and a wounded one the most dangerous creature utterly fearless and full of intention to get its man. I love watching buffalo. We saw a man down in the flats and came up with him. It was a Game Guard doing his job and doing a little for himself as well. I like the way Vaughan treats his Africans, and he is an old-timer. This African had 20 or more large fruits, ovoid, the size of a Jaffa orange. They were from the baobab trees. The coating of the fruit is like fawn velour and when you have cracked open this outer shell, the seeds inside are like white sugar candies. If you eat these, the outer coat of them is sharp, sweet and pleasant like cream of tartar. The natives use these seeds to flavour their beer. The Game Guard was also carrying fish spears and when Vaughan offered him a lift the five miles back to the game guard's camp, he said the other Game Guard and his own son were down in a pool with the fish. Out they came with several large barbels or catfish, all smiles — on the faces of the natives, I mean — and they enjoyed the ride back to their camp of two huts and a spare. We carried on and the native we had in the back pointed to a little white flower which I have already found to come from a true bulb, and he told Vaughan that the bulb could be used as soap. We got out and tried teasing up a bulb or two and then using them with water as soap. Sure enough a rich lather developed and I was glad of this chance to clean my hands which were black with the ash of grass fires. We came out of bush country on to a plain of a few square miles. A few buffalo were grazing a fair distance off, but to the right of our track we saw two lionesses 100 yards away. They stopped and sat on their haunches to observe us, quite unconcerned. Then we saw three lions 200 yards off on our left and we decided to get nearer to them. Vaughan says he has never seen these lions before, and with considerable skill sidled the Land Rover this way and that till we were only 50 yards away from the lions at a small pool. They also sat down to observe us, as we did them. No fear, and we gently sidled away. Thirty miles back to Ngoma and after a pot of tea Vaughan went back to the lions with his wife and a young man called Johnny Uys, who had come up to meet Mitchell and me. When they came back at 7 o'clock they said what a wonderful time they had had. About 40 buffalo were chasing the lions and had them on the run. The buffalo then went back towards the plains and the Land Rover went nearer the lions. One came near the car and slightly later Vaughan threw one of the lionesses a bit of biltong he had in the car. She came up and ate it. Such is the confidence of creatures which are not accustomed to showing fear. Obviously lions are excessively inquisitive animals. This young man, Johnny Uys, is an interesting character: born in the bush and has a quite remarkable knowledge of the bush. He is handsome, very fit and with a tremendous energy. He is reckoned to know more about lions than anyone else in the Rhodesias and is the hero of many exploits with man-eaters and cattle-killers. This boy is very charming, and having come from poaching to being an ardent preservationist, the Game Department has made him a Game Ranger. He is heir to considerable estates and cattle ranches and will eventually have to leave to take over these places. Meantime he is interesting and I am looking forward to two days with him. There is a plant growing at the foot of many anthills called *Sansevieria*; it is allied to sisal, which is a cactus, but looked at casually it looks like some very strong almost giant sedge. This plant has a fibrous leaf as the sisal has, but it must also be palatable for the elephant chews it and then spits out the bolus of fibre. I took a photograph of the stringy rejects alongside the growing plants.

8 October 1956

Away by 8.30 o'clock to Lochinvar, 100 miles or so. Our first stop was Namwala on the Kafue River, where there is a District Commissioner and other administrative offices. The place is as uninteresting as most of these small towns or administrative stations. The joys of Northern Rhodesia are in the bush. We stopped to make a cup of tea and eat a slice of bread and cheese under a tree which gives heavy shade, *Parinari mobola*. I have seen these trees in many places and I particularly wanted their name when I found them on the western edge of the eastern forest of the Busanga Plain. Soon we came on to the edge of the Kafue Flats proper and cruised across them for miles and miles. How lovely they are! Far away in the mirage are large herds of zebra and lechwe and gradually they become real things taking on shape and colour. We camped at a point where the anthill zone comes close to the flats, called Chunga. There is a hut and a tank for the ranch cattle and one or two acacia trees. The lagoon is 100 yards from us and between us and them are the lechwe grazing unconcerned, and stretching to the left and west are hundreds more. The edge of the lagoon is lined with hundreds of pelicans, spur-winged geese, knob-nosed ducks, whistler and pink-billed ducks, saddle-bill and open-bill storks, marabou storks, and some wattled and crested cranes. Spoonbills, egrets and ibises are common; there are some cormorants and anhingas, white herons, blue herons (like ours), Goliath herons and squacco herons. The flats themselves show what you would expect, wattled and blacksmith plovers, coursers, pratincoles, pipits and larks, and on the anthill edge are the wheatears. Out there in front of us over the lagoon are miles and miles of flats and water, and the impression is of the great ocean. We have set our beds so that we can hear and see, and when we wake in the morning about the dawn time, this splendid array of wild life will gradually come to light.

9 October 1956

It has been as I said. We heard the lechwe grazing near us in the night, and as we look forth they are still only 50 yards away, but as the sun comes up they move westwards another 50 yards, gently and as if casually. A quarter mile east of us is a fishing camp of Africans living the shieling life. Last night they were singing late and well. The ranch allows them to camp here so long as they behave themselves, and this they do. The lechwe, therefore, accept them and graze to 100 yards range. Yet over so much of the flats this lovely gentle antelope is being harried and their numbers reduced by poaching and lack of firm policy by the Government. It may well be that one of the best things I shall ever do in Northern Rhodesia will be to get the lechwe acknow-ledged as being the animal best fitted to use the flats, get aerial patrol established and a conservation policy accepted. The lechwe is the quietest of the antelopes and almost domesticates itself if given peace. This is what has happened on the Lochinvar Ranch of well over 100,000 acres, and the trouble now is that because of the harrying elsewhere the lechwe crowd on to Lochinvar in large numbers. It is reckoned just now that 16,000 of the surviving 25,000 lechwe are here. Conservation-minded people want to buy Lochinvar for the splendid wild-life sanctuary it is, and Sir Alfred Beit has offered to start the £50,000 price with £25,000. All well and good, but unless this is backed up by a proper Government policy the idea will not work. We must get a 5-year total close season on the lechwe, vigorous control work against poaching and see the lechwe spread over their former haunts. The problem of parasitosis through crowding is becoming serious, I think. The wretched Government, having prohibited the four communal hunts or *chilas* from taking place, thereupon allow three of them this year. The Game Department were present as observers and counted the toll by this chasing and spearing to be over 3,000 head of lechwe. Naturally there is no selection and many gravid ewes and small lambs are slain. It seems to me that if we could get a 5-year absolute protection backed by aerial patrol, which would get two new accessions of breeding females to the stock, we might change the *chila* tradition from several hundred of men spearing about to a similar number rounding up chosen herds of rams into a large corral, the choosing of herds having been made easier by the constant aerial patrol which would have already yielded such valuable data on movements of the herds. I would think the orderly slaughter of 3,000 surplus rams passed through corrals and shedding chutes more humane than indiscriminate killing with spears. The social occasion of the round-up would remain and could probably be elaborated to equal the present *chilas*. This morning we drove four miles along the flats towards a string of *mulemas*, which are permanent swamps within the general scheme of the river and flood plain area of the flats. We left the car and waded several miles, at first little more than ankle-deep, but finally we were in the *mulema* complex proper and found ourselves waist deep in *Polygonum* and grass stems and roots. Occasionally we could get on to a floating island of buffalo grass strong enough to support us, and sometimes we would find ourselves going through two layers of submerged vegetation and be up to our shoulders. Getting out or along can be quite a job. I made Johnny Uys laugh so much that he would snatch off his broad-brimmed hat and throw it down on the floating *sud*: the reason being that I was trying to develop a gait which would get me going on *sud* not firm enough to support me for more than a moment. I would get on well for several yards, swinging my arms round and backwards, and taking a quick off at each step, but then plump down I would go to the chest level. Finally one comes to beds of reed mace and clumps of ambatch. This shrub has a multi-pinnate small leaf and a stem thin at the top and thick at the foot where it is below water. The stem contains a pith of extremely light nature. The natives use it for floats on their fish nets. As we came away from the *mulemas* where some of the water birds are thought to breed, we drove through many hundreds of lechwe and zebra. With the decline of the fringe-feeding lechwe, the zebra have greatly increased. Out again in the afternoon through the anthill country of the ranch. This zone has been heavily overgrazed by the game because of the sanctuary quality of the place in the midst of an area of relentless native hunting pressure.

10 October 1956

Waking again to this wonderful spectacle, which is my last day in the bush on this tour. We left after breakfast, I reluctantly, and came through the ranch to the boundary with Native Reserve (Tonga). The change was striking. The Tongas are cattle people but not good ones, and they have beaten up their bush range (as distinct from the flats range) to such an extent that there is extensive erosion, both sheet and gully. The Administration will take no action in reducing cattle stocks to what the range will take. I cannot understand this irresponsible attitude. The Veterinary Department has pressed for action for years and at last got an order signed by the Governor. But the Provincial Administration sat on it for eight months and now only very half-hearted action is considered. Lunch at Mazabuka with the irrepressible Johnny Uys. I respect and admire this boy who is so very much a man. Then back to Chilanga, where the first thing I heard was that Parnell has been confirmed as Director of the Department. I am delighted. Out to Makeni for a delightful dinner and talk with Ronny and Erica Critchley. Ian and his wife and Bill Mitchell were my companions.

11 October 1956

This should have been a good day but it was spoilt and clouded by tragedy. Peter Maclaren, the head of the Fisheries side of this Department, was killed by a crocodile in the Kafue River at 1 o'clock this morning. I saw Peter three days ago at Namwala when he came in for a cup of tea. Apparently he was in the river looking at some nets. He was a man quite absorbed by his work and I can quite see him ignoring risk if he wanted to know something about the behaviour of fish at that time of night. He was a good man all round and he leaves a wife and three young children. I am much upset, but how much worse for Ian Grimwood who has had to go and tell his widow, and take the widow to Namwala for the funeral this evening. Peter's Fish Guard fought the croc and recovered Peter, but he died soon after getting him ashore. Bill Mitchell and I went to do various jobs in Lusaka, and lunched and had tea together. I have enjoyed being with Bill because he is not only a biologist but a fine naturalist, and God knows these are scarce.

* * *

Frank arrived home to what was perhaps the greatest crisis of his life – the death of his wife at the early age of forty-four. Averil's health had been in decline during his absence on these early forays in Africa. The Journals were ostensibly written to her as a continuous narrative letter in which he occasionally expressed longing for her, but little or no concern for her. Beneath Frank the pragmatic man of scientific discovery, was Frank the romantic – the mystic. He spoke to me on occasions of Africa being a great seductress, leading him on and on into her wonderland, and away from the realities of life at home. Livingstone knew it. Most of us, who have left home in temperate lands to work in tropical Africa, have known it.

Frank was in the grip of Africa and was clearly conscience-struck in the few references which he did make to his wife, and in the dream which he had of living with her in the grass house in the Luangwa Valley in the bosom of the Africa which he had come to love. It was his desire that Averil should share with him the opiate of Africa, which life in the bush gave him. Tragically, that dream never came true, for, on returning home to Shefford Woodlands in mid-October 1956, he was to learn that Averil had cancer of the spine, and had only a few months to live. She died on 16th February 1957.

The trauma of her passing was accompanied by the winding up of her estate, and of making provision for the care of their children; the attention to his senior lectureship at Edinburgh University, and to the Red Deer Survey of the Nature Conservancy, both of which had been starved of his energies for over six months; and many meetings and visits relating to his various interests. He was pressurised from all sides to meet his commitments, but no doubt benefited from a wave of sympathy when his wife's illness and death became widely known.

Before she died, Averil made legal provision for the care of the children. There is no place here for the detail of these arrangements, other than to say that the welfare of the children, until they came of age, was not placed in Frank's hands, but in those of trustees. This was primarily done for tax reasons; however, from an account of Averil's words in the last days of her life (given to me verbally by one of the trustees), there is an impression of estrangement; even had she been in the best of health, the grass house in the Luangwa Valley would probably have remained an unfulfilled dream.

Yet Frank's grief was real to me. In her last weeks he wrote to his friend Douglas Grant of her great courage and valiant spirit. Some five years later, when we were alone together at Shefford Woodlands, he spoke to me of her death with deep emotion, and presented me with an old English wine glass with a tear drop in its shank − just to commemorate that moment of grief shared with me.

CHAPTER FIVE

The Bangweulu Swamps, Mweru Wa Ntipa, Sumbu, Nyika Plateau, and East Luangwa

July to October 1957

IN THIS THIRD and last visit of Fraser Darling to Northern Rhodesia he aimed to complete his coverage of the major habitat types. He was now familiar with much of the northern catchment of the Zambesi, by his studies in west Luangwa, Kafue, and Barotseland (now Western and North-Western Zambia). This had given him a great deal, but not all, of the information he required. Though he had flown over some of them in a state of air-sickness, he had yet to see most of Northern Zambia with its vast lakes and swamps, and mountain habitats, situated on the south-eastern drainage limits of the Zaire (Congo).

28 July 1957

After two nights as the Parnell's guest I set forth this morning at 9 a.m. in the big Chevrolet heading for Serenje Boma. The road is the Cape to Cairo highway, dusty but wide and not very ridged once you get up to 60 m.p.h. We did the 270 miles in just five hours, scarcely a slowing down for anything. Met Estcourt the Game Ranger at the Boma and he was ready to do the 64 miles into the Kasanka Reserve if I was. Estcourt a fine figure of a lean young man who knows his Africa. I said yes and changed into bush clothes while the Land Rover was being packed. Estcourt's wife had just gone to Broken Hill to have another baby, the first being an 18-month old Christina, who is now going to be parked on the District Officer on the Boma. The whole journey northwards so far had been through *Brachystegia* woodland, occasionally opening on a grassy *dambo* or, as we climbed gradually to about 6,500 feet, the country would open a little to see the limitless woodland and granite hills in it here and there. At one point I could see eastward to the Muchinga Escarpment, beyond which is the Luangwa Valley. Leaving Serenje at 4 p.m. we went forward on small tracks made by Estcourt's Land Rover in the last few years. Woodland and *dambo* all the way but we were going westward now and downhill to the Kasanka, which is a swampy area with very low wooded ridges and some quite considerable lakes. It was dark when we made camp at 6.45 p.m., but there was a beautifully thin new moon and the soft brilliant starlight of Africa.

We have two African servants in the back of the Land Rover on top of the bedding and food. They now sprang to life and made a fire and set up our beds. Then the evening meal, with the crickets and the frogs as orchestra. We have no tents because it is now the dry season and we shall not need them.

29 July 1957

The most comfortable night since I left home. When I woke in the night the silence was complete except for the occasional foghorn of a hippopotamus in the swamp. No lions heard, yet there are several near here and it was while I was here first that near this place four natives camped on the edge of the Reserve with intent to poach elephant. A man-eating lion killed three of them that night and only one escaped to tell the tale. There is a long history of the lions being truculent in this neighbourhood. We left camp at 7.30 a.m., but before that we had gone to the edge of the swamp and climbed up a crazy ladder into a high tree where Estcourt had made a little platform. We could see a long way over the swamp and 8 or 10 sittitunga were in sight. These shy and rare antelope are swamp dwellers, never coming on to the truly hard ground. They are dull brown and have arched backs, and lift their feet straight up as they move about in the water. The Kasanka Reserve is a small one, only 180 square miles, and its principal animals are the sittitunga and elephants. The swamp gets a lot of Congo elephants on migration and there seems to be a resident herd of around 150, broken up into family groups of 20 to 30, and several bulls who live alone or in pairs. We drove through 20 miles of woodland to a great open swamp several miles long and more than a mile broad. We got there around 10 o'clock and walked along the edge among the high anthills which carry a few trees. This was just the time the elephants came out of the woodlands and waded into the green swamp to eat the lush grass. We were exactly right in time to see this happening without getting in the way of the elephants. A herd of ten cows with nine baby elephants and a couple of well-grown young elephants came out of the woods on our right and walked at right angles to us less than 100 yards away. The wind was just right for us and allowed Estcourt and me to get to an anthill and watch them grazing belly deep in the swamp. I took two photographs. There were several elephant wallows and rubbing trees along the edge of the swamp, and in the bordering woodland a good deal of *masuka* bush, the fruits of which are much liked by the elephants. There are many *Bauhinia* trees also, which bear brown pods of beans much eaten also by the elephants. The animal shakes the tree and the pods fall if they are ripe. The range in this reserve shows no signs of over-grazing or over-browsing. In the 54 miles of bush travel today and several miles of walking, we have seen 59 elephant, 40 to 50 sittitunga, 3 roan antelope, 100 or more puku, 3 bushbuck, 1 duiker, 4 waterbuck, 1 hartebeeste, several warthogs, 4 reed-buck and a dozen or more hippopotamus. We also had a treat this morning in getting to within 50 yards of a herd of 14 buffalo bulls. Two of them were quite magnificent creatures with enormous horn bosses. I still think the great buffalo to be the most beautiful of African game. The movement of the elephant is another lovely thing.

We got back to camp at 2.45 p.m., had some tea and then crossed a lake in a dug-out canoe which Estcourt keeps here. It was well over a mile across, with several great rafts of papyrus sedge. Spur-winged geese were in pairs here and there along the edge, with broods of youngsters. Cormorants, herons, egrets, jacandas and haa-de-daas. Then a walk to another section of the Kasanka River swamp, and by wading out to some anthill islands we were able to get some close views of the sittitunga grazing. Back to camp by 6.15 p.m. Then a bath and sitting down to write this. We came through some areas of *Chipyea* today. These trees indicate better soil. The game knows this and occupies such areas. The African tends to leave them alone because the game are there, but in regions where there is no game, the *Chipyea* areas tend to be farmed native fashion, i.e., gardens for finger millet, maize or cassava.

30 July 1957

Away by 7 o'clock, all packed. We have travelled in the open Land Rover with the windshield flat, and my face burns tonight. We travelled through the Kasanka to the

The monument marking the place at Chitambo's village where David Livingstone died in 1873, visited on 30 July 1957.

Wiassa Lakes, where the water was still at a high level. The 50 per cent. more rain last rainy season has raised the water table everywhere. Apart from a few hartebeeste and two duiker we have seen no game all day. The bush was in very good condition and much was unburnt. We came to a track ending in a T and by turning left for 15 miles came to the monument built on the spot where David Livingstone died in 1873. He remains one of the few heroes I have left, and I was glad to pay my homage. Just the monument in the bush and nothing else. Then back for 40 miles to Lake Wakawaka, where we met Alan Savory, the young Provincial Game Officer who is to be my companion for the coming month. I liked him.

31 July 1957

Forgot to mention yesterday that Laurie Estcourt showed me a hyaena burrow in the bush. It was the largest he had ever seen and I should imagine it was most unusual. There were eleven holes altogether and many cartloads of earth of red colour had been thrown out. The whole system covered quarter of an acre. Most of the holes showed that the animals were down there, for many flies buzzed within the mouth of the hole and there were tracks in the earth outside, much as we found with the badger sett at Cole Park.

We were away soon after 6.30 this morning, making for the Lavushi Manda country on the Lukulu River. We have Land-Rovered 106 miles today, about 30 being bush-bashing and perhaps 40 on the Great North Road; the rest on tracks which are pretty good. We went first to a very rocky kopje of gneiss rock, a very fine hill called Nsalu. Half way up was a wide-mouthed cave like a quarter of a sphere, on the walls of which were rock paintings of geometrical designs and successions of finger marks. They are probably very late Bushman paintings, perhaps as late as A.D. 1800, but possibly going back to the first century A.D. The paintings were in red and buff: no animals were portrayed and the work was nothing like so delicate as those I saw at Kasama last year with Benson and which I sketched in my diary then.

When we reached the Lavushi Manda we were in a great area of poorish *Brachystegia* woodland among kopjes and ridges of highly stratified undifferentiated schist. There were cactuses in the rocks and some dried up plants called resurrection flowers. If you break off these dried and withered twigs and put them in water they burst into green leaf within 24 hours. The Lukulu River is narrow, deep and very swift, rather frightening I thought, and its water was slightly milky not clear, as are so many streams up here on the plateau. It holds many crocodiles and some hippos. Alan Savory and I went for a slow walk for three hours in the afternoon, just along the river. The strip of riverine vegetation is very narrow indeed, perhaps nine feet, and then the *Brachystegia* with its fire complex of vegetation. Alan tells me that what we have called *Bauhinia* is now called *Piliostigma thonningii*. The willow-like tree or shrub along the river may be *Ochna* but can't be certain. The hippo grass on which I remarked last year is probably *Cynodon*. Following down the river we found it went through a rocky gorge, then opened out into a wide pool between 100-foot cliffs and went over a natural weir and a tumble of rocks, very spectacular indeed. We identified a small smooth-leaved shrub in the rocks as *Marsdenia umbellifera*. It has white milky sap and round green fruits about the size of a small apple. There was also a fig of some species in the rocks and it also had white milky sap. A small-leaved riverine shrub which looked something between a *Cotoneaster* and *Lonicera nitida* was *Cadaba stenopoda*, small oval smooth evergreen leaves. We have seen no game except a single hartebeeste all day. But we can tell there are a few waterbuck round here and I just saw the back of what I think was a duiker. This is not great game country and I doubt if it ever was.

1 August 1957

Away by 7 o'clock this morning. We were heading back to the Great North Road, then to Mpika and onwards up the Kasama road to the Mayuke turn. From there another 50 miles down a track to Mayuke to Lulilinga River. On the way, before Mpika, we went off the road to a granite kopje, Kachifukwa, where there are Bushman caves, the ceiling blackened by smoke, the floor covered with bat guano now reduced to an extremely fine powder which kicked up and got into the nose. The cave was 30 yards deep and though there was a strongish breeze outside the cave was still. There were two very small geometrical paintings of much finer type than in the caves we saw two days ago. It was scarcely pleasant stopping at Eustace's house at Mpika when he was not there. He is going on leave and left three days ago. The bush after the Mayuke turn and when we had got down off the ridge was particularly good in shrub and field layer. But 30 miles down we came to a White Fathers' Mission where sheep had been imported and agriculture attempted. The people were not particularly healthy – pot bellies endemic and a good deal of ophthalmia. The countryside was devastated and hard to look at, yet I doubt whether the White Fathers would ever see it so. On again and we came to a great plain, treeless, burnt, and with the small pointed anthills that I have come to associate with such country liable to inundation. There was a roan antelope far away and in the six miles of it we saw 20 to 30 oribi and 2 female reedbuck. It could carry a lot of game but it is down to these tiny oribi. Mayuke is a rather poor village. Our arrival created a lot of interest and we soon heard that Peter Jackson and Jim Soulsby of the Fisheries Department had arrived with the launch *Hydrocyon*. Alan and I are to travel through the Lake Bangweulu swamps with Peter and Jim in this little ship 32 feet 6 inches on the keel, aluminium, with two 43 h.p. Diesel Perkins engines and tubular propellers. Cabin forward and plenty of room aft, shallow in draught. We fed on the launch but had our beds put up ashore. Plenty of mosquitoes but we slipped across to bed quickly and I had a very good night. Very little game seen today, a few duiker and a couple of hartebeeste, and then the oribi on the plain.

2 August 1957

Away in the launch at 6.50 a.m. Through the reeds into the Chambesbi River, which is just a channel between the papyrus swamps. The launch does twelve knots easily and we were soon in Lake Chaya, the current with us all the way. The swamps of Lake Bangweulu extend over an area bigger than Wales, and landward of the swamps lie grassy plains inundated for several months of the year. Islands occur in the swamps, inhabited by the Unga tribe, a primitive and pretty low sort of African who were driven into the swamp by the Bemba 200 years ago. They intermarried with the Batwa, who were the still more primitive Bantu people occupying the swamps. This doubtless saved the Unga because this swamp life is fairly specialized. First they are a fishing people and that means skill not only in fishing but in the manufacture of gear. The swamps show a pattern of many narrow channels and sometimes the Unga cut channels through the papyrus, reeds or rushes to make short cuts here and there. The universal mode of travel is the dug-out canoe. There are one- or two-man canoes, and a few much larger ones made from very large trees. The life of a canoe is about 15 years and suitable trees are getting scarcer as pressure is considerable and the population is increasing. The canoes are pencil-like in shape, beautifully streamlined and heavy enough to have momentum when being poled through the swamp growth. They have about two inches of freeboard and sometimes less. Paddles are carried for deeper water. The poles are the stems of leaves of the *Raphia* palm, very light and smooth and strong. A slightly forked end is often spliced and wired on to the foot of the pole so that it does not dig in too deep in the mud, and each canoe carries a small baling scoop for keeping the canoe clear of water. Two men can push a canoe along at 15 or even 20 miles an hour if they are going flat out, but the usual pace is 4–6 miles. Balance is quite exquisite and the use of such craft means that the water of the swamps is always calm. There are both deep and shallow lagoons in the swamps in which the Unga set their nets. In general the Unga are in water not more than waist deep and as their canoes are made of wood heavier than water, it does not matter much if they should ship water and overturn and sink, but they are much more careful about going out on the Lake proper where storms can blow up in short time. The Lake itself has very few fish anyway, except round the shores. As we sped along the wider and deeper channels in the launch *Hydrocyon* we saw many bream, *Seriunochromius* tiger fish, *Hydrocyon*, and once about thirty large catfish, *Clarias* (No!) in a deep hole in the channel. Two of these fish must have been five feet long. The characteristic fish of the shallower swamp is *Tilapia*, a vegetable feeder, and the muddy fish [blank in MS].

Within the swamps of lake Bangweulu is a large alluvial flat known as the Unga Bank, on which numerous Unga lived and grew rice and cassava. There was a big flood in the Chambesbi River in the mid-thirties which caused a diversion of its course and much of the Unga Bank was inundated. The Unga had to go eastward into Bisa country and feeling between Unga and Bisa is not good. It was partly to remedy this situation that Frank Debenham, professor of geography at Cambridge, came to do a survey in 1949. He and his assistants advised a cut through the great papyrus blockage, to bring the Chambesbi River south of Nsalushi Island, so that the Unga Bank would drain clear again. The cut has been made and it works, but the current is 3-4 knots and the bed scours so much that the distal end of the cut is silting. As far as the Unga Bank is concerned, the Unga seem to be pulling out, preferring the Bisa shore, and such as remain grow only cassava and no rice. The miserable Provincial Administration does nothing about it.

I had plenty time to look at papyrus swamp vegetation and animal life. The dominant plant of course is *Cyperus papyrus*, the most characteristic swamp plant of Africa, with its feathery plume which one recognizes so commonly in ancient Egyptian paintings. The main run of papyrus is 6-8 feet high, but for a few yards it is much higher, 10-12 feet, and carries a much more varied flora. The actual fringe of the channel is often the hippo grass *Vossia*, and there is a plant I call *assegai* weed, about 3 feet high, which seems to be liliaceous and carries blue flowers, though I saw only two or three of the flowers in the course of our 200 mile voyage. There is a beautiful reddy-purple convolvulus that climbs up the papyrus, and there was a good sprinkling of a marsh hollyhock growing to 7 feet high, a purple flower and very spiny on the stem. When I picked one for Alan Savory's press, fine hair-like spines were left in my fingers. There was a yellow *Epilobium* and occasionally a larger red *Epilobium*. The sub-aquatic fringe vegetation included a many-petalled blue or blue-white water lily and a yellow three-petalled *Potamogeton*. A similar white *Potamogeton* in the same habitat may have been just a colour phase of the yellow one. *Arum* leaves were a constant sight in the papyrus, and a little lower a fern rather like our common one at home. A still smaller fern was at the bottom of the profile. We saw occasional small-headed composites, and Alan found a small hair-like leaved yellow labiate. The edges of the papyrus swamps bear one's weight, but farther in and especially when you get through the few yards of higher denser growth, you go through.

Weaver birds build their nests in the plumes of the papyrus. I also kept hearing a bird with a song like a nightingale – marsh warbler. No one could tell me about it and I must look up Benson's new check list. The bird life of the swamps is not profuse as it is on the Kafue Flats, nor so varied. A small cormorant seemed the commonest bird of that size. Swallows were common and seemed to accompany us, taking advantage of the insect life we disturbed. Many pied kingfishers and one or two crested ones. The heron was common, so was the smaller buff-backed heron, and we saw two or three Goliath herons, and occasionally a white heron. We saw three pairs of wattled cranes, one saddle-billed stork and many open-billed storks, the bird which feeds on the big snail *Pila*. The nearer we came to the Lake we saw the grey-headed gull and white-winged terns. Where there were lily lagoons there were always pigmy geese and the jacanas. The black crake appeared when there was plenty of *Vossia* or *Phragmites* to carry it. Cuoculs were common in the higher papyrus and *Phragmites*, and occasionally I saw a purple gallinule low down in such stuff. Once or twice carmine bee-eaters flew over the swamp near us. A swamp warbler was constantly heard but not often seen. Knob-nosed ducks, yellow-billed and whistling duck occasionally rose from the lagoons and we saw two or three spur-winged geese. One of the loveliest sights was of marsh harriers or pallid harriers covering the marsh.

We saw no crocodiles during our voyage. There are relatively few in the Bangweulu swamps because land is scarce and is heavily populated by human beings who dig up and eat crocodile eggs whenever they are to be found. Neither did we see hippos, which are also scarce. There is not enough quiet hauling ground for them and the tribes have hunted them hard also. The means were a multiplicity of canoes and spears carrying floats. Here and there along the sides of the channels I saw places where otters came out and left scats. The vegetation was changed to a little mat of green shorter grass. Sometimes the swamp would firm up a little and the hydrophytic leguminous shrub ambatch *Herminiera* [in margin: *Aeschynomene elaphroxylon*] which I found last year in the Kafue swamps.

The first night we fetched up at Ncheta Island, where there are several villages of Unga, the one we called at being the largest, where the Chief lived. We arranged to hire canoes from him on the morrow and strike south into the black lechwe habitat.

The chief's 'pencil' canoe with interpreter and polemen in the swamps of Lake Bangweulu on 3 August 1957.

The whole village came to the landing channel to see us, and an ugly bunch they were. Many of the women had such bad lordosis curvature as to deform them. The Chief was a very big strong-looking man, pretty crafty as well. He said he would come with us tomorrow and I think that is to see just where we are going. The Unga are ill-disposed to the white man and particularly to the Game Department. These are the people who are hunting the black lechwe to extinction.

The place is alive with malarious mosquitoes and as the launch won't hold us all, Alan and I have to sleep ashore by the village. We have our nets but even so have been nearly eaten alive.

3 August 1957

Away by 7.30 a.m. The Chief and his interpreter in a 'pencil' canoe and two paddlers; the four of us in the largest and roomiest dug-out I have ever seen, with three paddlers, or more correctly, pole men. We were even able to sit comfortably in Government-issue canvas chairs. Even so, when we got back at 4 p.m. I had had more than enough, for the glare from the water was very tiring and there was so little one could do. We struck southward to Itulo Island, where there was another community of similar sort, but they grew a kind of bunched onion, like 20 or 30 spring onions growing together, and they also had tobacco. Otherwise just cassava. It seems that once the African gets a taste for cassava he is like the Irishman and his potato.

We were now in a shallow marsh of more open plant growth, mashenka rush being the dominant, with odd patches of papyrus and *Phragmites*. Lilies and *Potamogeton* grew among the mashenka rush. We pushed several miles south of Itula Island towards Myunge Island, by which time anthills had begun to appear, the large kind which create the islands on which the lechwe can come out of the swamp and rest.

Actually, all the anthills we found were either occupied as fishing camps or were used as latrines by the canoe men. This is a moment to remark on the human ecology of the swamps. We passed canoes every few miles and near the islands much oftener. There were even fishing camps set up on the floating papyrus, the canoe men using them for sleeping, cooking, and drying the fish until a full load was made up. The anthills were obviously important in the open swamps as the only possible dry ground.

The Chief charges men of other tribes fifteen shillings to fish his area and occupy an anthill. We called at one and the Chief took advantage of the moment to have a good feed of mealie meal and fish at the fishermen's expense. He also got a present of fish, which he added to the good quantity of onions presented to him at Itulo. And when we got back to Ncheta he allowed us to pay his pole men as well as our own — two shillings and tuppence the day each. And Peter Jackson gave him two bottles of beer, so the Chief had had quite a good day. Alan and I were so eager for exercise that we went ashore and went almost as far as we could on Ncheta. We went through four villages and were greeted 'Mapulene mokwai' only by old men here and there. The young ones looked sullen and made no move. We notice that the dogs here are all large and well-fed and we have no doubt lechwe hunting is still going on. This antelope will not be saved until there can be aerial patrol and poaching stopped. Once more, it is not hunting merely for their own pots but for export by bicycle carriers to the Copper Belt 80 miles away. The fishing of the Bangweulu swamps is all for African consumption on the Copper Belt. Last year over 1,000 tons of dried fish went out. This is a considerable industry which can be much increased if need be, as the potential fishery is very large.

4 August 1957

Away by 7.30 a.m., our beds lifted from the island and all complete. Took the aerosol into my mosquito net last night and killed all the mosquitoes that got in when I did. Consequently I had a good night. Down current now to Lake Chali, several miles long and really a very large lagoon in the swamps. Then into narrow channels in high and dense papyrus until land ahead of us showed clear as a 100-foot ridge with many trees. The Luapula River runs under this ridge, and here we came against the current coming from Lake Bangweulu. The ridge was the Kapate Peninsular, behind which is Kampolombo Lake. We found this way and that up the Luapula until we came into the open water of Walilupe Lake, then under Mhawala Island and into Lake Bangweulu proper. Sixteen miles across to Samfya where the launch moored and we stayed the night with Peter Jackson.

The Luangwa Valley – 'We come to the Muchinga Escarpment and see the vast Luangwa Valley laid out before us' on 3 March 1956 (see p. 18).

The *ulendo* emerges from the riverine forest of the Mutinondo River into the grasslands of the Chifungwe Plain on 9 March 1956 (see p. 28).

Crossing the Mutinondo River on 9 March 1956. 'The water came only to my armpits and the current was never so strong that I was in danger of losing my feet' (see p. 28).

The Chifungwe Plain – 'magnificent open rolling scenery of grass about 7 feet high', about the size of Salisbury Plain. Crossed by the *ulendo* on 9 and 10 March 1956 (see p. 28).

A baobab tree badly injured by elephants near the Mupamadzi River on 11 March 1956 (see p. 33).

Deep tsetse country – the flood plain of the Munyamadzi River near Nabwalia's village, scene of the elephant hunt on 14 March 1956 (see p. 38).

View of the Luangwa Valley from the Muchinga Escarpment looking to the border hills of Nyasaland (Malawi) which was called 'Poles' Peep' following tree clearance on 20 March 1956 (see p. 47).

Frank Fraser Darling in *mushitu*, relict high forest, at Danger Hill on 23 March 1956. The structure of the ancient forest is well seen with the trunks of high canopy trees and dense understorey (see p. 50).

The Chilabula Falls near the Roman Catholic mission in the Kasama district on 26 March 1956 (see p. 53).

Ronnie Critchley's Barotse cattle at the Blue Lagoon Ranch on 2 April 1956. They were said to survive better than the improved Afrikander breed (see p. 58).

Lozi tribesmen making a dug-out canoe from an acacia tree in the Kawambe-Ila country, Kafue River, 4 September 1956 (see p. 72).

An iron ore furnace made from ant-hill clay near Lake Kaka on 20 August 1957 (see p. 132).

Mzima Springs in Tsavo (West) National Park, Kenya, visited by Fraser Darling on 23 August 1956, and the subject of a test case in conservation, as they were in 1965 (see p. 162).

Hippos in the clear pools at Mzima Springs, Tsavo (West) National Park, 23 August 1956 (see p. 162).

A black rhino and calf *en route* to the Tiva River on 21 August 1956 – 'Once a mother rhino with a calf charged the Land Rover . . . but David was stepping on the accelerator' (see p. 158).

The Tiva River near Tundani Rock, Tsavo (East) National Park, on 21 August, where elephants and rhinos dug in the dry river bed for water at night (see p. 159).

5 August 1957

Spent the day at Peter Jackson's desk writing my journal and pulling the guts out of Debenham's *Study of an African Swamp*. Alan and I left Samfya at 4.30 p.m. for a 50 mile drive to Fort Roseberry to stay the night with Jim and Elizabeth Soulsby. There is a heavy African population here and the bush is looking the worse for a too heavy succession of *chitemene* gardening. The Soulsbys are a very fine young couple, both handsome and with their heads screwed on.

6 August 1957

Fort Roseberry is a very small town which is going to become a boma. At the moment there are three stores and a few houses and nothing else. Away by 10 o'clock heading for Lake Meru and the Luapula Valley by way of Kawambwa, which is a boma and a very pleasant one. We passed through a lot of excellent bush country of mixed trees with a good shrub and field layer. Occasionally as we came to *dambos* there were small *mushitus* obviously suffering from the encroachment of grass fires. At 3.10 p.m. a road gang of Africans asked us to stop, which we did. One of them had been bitten in the big toe by a snake at noon and he was now a very sick man, his foot and ankle swollen with much pain and his head throbbing. Alan whipped out his First Aid case and we had soon given him a 10 cc. injection of anti-venom intramuscularly in his buttock. He was much better within five minutes and meanwhile we had found the small grey snake 15 inches long that had bitten him. We wrote a note to go with the man and the snake to hospital. Reached Nchilengi on the shores of Lake Mweru before 5 o'clock to embark our gear aboard a 50-foot ship built on the style of a Greek caique but carrying upperworks and sun deck added later. She is owned by the Fisheries Section of the Game Department and is called *Ipumbu*. Alan and I have her to ourselves entirely to do as we wish for a couple of days. She carries a supercharged 53 h.p. Diesel engine which pushes her along at 8 knots. A Fish Ranger called Kleinschmidt explained the engine to us, gave us an African crew of three and we took aboard our own two servants. We got under way at 7 p.m. with a good moon and we steamed for five hours to reach Chisenga Island which is in the great estuary of the Luapula River. We cast anchor a hundred yards from a village. I did not sleep well for some reason though I had no fault to find with my bunk.

7 August 1957

Ashore by 7.45 and walking across the island, a distance of 6 to 7 miles. The first two were through almost continuous village, for the Luapula fishing ensures a high population. After that we got into a plain of good grass with shrub-covered anthills and dispersed *Securidaea* bushes. We soon found herds of puku and during the day thought we had seen over 300. We did not attempt actual counts because we knew we could not see all the animals present in the long grass. The puku looked well and there was a goodly number of yearlings and a few newly dropped lambs with the ewes. Then we found red lechwe among the puku and ultimately came to almost all lechwe, i.e., nearer the swamps proper. At one time, looking over the plain, it was possible to see about 300 mixed puku and lechwe at the same time. The animals were not unduly shy but they stayed at 200 yards or more, i.e., beyond the range of a muzzle loader. We counted four wounded beasts – two lechwe and two puku – which means the animals are being poached, probably from the river by fishermen in their dug-out canoes. Chisenga Island is not a reserve, but a Chief's private hunting area, which means that nothing is to be shot without the Chief's express permission. He himself shoots very little and is willing and even anxious to breed up his stock of game. There are no lions on Chisenga but a lot of leopards. There are plenty of wild pig, which cause some damage to the gardens. Food chains on Chisenga are fairly simple – leopard preying on pig and to lesser extent on puku and lechwe. We examined four kills in the course of our walk, all of male puku, one being a youngish adult male, two being males in prime, and the other an old one. We determined this from studying the condition of the premolar teeth in the lower jaw. Gradually the bush thickened and became almost dense. Now there were neither puku nor lechwe. There were quite a lot of acacias *A. albida*, often carrying the climber, family *Curcurbitaceae*, which had ovoid green fruits the size of elongated cricket balls, with papillae looking like spines on the marrow-like skin. These fruits carried a mass of emerald green seeds encased in green jelly. They were most refreshing to eat, with a faintly acid flavour. When one spat out the seeds, they were small and cream in colour, much like melon or squash. One of the acacias carried a vulture's nest, and on a branch a few feet below the nest was a newly fledged young vulture who was faintly worried by our coming. What a beautiful large eye! Brilliant blue starlings flew among the bushes and occasionally some small green bee-eaters. When we reached the water – really the swamp at the other side of the island – we found the shore much uprooted by pigs. There was a lily of some kind growing in fan profusion just here. The stem was 6 to 8 feet high looking like a stalk of dead grass, and the several white flowers were carried as a panicle at the top. The root was a tuber the size of an egg with roots coming from it. Festoons of a purple convolvulus-like flower covered some of the shrubs in this kind of bush and made little tents almost under the bush. A large bat flew out of one of these, crossed a few yards of water and settled in a large bush. It had bright gingery hair on the wings. The walk back was in the heat of the day and took two and a quarter hours steady going. Was glad of a bath and shave. The black from the burnt ground goes up to one's middle.

8 August 1957

A most interesting day. We drew anchor at 6 a.m. and made for Kilwa Island. This island is many miles long and there is a hill at each end. The nearer hill actually formed a red cliff of 80-100 feet against the lake. I am getting on too fast, because before we had left the Luapula River we managed to get near enough to an ambatch forest to drop anchor and investigate the habitat in the dinghy. Ambatch as a forest is something new to me. Here this curious tree grew to 30 feet and over in height, with trunks as much as 18 inches diameter. They grew straight out of the water and it would be difficult to get a canoe along within the forest. The only other growth was a fleshy-leaved rosetted plant that floated on the water and had no moorings. The water was more than five feet deep, but how much more I don't know. The foot of the trunks at water level and above and below was fringed with fine roots making almost a fur round the trunk. This tree is papillionaceous and has a handsome yellow pea flower 2 inches long. The interior of the trunk of the ambatch trees is pulp, full of air which helps to float the tree. We noticed the trunks were fairly easily pushed around. The growth is so dense that the forest is fairly dark inside. Bird life in the canopy seemed to be prolific and included the nightingale-like bird I have heard in the swamps. The edge of the forest is bounded by *Vossia* grass and sometimes some papyrus, and at this edge I saw a small black-crowned and backed bittern with delicate chestnutty underparts and white eye stripe. Then away to Kilwa and half way along the island where it is low and wooded we picked up a *kupassu* (chief's messenger). He was to guide us to some caves at lake level under the cliff at the other end of the island. Here we had good shelter from the wind and were able to examine the caves in comfort. Here were limestone cliffs and the water had made caves at the foot. The cliffs were overgrown with lianas which we had to get through, then a tumble of boulders and then the caves, which were waist deep in water. I didn't much like wading into this place. The caves went in a good 50 yards and were narrow and 12 feet high or a little more. The *kupassu* and the game scout followed us gingerly, but when I trod on some living thing and it wriggled strongly away I cried out, for I hate that kind of thing. The scout hurried back to the entrance and the *Kupassu* went back a few paces, but by this time Alan's torch had revealed thousands of bats hanging from the roof farther in the cave. So we went forward and heard the squeaking of the bats, and the *kupassu* overcame his fear. We reached the head of the cave and were amazed at these close masses of bats, scrambling and scratching over each other like a great swarm of bees. When one would try to disengage itself and fly down the cave, it found it difficult to do so because some other bat was holding on to it. The squeaking was terrific. We reckoned there were two species in there anyway, a large and a small. The head of the cave was hot and stinking of bats and we were glad to go back to the entrance and be sure there was no crocodile around our legs. As always when I come out of caves, the sunlight seems lovelier than ever. We now tried to get up on to the hill above the sedimentary limestone cliff and had to scramble through lianas and thorns. Every wretched plant seemed to have prickles of some sort, and what was worse, buffalo beans were everywhere. These long curled beans are covered with dense fine brown hairs, which if they get on your skin, cause intense irritation. Alan, being in shorts, was soon affected. It came to me much later when I began to get the tiny hairs on me from my trousers. A legewan, that is a 3-foot monitor lizard fell out of a tree near us and scurried into the undergrowth. The lizards had made the beginning of paths through the long grass and prickly herbs and I was glad to use them. Alan would come no farther because of the buffalo beans. This island has no game on it except some sittitunga in the swamps. But soon after getting up on to this dry rough slope I found some droppings of a smallish buck. The game scout and the *kupassu* held to it that these were sittitunga droppings but this is a most unusual place to find them and I still keep a little reserve about it. I was glad of this turn ashore because there had been a strong breeze on the lake and we had been coming beam on; I felt pretty sick. Now going across the lake to Nchilengi the wind was much less and on the quarter. We reached Nchilengi at 3.30 p.m. and were away again in the Land Rover by 4.30. We camped 30 miles on in the good bush we had noticed on the way. A horde of monkeys had run across the road a few miles back and among them was a tiny agouti-coloured buck which Alan recognized as a gruysbok, which though recorded in Northern Rhodesia is rarely seen. It was as small as a hare.

9 August 1957

Away early this morning on a very rough bush-whacking ride to Lusenga Plain. By the time I got there I felt rotten with the pain between the ears which I get after a lot of bumping in the Land Rover. I also felt liverish from yesterday's rolling in the *Ipumbu* without being properly sick. We passed through rich *Isoberlinnia* bush with a good deal of *mushitu*-like growth. The tsetse flies were terrific and we were all well bitten by the time we reached the Lusenka Plain. This is really an extra large *dambo* between three and four miles across and perhaps seven or eight miles long. Someone at an earlier time had thought this 'plain' (wrongly termed) was a good game area for plains game, but in fact the depression – for that is what it is – is covered with poor fine grasses and there is a soft spot along the middle and at places round the edges where there are *mushitus*. Forgetting about the 'plain' as a game area, it is still worth keeping as a Reserve because of the rich bush round it and for the fact of the *mushitus*, which should be preserved anyway. As I say, I felt rotten and mixed myself a stiff 'bush cocktail', that is, an orange drink with Eno's Fruit Salt. I drank a good pint of it and a heaped tablespoonful of Eno's, then had a bit of breakfast before Alan and I and the Game Scout set off across the depression. It was stiff going and I wondered if I would ever get, but from camp I had noticed through the binoculars an elephant and some buffalo on the far side, and I wanted to see the range on the edges of the *mushitus*. The flowers in the *dambo* were brilliant, quite small ones for the most part but they were so many – pastel blue, dark blue, and white. The *mushitus* appear to have grown on springs of water which feed the *dambo*. The *mushitu* forest here are high and wondrously green but are a little more open than those I saw at Danger Hill last year. This tendency to more openness must be caused by elephants which here go into these remnants of equatorial rain forest and browse among the great variety of plants. We were interested to notice that on the lower edge of two patches of *mushitu* forest there was a dense growth of 12-15 foot high ambatch. We found our elephant standing in the bush just at the edge of the *mushitu* and stalked to 60 yards. We then heard movements in the *mushitu* and saw some bamboo moving which indicated there were several elephants there. Eventually we saw a little toto and its mother and another young bull, and decided not to disturb the herd, for our wind was blowing towards the *mushitu*. The old bull was indeed an old one, very hollow around the forehead and high on the chine. He had a nice tuft of bracken in his trunk as a fly switch, and swung this around his head so delicately. We moved parallel with the edge of the *dambo* and on the other side of the *mushitu* saw a cow elephant with a very young toto and a little bull which was evidently her previous calf. They were suspicious but not bothered overmuch. Their trunks would go up feeling the wind and the ears came forward; then they moved gently into the bush. On again until we reached 1,000 acres or more of burnt ground on which the grass and sedge were growing a brilliant green. Five large antelope we had assumed to be roan turned out to be cow sable, all heavy in calf and very dark chocolate brown in colour. This was a nice find, for sable antelope are scarce up here. Among them was one solitary little warthog strolling about in his cocky little way. What nice things they are! The volcanic rock comes just to the surface of the *dambo* here and there, forming a pan of water which the buffalo evidently enjoy. The walk home across the *dambo* was much easier than going out. I had got better myself all the way, and as Alan wanted to talk to his Game Scout in this area, I walked on at a good pace and enjoyed it. Had another 'bush cocktail' when I got home, then a pint of tea and the world seemed good. It was now 6 p.m. and time for a bath. I get black up to the middle from the ash on burnt ground.

10 August 1957

Breakfast before we left this pleasant camp on Lusenga Plain. No game had come out on to the great *dambo* by the time we left at 7.30. It is not a good feeding ground. We bumped the 20 miles back on to the road and came along to Kawambwa Boma through this good bush. We were hospitably entertained to lunch by Gordon and Mrs Gathercole: he is the Tsetse Supervisor in this district. The fact that our trailer was giving some trouble gave me time to write a letter or two. We left at 2 o'clock and did 50 miles over atrocious roads to Lumangwe Falls on the Kalanguishi River, where we made our camp for the night. These falls are so little known, yet after the Victoria Falls on the Zambesi these must be the most sublime in Africa. They are about 150 yards across and 100 feet deep. A large body of water comes over and turns to whiteness, and below are the most terrific turmoil of rocks and angry pools. Vast clouds of spray sweep away down river and create fantastic rain forest conditions on the steep banks below the falls. Were I a botanist of any quality I could spend a day or two itemizing the wonderful plant life, so different from the dry wooded plateau conditions so very near at hand. The roar of water is tremendous and awe-inspiring. This evening it is full moon and Alan and I have gone to the edge. However breath-taking they were in daylight they are even more sublime in the light of the full moon. The great white waters are wraithlike, the foot of the falls is even more horrific in mystery and unapproachableness. And a perfect lunar rainbow came out of the depths and arched over the river. These great falls go unseen from month end to month end, and we wonder how many people have seen them as we have tonight in the light of a full moon. Camped here less than 50 yards back from the edge. I am wondering whether the roar will keep us awake or make us sleep sounder.

11 August 1957

Away fairly early this morning for the Mweru Wa Ntipa, which is a huge depression occupied by lake and marsh. The lake varies in size in approximately five-year cycles, and may go bone dry. There are *Tilapia* fish in the lake and in such quantity that insofar as Lake Bangweulu gives an average fish weight of 9 lbs per 100 yard net and Lake Mweru 29 lbs, the Mweru Wa Ntipa gives 123 lbs. Nobody quite understands why, when the place occasionally dries up. Also there are immense numbers of crocodiles here, more than almost anywhere else. One big *dambo* leads out of the southern end of the marsh, called the Mofwe *dambo*. The lake and marsh are the breeding grounds of the vast hordes of locusts which occasionally break out and devastate much of Central Africa. The International Locust Organization is centred at the north end of the lake and by spraying and what not have pretty well got the locusts under control. If they let up their efforts the locusts would be as bad as ever. Naturally, the organization is expensive. It is now thought that by damming the Mofwe *dambo* the level of the lake could be stabilized at its high level; the result would be – or rather is expected to be – that the fishing would continue year in year out without interruption by dry times, and what is more important, the area would cease to be the biggest potential breeding ground of the red locust in Africa. All well and good, but the Mweru Wa Ntipa area is also a game reserve and the question is how will the flooding and permanent level affect the feeding of a very large herd of elephants (some say 2,000) and a lot of buffalo? My feelings are that such game as roan and zebra that now use considerable areas of the marsh as grazing in the dry season, and possibly the buffalo, would be affected, but the elephant would probably be helped, for they use the marsh for feeding on ambatch and long grass in water up to 4 feet deep. Water in the marsh gives them sanctuary. We came to a locust camp called Mzeshi where there are two or three African families living near a hot spring which flows down into a small *mushitu*. We camped here and sallied forth in the afternoon along a track made by the Locust Organization. This track runs along the edge of the marsh and the edge of the mateshi bush which stretches inland for several miles. This whole area is different country from what I have been through so far this tour. You can't call it low veldt of the Luangwa-Zambesi Valley type, but it is not the plateau woodland and swamp either. *Brachystegia* almost disappears and there is much more *Combretum*, *Acacia*, and *Terminalia*. The mateshi bush is a formation on its own and is always reckoned good for elephants. Mateshi bush is dense but honeycombed by elephant paths, and the species are several, *Combretum*, *Pseudoprosopis* and *Baphia*, among others. At the edges are *Zysiphus* and *Terminalia* and if the mateshi has been burnt out, *Terminalia* becomes the dominant bush or small tree. Some people think of mateshi bush as degraded woodland habitat, but I doubt it. I should say it is a fire-tender formation kept as it is by elephant, and to a lesser extent buffalo. (We also saw another gruysbok in the mateshi.)

The set-up here is perfect for elephant and buffalo, the marsh with its rich feeding of grasses and ambatch in both wet and dry season, and the mateshi bush for dry cover and chosen foods of lesser sort. No wonder there is such a good population. But both elephant and buffalo are very shy because the Africans at the south end of the marsh are completely ignoring the game reserve laws and are hunting hard. What is worse, this locust track gives a perfect line of mobility on which it is possible to interrupt the elephant and buffalo as they come out or go into the marsh. Furthermore, it is probable that from here there is an ivory racket being run into the Congo or to Lake Tanganyika and by water to East Africa. As I say, we travelled this track this afternoon and found the remains of a dead elephant. Next we met an African with big awkwardly filled sack on the back of a bicycle. Alan ordered him to open it up. He opened it a little and said, 'Pig', and so were the two top pieces – warthog. But underneath, all the smoke-dried meat was buffalo. 'Where did you get this?' He bought it. 'How much?' The man was getting troubled. 'Where did you get this?' He took a sack of cassava over to the Congo and exchanged it for the sack of meat. This was very unlikely. Alan confiscated the meat, took the man's name and the number of his bicycle and told him he would have to go to court. That evening Alan got suspicious of the two Game Guards of this reserve who lived at Mzeshi but were away today. Alan went to their huts and told their wives he wished to come in and inspect. This he did and found a lot of buffalo meat. Our servant William has also been scouting and reveals a fairly well organized racket. Alan says he must now catch the Game Guards and take them to jail, and the man who was carrying some of the same buffalo must go too.

12 August 1957

Alan sends William off scouting, and with two local boys we set off to cross the marsh to Nconta Island, a long narrow ridge of perhaps 150 feet, running north and south about four miles away across the marsh. It is rough going, first up to the waist in water through ambatch and *Vossia* but only for half a mile, then into ankle-deep stuff pitted with elephant footmarks into which one stumbles perpetually. Then on to dry but soft ground in which the elephant marks are not more than 6 inches deep. The grass is 7 foot 6 inches high. We go forward working by the sun, making a good pace which I find pretty tiring. We cross the marsh in an hour and a quarter and find ourselves on Nconta Island, which should be covered with mateshi bush but is tending to lose it because of burning. The Africans are obviously burning and doing as they like. We climbed the hill and looked northwards over the marsh. A large herd of buffalo were two or three miles away and Alan and I counted independently and said between 250-300. We crossed the island to the west side, less than a mile, and walked round the south end. A herd of fifteen elephants was out in the marsh, but dotted here and there in the marsh were fishermen, nearly a score in all, and in the Reserve where fishing is not permitted. What could one expect in the way of proper care when the Game Guards are traitors? We had seen only one roan antelope in the dry part of the marsh, rolling fat on the good grazing, and there were a few zebra on the island. The place ought to be thick with game. Back again across the marsh and then a potter in the fringing vegetation before driving back to camp for tea. Now William came in with a Game Scout and Game Guard, bringing the two Game Guards who are helping run the poached meat racket. What was bad for them was two bundles of eland meat and the animal's head which William and the Scout had found. The disgraced Guards said the eland was a lion kill and the buffalo had been killed by poachers and they had merely taken the meat. Alan did not find it difficult to upset their lying. The African constantly employs lying but is not clever enough to sustain it and constantly lies himself into confession of his guilt. Alan stripped the two Guards of their belts and hats, and made arrangements to take the whole party of malefactors to jail at Mporokoso tomorrow. Alan and I broke camp and came nearly twenty miles into the Reserve, down the locust track northwards. We camped under a large fig tree. Mosquitoes rather bad and particularly so when we went down to bathe at some elephant wallows below the track. The water situation is not too good, no clean water either for washing or to drink. Alan shot a guinea fowl yesterday and we have had it roasted tonight. Our cook almost cremated it and we did not enjoy it as much as we had hoped. Our cook is really a shocker and so dreadfully slow. How lucky that I have such a good set of guts to put up with the standard of feeding of the last fortnight! How wonderful it will be to get back to Wigforth's apple pies and some prime venison steaks washed down with a bottle of claret!

13 August 1957

Alan away early this morning to take the men to jail. I shall be here alone for a couple of days and I hope to get some writing done, a chance I am not likely to have again. I wrote solidly until 1 o'clock and had a bite of lunch before leaving at 1.30 with William and the Game Guard, who had by now reported himself to me at the camp. We went eastward across a dry depression of acacia scrub much used by elephant and buffalo. We came on to the edge of the marsh and walked along to a point of mateshi bush on slightly higher ground. In effect we were going along the high road used by the wild life of this area. Not only were there the trails of the big game, but I could see the spoor of genet cats, cervals, a hyaena and a leopard. Last night, or rather early this morning, I could hear a leopard's voice. Here along the edge of the marsh were many birds I now know well — spur-winged geese, knob-nosed and whistling ducks. There were two or three sandpipers, a flock of pratincoles with bright red on the upper beak and a charming red line pendant from the side of the face and passing under the throat. I think this bit of bare land among stones between bush and marsh was their nesting ground. The birds made a bit of fuss, enough to startle an 8-foot crocodile into the water, which must have been about 6 feet deep a few yards out before the giant sedges, for a hippo popped up its head and made a bit of a fuss. So it might, because I had seen 30 yards further on two baby hippos lying together ashore in the mud. I got right up to them before they toddled into the water, more pained than frightened. They were about the size of a moderate pig. Which reminds me that we also saw five wild pigs rootling at the edge of the marsh. They are not so attractive as the warthog. There were 17 puku on the point and as they ran off into the bush another long, low, grey creature with a very long tail bounded away also. It was larger than an otter and I wondered if it was some very large mongoose. William, who is a Bemba, said they called it 'vaow', so I must ask Alan what it was. The water's edge had not quite finished its vividness in display of wild life. There was an elephant 70 yards out in the marsh eating ambatch. We walked openly past him and he took no notice, but when we were well past he got our wind and moved slowly farther into the marsh. We walked on and spied the horn bosses of buffalo at the edge of the scrub 70 yards away. There were five bulls watching us. When we stopped and I looked through the binoculars they moved into the bush at the run and then it was obvious there were a lot of them. I asked myself, 50 or 100 by the amount of sound? I said 100 to myself. We came up to the edge of the bush a few yards farther on, when suddenly there was a thunder of galloping and through my binoculars which help to see through thick bush I could see the herd coming towards us. William and the Game Guard urged me to run, which I did, at right angles. They shouted and it turned the animals so that they ran parallel to us and out into the open and away. It was a herd of bulls, cows and calves and I counted exactly 100 though I think there were a few more. Home for a large pot of tea.

14 August 1957

Away by 8 o'clock with William as carrier and the Game Guard also with his Lee Enfield .303 (a relic of 1914–18 but still a nice rifle) for a three-hour walk into the mateshi bush. We went for a good way up a lightly shrubbed flat between two ranges of mateshi-covered hills, then struck left up the hill to the ridge. The elephants had punished the strip of perhaps five yards between the flat and the stony hills. The mateshi bush was pretty dense and inside one ever-green bush the Game Guard pointed out the lying-place of a gruysbok. Unfortunately he cannot speak English at all, and William's is sketchy, so I have to use a lot of sign language. It is one of the striking things in Central Africa how few Africans have even a smattering of English. When we reached the summit ridge, really a flat boulder-strewn tableland 500 yards wide, I found it had been burnt, much to its disadvantage. It is quite evident to me that mateshi will not stand burning. The bark of the shrubs is not of fire-resistant type at all. There was sign of elephant and buffalo everywhere, but we saw none. At this time of year the mateshi is completely waterless. Gradually down again to almost marsh level, to what might be considered light acacia scrub with much long grass. Coming back to camp through this we saw two young female reedbuck.

A shave and tea and settling down to writing. Tsetse are bad, and though our camp is well up in the dry away from the marsh the mosquitoes cannot be ignored, even in the heat of the day. Last night I had ordered dinner for 6.30 and thank goodness I did. The mosquitoes were thick and I could scarcely eat. As soon as I had finished I retired beneath my mosquito net and read the Meinertzhagen *Diary*. Mosquitoes make such a mess of me and candidly I feel quite ill from the multiplicity of bites, each of which leaves a big red lump. Fleas and mosquitoes both have this effect on me, of producing what seems to be a light fever. I have another full week of this before getting to the more delightful sand beaches on the shores of a clean Lake Tanganyika.

Here, we start washing in water you can't see through at all, and even boiling for drinking seems to remove but little of the mud in suspension.

This whole area carries a good head of guinea fowl and of francolins (partridges). One sees a score or more francolins each walk out of a few miles.

Alan got back at 7 o'clock tonight, bringing his wife. Both of them are twenty-one. Alan is such a developed young man, and in the responsible post of Provincial Game Officer I think he is doing remarkably well. He knows his Africans inside out and is good to them. His standards are very high and one wonders how far the rough and tough of petty politics and jealousies within Government service will make him cynical. Already he is well hated because he knows more and does a better job than most chaps in the Department twice his age.

15 August 1957

We left this mosquito-ridden spot by 7.45 this morning and came along Locust Organization roads for 100 miles to the shores of Lake Tanganyika. We passed through a good deal more of the Mweru Wa Ntipa reserve through some excellent bush with *Borassus* palm quite common. Then outside the Reserve, still in beautiful bush, there were new villages which had come there because of the Locust Organization tracks. In other words, they are there for the killing of the big game. The District Commissioner should never have let these people in to this site, but this is how things run in this Territory: the Administration cares nothing for the animals and definitely hates the guts of the Game Department, so no chance is lost of spiking its guns. We came 72 miles to a Locust Organization Post where their chief mechanic lives and where we also met their road construction man. This latter spoke of a hill ridge within the Wa Ntipa Reserve where there is a water hole and where there is a great deal of game because of the water. He is going to run a new track to the water hole and onwards; in other words, bend it to the water hole so that a Locust Camp will be able to get water from it. This of course will let the African in with his bicycles and the game will soon disappear. All this is projected without any reference to the Game Department or to the fact that the area is a Reserve. It is apparent to me that the situation in this area is completely out of control. The tribes are definitely awkward and are looking on both the Wa Ntipa and the Sumbu Reserves as being their private hunting grounds. With the muzzle loaders and black powder which this fool Government allows them ad lib. they will soon finish off the natural resource of game animals. And there will be nothing to take their place. If Britain continues her idiotic policy of handing over to the Africans as quickly as possible, the tribes of Northern Rhodesia will be in free barbaric control and there will be bloodshed very soon. If this is to be the only way of reducing the numbers of people and keeping them in a condition of perpetual warfare the animals may get a chance. The Administration refuses to see that the big animals are a natural resource to be treasured and wisely cropped as a meat supply for the African.

The view of Lake Tanganyika as one comes down the wooded hills is quite magnificent, lovely blue water and hills. We have run along the Congo border for many miles and now we see Northern Rhodesia and Tanganyika on one side and the Congo on the other side of the lake. Northwards it stretches two to three hundred miles. We stopped at an African store kept by a German called Paul Mathas. This man was originally a monk of the White Fathers, who have many missions in Central Africa. He is square-headed and shortsighted, but full of humour and good brains. He gave up being a monk and married an African woman and now has a brood of little coloureds. He is still a devout Catholic and his children are being well brought up. In so many ways I feel this is what the African needs if he is to come out of barbarism at all — an infusion of white blood. The Africans of Central Africa are pure Bantu and have no Nilotic or northern Arab strain which might make them fitter for accepting the truer values of civilization. I met a 'coloured' man the other day who is an under-ranger of the Game Department. He is doing splendid work altogether and I found him pleasant to meet and talk with. He is the result of a District Commissioner's misdemeanours in this northern province. Mathas gave us a most hospitable lunch and was a delightful host, but the African wife did not appear at all. We hired a native-made boat from Mathas with a 10 h.p. outboard engine. The boat was not impressive and it leaked, and squalls blow up quickly on Lake Tanganyika. We had twenty miles to go from Sumbu to the base of a long peninsula of the lake. The wind did blow up a little crossing Kamba Bay and we had to go slow, for the boat is not really seaworthy at all. Schools of fish broke the surface of the water here and there, and flocks of small grey gulls dipped at them. These gulls had a habit of flight very like the fulmar. The wooded country of the lake shore is wonderful bush for game, but it is being heavily poached from the water by the African. It is scarcely credible that the Game Department has no boat at all on the lake and is content to risk the lives of its officers in crazy hired outfits like the one we were in — flat-bottomed, no proper ribbing, no keel or king plank, and no lines.

Anyway, we got to the narrow neck of the peninsula, the neck being a sand dune 500 yards wide. The Game Department has set up a camp here, but it is right in the road of the game coming and going on the peninsula. We went over the neck to a glorious sandy bay nearly a mile round called Kasaba Bay, with a great width of the lake out before us. No mosquitoes here and apparently no tsetse. We bathed in the lake, and for me it was heaven. Lovely clean water of just the right degree of warmth, the sand underfoot, and the fact of getting really clean. My numerous mosquito bites irritated much less after this wonderful leisured bathe.

16 August 1957

This spot is such a heaven on earth that I have done little but stroll about the isthmus and bathe in the sandy bay. There are plenty of crocodiles here; you can see them swimming on the surface if it is calm, but they do not have the reputation of being man-eaters. All the same I do not go too far out. There are certain things I prefer to be careful about, and the crocodile is one, especially as the water is not my natural element. Another thing is snakes. If one doesn't know them intimately — and I don't — it is better to leave them alone. Alan Savory's wife, going to the chimbuzi last night gave a yell, and well she might – there was a black cobra on the seat. Alan was all for abstracting it and letting it go, but I said no; if it was wandering around a camp it might go into the Game Guard's hut and bite his baby or something like that. It is quite a different thing leaving a poisonous snake alone in the bush: he is where he belongs. So reluctantly Alan fetched his .22 rifle and shot the cobra through the head.

A grass hut has been built round a very large *Acacia albida* tree. This is a leguminous tree, the fruits of which are curled beans of biscuity texture when ripe, rather like French croissants. Elephants, buffalo and other grazing and browsing animals eat the beans which are of high feeding value, and the hard seeds themselves are freed and spread in the animals' dung. Another legume here, *Cassia abbreviata*, has tremendously long pods, two feet and over, and each one contains hundreds of hard seeds, each covered with a layer of what looks and feels like dried jam. The birds eat these beans, or rather the seeds inside the beans, and spread them that way. This sandy isthmus might have been inhabited in the long past, for the vegetation shows signs of fire. Just south of the isthmus there is a stretch of *Terminalia* woodland typical of a fire climax. Yet on the peninsula and well south of the isthmus the bush is a pure mateshi association and very dense. Apart from this dune sand of the isthmus the country is sandstone, and the shore line is a tumble of boulders. The sandstone is reddy-brown and well compressed. I went to bathe this afternoon by myself and took a walk along the whole length of Kasaba Bay and round the boulder shore to another bay, only the head of which was sand. Mateshi bush comes right down to the boulders. At one point along the sand shore I saw a small snake curled up just where the wavelets could touch it. I threw some small shells on it but it did not move. So I got a long bamboo from high water mark and tickled it to see if it was alive. It was, and after looking surprised wriggled into the lake and swam away quickly. Then it put its head above water for a few yards and looked around, but it did not come ashore again. There are four or five bushbuck on the shore on the west side of the peninsula, and on the edge of the *Terminalia* and the mateshi this afternoon I saw a group of eight waterbuck. They were of the dark-behinded variety.

17 August 1957

A quiet day, strolling around, soaking up impressions of the habitat and having two glorious bathes. I am taking advantage of these few days as a holiday and a rest. Alan went fishing this morning and I went out with him in the late afternoon. Result: two large fish, a yellow belly and a Nile perch, which provided an adequate meal for the three of us tonight. I could eat more than we are in general getting, and we eat only twice a day anyway. But all this must be good for the figure; nevertheless Wigforth's apple pies are a delicious dream. We are getting no game meat because we are spending so much time in Reserves, and boy Alan takes his job very seriously and is dead against shooting meat for the pot because of the example to the natives. So apart from the two tough guinea fowls and this fish, everything comes out of a tin. If this young man goes on feeding irregularly and in the manner he does, he will soon ruin his tummy.

This little bit of acacia woodland near camp holds some quite magnificent *Acacia albida* as individual trees. The elephant spend a lot of time under them and the buffalo come grazing the pods. Often a *Trichilia* grows in association with these large acacias, perhaps taking advantage of its shade. The *Trichilia* has a very dark green foliage, contrasting with the light green of the acacias. I have been into the mateshi and am convinced that it represents a climax and that it is fire tender. It is extraordinarily pleasant to be in, with its elephant paths. Quite a quarter of a mile from the lake we found a crocodile skull – a very large one. I abstracted a canine tooth to bring home as a curio for the boys. The teeth are hollow all the way to the point, and new teeth grow up inside the old ones to replace them as they wear out. The elephant, I noticed from a skull in the bush the other day, has another good way of keeping its teeth going. There are three large molars on each side in each jaw, i.e., twelve in all. They grow in from the back and work forward and as the front ones wear out, portions are shed off the front of the tooth, a most ingenious and convenient arrangement. There are about a dozen very large crocodiles in the bay by the camp. They come out to bask in the sun about 400 yards away but are no trouble to us. We have seen none in the great sandy bay on the other side of the neck of the peninsula, but there was a hippo there this afternoon, 100 yards out. No trouble, but he did throw himself up once and open wide his jaws as a demonstration. There are bushbuck near the camp, four or five of them, and moderately tame. The elephant and buffalo do not come near the camp in the day – they are hidden in the mateshi, but they are less than 100 yards away each night. Gruysbok in the mateshi here.

18 August 1957

Did very little but bathe and walk about in the acacia woodland. There are several bushbuck about camp, relatively tame. Farther round the bay the big crocodiles were lying out and I stalked them in desultory fashion. I got no nearer than 35 yards, the bushbuck giving the alarm. It blew very hard in the night and there has been quite a lot of wind all day. Finished reading Meinertzhagen's *Kenya Diary 1902–06*. What a show-up of the Colonial Office it is!

19 August 1957

A big blow in the night and a lot of wind this morning. We left our camp at 10 o'clock in the crazy boat, complete with the Game Guard of this area and his family of wife and five children. They had been camping on the isthmus also for a few weeks. The children, from about 12 down to 15 months, were so well behaved and quite delightful. We got out of our bay and after a couple of miles turned into Kamba Bay in order to hug the coast all the way. The seven miles across the bay seemed too much of a leg of open water, and so it was because a wind known locally as the *kapata* blew strong. We had no more than an inch of freeboard at the stern as the swells came behind us. I didn't like it at all and took over the steering myself. The wretched boat has no keel and yaws about unless your hand is constantly at work on the tiller. If we move the load forward in the boat the propeller comes out of the water, and when we are trimmed deep at the stern we are in danger of being pooped. All the same, thank God we weren't going the other way, otherwise she would have pounded as she did the other day, and the water spouts up between the planks. Anyway, we ran her up full tilt on the lee shore, which was the sandy beach of Kamba Bay at its head. Then the animate cargo went forward quickly and I went over the side to get the boat up the sand a little more, and I took the Johnson 10 h.p. outboard engine off the stern. We shipped practically no water over the stern and we kept all our stuff dry. I was delighted to get ashore here and explore. We had come up the bay between hills of fire-climax woodland, and now in a two mile wide strath at the head of the bay was a flat expanse of alluvial sand and mud growing a typical African parkland of *Acacia albida, Ficus* and tamarind. I carried the native 3-year old across a little marsh to where we made the fire for some tea, and the native wife made a dish of mealie porridge for her brood. We noticed a honey guide calling above us, so Alan told the Game Guard to follow it and get the honey. This bird, which I have now heard so often, has a note like rubbing a wet cork quickly up and down a bottle. It comes to man and to a badger-like creature called a ratel and makes this call to persuade the man or ratel to follow it, when the bird will fly on ahead for a quarter to half a mile and point out a wild bee's nest. It is now the man's job to get the honey out and give the bird some of the wax which it likes very much, even though it is improbable that the bird is able to digest beeswax any more than we can. Today the bee's nest was only 100 yards from where the bird called us. Alan whistled back to it to let the bird know we were going to follow it and it called most excitedly. The nest was in the base of a large acacia tree and the Game Guard had quite a lot of work with his axe to get a hole in to the honeycombs, especially as a boulder had been grown in to the base of the trunk inside. The bees were kept quiet by a wood fire having been lit to windward and the smoke blew over the man while he was at work. Eventually he got a nice lot of honey in a basin and some comb was put out for the delighted honey guide. The taste of the honey was exactly like our stronger honeys, but you have to go a little carefully in eating it, i.e., if you are uncommonly greedy. Alan and I went a good walk up the strath, seeing most of a hundred puku, and seeing evidences of plenty of buffalo and of elephant. The elephants were piercing the *Acacia albida* bark here and there and one or two trees were completely ring-barked. There were two or three merula trees, *Stertocaria* or mongwa up here, the fruits of which the elephants are very fond of, and it would appear that the elephants take care of these trees, because they are never broken or damaged by the elephants. (They are sometimes.) The position of each merula tree in a stretch of country is well known and remembered by the elephants. Alan has had a constructive idea about elephant control, that where merula trees exist in areas where gardens are raided by elephants, the merula trees should be cut out so that the animals will not come in to them and damage a lot of other stuff incidentally. We went on the rough in the strath where there was a lot of *Vossia* and elephant grass well trodden down by the elephants and buffalo. I look upon this mass of grass trodden down as a measure of natural conservation of the soil beneath and not as waste and the need for fire as so many interpret it. This mat protects the soil physically in that the heavy feet do not go through and make those elephant foot holes which are always possible starting points of erosion, and organically in that there is some chance of some of the surplus grass rotting and getting into the energy cycle. There was one long erosion gully running down the strath to the lake, but if it does not get worse it probably serves some purpose in varying the habitat. I came on a small flock of stone curlews on the edge of it, and a Goliath heron was also fishing in it lower down. We found old village sites not far from the lake, and farther in, two miles, near a *mushitu* of perhaps four acres. There were old cassava mounds also, showing that occupation could not have been so very long ago. One hut circle had the fireplace intact with the three bosses of clay, 10 inches high by 8 inches diameter, serving as hobs at the edge of the fire. The *mushitu*, in standing water and obviously penetrated by the elephants, called for exploration, but we had no time. A little lower down on the west side of the strath was a 100 acre swamp of giant sedge in which three bull elephants were keeping their feet cool. We went no nearer than 200 yards as time was short. The lake had calmed and we got away by 4 o'clock, making a good pace of perhaps six or seven knots. We saw nothing on the shores of Kamba on the west side except a sleeping hippo. The steep hillsides are just standing trees of even age with no undergrowth, a typical fire savannah. Round the point on the lake side, Cameron Bay as it is called, there were elephants and bushbuck near the shore, and the mateshi bush was still intact. It was dark when we reached Sumbu and deposited our native family. The high wind which bedevilled us today and which has blown in the night these three days is known locally as the *kapata* and has the effect of sending great shoals of small sardine-like fish to the southern end of the lake. The larger fish follow them to feed on the shoals. As we came into Sumbu, the native fishing fleet was just going out in their narrow 14 foot boats. A bowsprit on each boat carries a lighted incandescent paraffin lamp, the light from which attracts the shoals of small fish, which are then scooped out with wide-mouthed nets of mosquito netting. There were over 50 boats being paddled out, making a pretty sight as they fanned out over the lake. We camped.

20 August 1957

Elephants came to within 25 yards of my bed last night to feed on the biscuit beans of the acacia, but they made no fuss, nor did I. Had quite a long talk with Paul Mathas before leaving Sumbu. He is so intelligent and typically one of those liberal Germans that one tends to forget exist. Would that there were more Britons of his calibre in this northern area.

Twenty miles on a Locust road and then bushwhacking through poor bush to a very large *dambo*, at the northern end of which was Lake Kaka. There was open water in the middle (thanks to a herd of hippo) but most of the lake was papyrus and sedge. The Game Department calls this great *dambo* a plain, but this is quite wrong: in this country a plain, i.e. almost treeless grassland, is somewhat raised and lacks trees because of drought conditions at some time of the year. A *dambo* is free of trees because of waterlogged conditions for the greater part of the year. This particular *dambo*, within the Sumbu Reserve, gathers game from the surrounding terribly poor bush at the end of the dry season, because it offers the only water in the area, but such a momentary concentration of animals does not make it a great game area. I saw nothing in the *dambo* itself except a solitary hartebeeste, but the *dambo* edges showed severe signs of over use at some time of the year. Anthills were bare with the skeletons of bushes overbrowsed. I have rarely seen anything as bad in Northern Rhodesia and I put it down mainly to the poverty of the surrounding bush which is the remnants of over-firing and also, in this instance, of cutting at some period in the past. On the side of the *dambo* by which we had reached the lake I had chosen to walk and follow on to the camp, and at intervals of a quarter mile I had found the remains of three smelting furnaces built by anthills, the anthill earth having been used evidently to make the large clay blocks of which the furnaces had been built. The inside of the blocks remaining were vitrified. A heap of slag was near each furnace, very heavy stuff, as if all the iron had not been smelted out, and there were the remains of clay pipes of an inch and a quarter bore and in some of these short lengths were cores of impure iron as it had run from the furnace. We had a midday breakfast and then Alan and I set forth across the *dambo* through a bit of marsh and a bright green stretch of *Syzigium* woodland to the line where *dambo* edge and mateshi met, an area of indeterminate woodland. Here we found many furnaces about 100 yards apart, intact or nearly so. We were quite excited. Each was about 10 feet high and 6 feet across at the base, with holes at the bottom like Gothic arches. Sometimes there would be a peephole at eye level. The furnaces had fired red. There was the heap of slag and the 4 to 6 inch lengths of clay pipe lying here and there on the slag heaps. It was interesting that the elephants had not pushed down these old structures, which should now be looked upon as of archaeological value. I must try to find out something as to when these were used and when they stopped being used, and where the ore came from. There were no signs round about of any ore extraction. The hills were of the sandstone I have already described. In the woodland of this furnace zone we saw a herd of sable antelope with some very young calves. The bull was a magnificent animal, the blackest I have ever seen, and his horns were phenomenal. Alan said he had never seen one like this before and that we were probably looking at a record head. What a good thing he and I are not interested in record heads! We had better shut up about the existence of this beast. He was curious and came several yards towards us to give us splendid views of him. Of all the antelopes, I think the sable is the most spectacular and even thrilling in a queer way. He is so splendidly built and full of tight power. The furnaces and the sable quite made the day.

21 August 1957

A grazing hippo came into camp in the night but didn't bother us. I listened to his galumphing and snorting sounds. By early morning they were all back in the lake, sounding their great *basso profundo*. A long round of bushwhacking today in the Land Rover to a huge *dambo* called Tondwa. Again the Game Department miscalls it a plain. We passed a nice group of five roan antelope with two calves on the way. All we saw on Tondwa were four zebra, but there was plenty of sign of buffalo. Indeed we have a job to find water that was not just a soft spot trodden up by the buffalo, when we stopped for breakfast under some *Syzigium* at noon. The tsetse flies were really dreadful today and it was very hot. Had quite a bit of excercise going ahead of the Land Rover to find the best way. I watched two Bateleur eagles which were calling as they flew and soared around; as the call was made the bird lowered its feet. Back to the Locust road after three more hours of bushwhacking, and then along towards Kampinda, which is on the escarpment on the western shore of the lake of Mweru Wa Ntipa. We stopped to camp at a disused Locust Post, near which a village had arisen. We were well inspected by the inhabitants, who gathered in a great ring. Alan shooed them all away but they were back next morning. We had seen six impala in the bush near here, which surprised us.

22 August 1957

Alan and I went down to the lake shore at Kampinda before breakfast this morning. The village on the escarpment is abject and filthy and the children did not look healthy. Down below was the chocolate-coloured lake, the shore also filthy and littered with dug-out canoes. Away to our right was a small island which looked like a volcanic boss or extrusion. It is dry and waterless and is a breeding ground of the Goliath heron. The island may be three or even four miles away and as the sun gets higher the wind rises. Alan has been to the island by dug-out and was almost capsized. He surprised two natives who had almost cleaned up all the young Goliath herons. This is in a Reserve, and the fishing camp at Kampinda is allowed as a concession from the Game Department. Once in, the Provincial Administration objects to the natives being kicked out, even though they use the concession to poach in the Reserve. Leaving Kampinda we made for Lake Chishi, which is almost part of the Mweru Wa Ntipa. It lies below a 400–500 foot escarpment of sandstone and is quite spectacular. There were marabou storks nesting in the cliffs, the young sitting upright on their hocks and looking like penguins. There are two *mushitus* below the cliff and between them is a fishing camp. Dug-out canoes only are used for this fishery of *Tilapia* and bream. In all this primitiveness the natives are using nylon nets. A young native down there was holding forth, in freshly laundered shirt and trousers, and bashing a Bible. He turned out to be a Jehovah's Witness, a group that pretty well stands for anarchy here. They call themselves *The Watchtower*. I didn't think he was getting far here among the fishermen. Back then along an atrocious road for three and a half hours to Mporokoso, where there ia a boma, through some of the poorest and worst-treated bush I have seen in the Territory. All this was round Nsama. Chitemene gardening was being done before the bush had recovered at all from the last round. It was poor sandstone country anyway, which would have been better left alone. We met two natives pushing bicycles loaded with baskets of pleasing design. There is a village near here which specializes in this craft. We stopped and bought some. I could not resist buying a most useful lidded basket for five shillings and I shall try to get it home, though that may be scarcely possible. It would be so useful in lots of ways. Called on the District

A fishing camp near Kampinda, Lake Chishi – 'in all this primitiveness the natives are using nylon nets,' 22 August 1957.

Commissioner in Mporokoso. Charming fellow and all that, who pretty well expressed his contempt for the Game Department and did not ask me to have a cup of tea, though it was the right time for one and it was badly needed. The Lake Chishi fishing camp was another concession of the Game Department, and now this D. C. is building a road to it for fish lories and wants the whole area excised from the Game Reserve. We left Mporokoso at 4.45 p.m. and I drove the 114 miles back to Kasama, which we decided was worth trying to make that night. In at 8.30 to Alan's little house, just twelve hours after we had left camp. Was ready for food and yet was past enjoying it. Food situation is bad, not enough and absolutely rotten cooking. I refuse to insult my guts with something badly fried and dripping fat, and usually half cold. It is elephant fat at that, which might be quite good properly used, but not this way. The journey to Kasama in the open Land Rover in the dark was very cold, especially as one came down and crossed *dambos*. Got a warmish bath by 10 p.m. and got clean, which in itself was quite hard work. Three days here now to get my writing done and washing done and so on before setting forth on the next *ulendo*, this time into the Rukwa Valley of Southern Tanganyika as guest of Desmond Vesey-Fitzgerald, whom I met here in Kasama at Birdie Benson's house eighteen months ago. Then away again to the Nyika Plateau in Northern Nyasaland. Five weeks gone and six to go. It is quite a severe physical strain which could be better borne if the food situation was in better shape. We need Wigforth on this expedition!

23–24 August 1957

Kasama. Getting up to date with letters and having stuff washed. Met a few of the Boma types, including Kerfoot of the Forestry Department, whom I have met before in Oxford and here. He is going to the East African Forestry Research Organization H.Q. near Nairobi.

25 August 1957

To Abercorn, 114 miles, today, to link up with Desmond Vesey-Fitzgerald for the Rukwa Valley safari in Tanganyika. Vesey as cheery as ever and very hospitable, inviting Gunn, Director of the International Red Locust Control Organization; Halcrow who is now a Special Commissioner in the northern province of Northern Rhodesia in charge of development; Lankester, an old-timer who was once in the Game Department, and one or two others, for sundowners. Very good dinner for Alan, Lankester and me afterwards. Had a good crack with Halcrow about the black lechwe situation round Bangweulu.

26 August 1957

Left Vesey's house in the pleasant township of Abercorn, a place which is being much upheld by the Red Locust group. We were soon into Tanganyika and more hilly country than farther south. The trees on the hillsides were coming into their spring foliage, that lovely red colour which later turns green. Then we got out of forest country into vast grassy downlands and Scottish Border-like hills. The soil had turned from red to grey. In the middle of this country, the Ufiba, an old Greek was growing coffee, altitude about 6,000 feet. The country then grew more mountainous, the rock appearing to be a mica schist, and we came to the edge of the escarpment of the Rukwa Valley. The valley is an extraordinarily flat flood plain with no drainage out of it. We had come to the best place to get down the escarpment, i.e. a fairly low point, but even so the drop is 2,000 feet and spectacular. The Locust people have made a road down it, a succession of little hairpins and they get down the 2,000 feet in three miles. We were then in a country of good acacia woodland where there were villages, finally into the grassy flood plain proper and along to Tumba Camp. This is more or less Vesey's headquarters; there are several rondavel huts and a native compound, for the Locust Organization maintains fire-watching, and there is spraying activity from time to time when a locust swarming becomes likely. Tumba is in an area of *Hyphaene* Borassus palms, some acacias and occasional small thickets of bush. There are elephant around and on this first evening we saw a group of 12 giraffe – 1 big bull, 5 cows, 5 calves, and one yearling female. Vesey has done much in this valley to conserve the game and has some hopes of a Reserve being established. When we reached Tumbwa we found Gerry Swynnerton, Game Warden of Tanganyika, and Tony Mence, a Ranger, already there. They will be of our party these days.

27 August 1957

Away this morning through the grasslands, 7 feet high, to the shore of Lake Rukwa. We passed along five miles of Vesey's firebreak and we saw 57 Hoare's reedbuck on the young grass. Most of them hopped back into the long grass as we approached but they were nevertheless fairly tame. The males have pretty horns, spreading and coming forward at the tips. Then we came into short grass country which was the chosen ground of the topi. These antelope are of the hartebeeste family, chocolate brown in colour and with horns lacking the distinctive kink of the hartebeeste. Where the grass was shortest the topi bulls had a sparring ground; each one had his own little worn patch where he played king of the castle. The one who would come and challenge him would show his horns and the two of them would go on their knees as the horns met and the two engaged in a shoving match. There was no serious fighting at all. During the day we saw many herds of topi, perhaps 1,000 in all.

The Rukwa Valley, being dead flat and alluvial and relatively dry, has an alkaline soil. This was most obvious when we reached the lake itself, where alkali tended to cake on the shore. The lake extends for the best part of 100 miles, but all this northern end is so shallow that natives from the valley edges come down with nets and wade through it not more than knee deep for miles. We saw some of them as giants in the mirage. The lake was a thrill for us because of a huge flock of flamingoes and another of two species of pelicans. There were also gulls (a small grey-headed type), spur-winged geese, whistlers, and many pratincoles. I saw a little stint among the sand-pipers. The rising and falling (and occasional disappearance) of the lake over the years creates instability of vegetational conditions and the existence of bare ground is a factor in the locusts' egg laying. Vesey also thinks that bare ground caused by fires is also a factor down here in the valley, and his practice is to keep as free from fires as possible. Indeed we exchanged ideas on a point he and I have come to, that a disturbed or degraded habitat is mixed up with plague proportions in animal populations. There are large areas of relatively short annual grasses in the valley, and because the soil is alkaline rather than acid the grass can well cure on the stalk to remain as palatable hay. This is eaten by the topi as long as there is water to be got.

28 August 1957

Out this morning a long way into the long grass *Echinocloa* and *Hyperrhaenia*, which looked like vast fields of ripe corn. We were working our way down towards where the river flowed sluggishly and where there was expected to be a fair concentration of game. We came on puku but never got into anything more important because we saw far behind us a plume of smoke. The party of us went back and sure enough a huge grass fire was on. All the square miles I likened to ripe corn were now a black expanse and the fire was obviously going to run for miles and miles. We went back to camp, had a hurried lunch and organized for fighting the fire. I cut many young palm fronds with my Ghurka knife and sliced off the great toothed thorns on the shank. These fronds make excellent fire brooms. We did long stretches of back burning but we feared rightly that the fire would sweep round and catch us in the rear. Much later at night the fire came from the other side of camp altogether and we were out for hours trying to steer it and keep it away.

29 August 1957

Out fire fighting this morning, most of the African squad back-firing to ring the camp effectively. But the fire beat the squad and a very few of us were left to stop it getting to the camp and the thatched roofs of the African quarters. We stopped it as it reached the corner of the first hut. We went back to lunch exhausted and dehydrated.

Forgot to say that on the afternoon and evening of the 27th we went to the edge of the opposite escarpment where there were granite kopjes above the great flood plain. The spot was called *Chem-chem* because of the warm spring which rose there. Coming through the long grass we had seen a lovely herd of 30 elephant cows and calves and a big bull moving out into the plain, and now from the kopjes we could see another such herd and one of about 100 buffalo. There were rock hydrax among the kopjes, the first time I have come across them. The foot of the kopjes against the flood plain (also called Nkungazi) reminded me of the narrow strip below the Mongu escarpment and above the Zambesi flood plain. The soil was rich and bore a varied flora of acacia trees and tamarinds and baobabs. Quite fascinating.

30 August 1957

Away to look over the extent of the fire – perhaps 100 square miles or so. A river bed would probably stop it now and it would burn itself out. The squad had beaten it on the other side the night before. I also went out on the burn myself to look at things; I saw marabou storks and vultures in a mob at two points and went to see what they had got; in other words, what had been caught in the fire. One was that amphibian, the legevan, and the other a small porcupine. It is surprising how few things are caught in an African fire. Francolins will not fly because birds of prey hunt in the van of the fire waiting for birds and small mammals to come out. Fortunately there seems always to be occasional clumps of dense green bush which do not burn and these become the havens of francolins and other birds. Doves and colies and kivelias were flying over the recently burned ground quite unconcerned and the night before Alan had seen two reedbuck squatting only 50 yards on the burn side of the fire. At one place where the fire was coming up to a burn of yesterday, the marabou storks were standing on the burn opposite the fire waiting for some poor creature to run out. We have seen several giant mongooses in these few days and it is certainly one of these I saw by the Mweru Wa Ntipa on the afternoon of August 13.

31 August 1957

Vesey, Alan and I left the Rukwa this morning coming up the Moze gap in the escarpment and over the Ufipa plateau to Sumbawanga, where there is a boma. Paid our respects to the District Commissioner and went up to the Mbizi Forest at about 8,000 feet on the very top of the escarpment. This is a relict high montana forest in patches of a few to 50 acres set among xerophytic fire-climax grassland. There was a good skirt of shrubs and herbs round many of the patches, which makes Vesey wonder whether fire is the immediate cause of shrinkage of the forest. Once inside, I was struck by the dense herbaceous undergrowth and the open stag-headed quality of the canopy. This enabled us to see several groups of the red colobus monkey which lives only in this sort of forest. Funny little black faces looked down at us from very high branches. I had used an expression earlier in the week of a forest being dead on its feet and Vesey had taken it up, saying he thought this expression applied to the Mbize. I think so too; it had the look of a post-climax. Most of the tree species, including a great *Euphorbia*, were new to me and I did not attempt to sort them out. We left Vesey regretfully when we got back to Sumbawanga and came back to Abercorn. Was very tired on the way, but Vesey had told us to use his house and a hot bath was awaiting us, and Dr Gunn, Director of the Locust Organization asked me to dinner for tonight and tomorrow. Delightful altogether.

1 September 1957

Away to Mpulungu on Lake Tanganyika where a young man called Wylie Watson was waiting for us with a metal boat with an outboard engine. He was to show me some tsetse control work, consisting of felling lakeside trees to the 100 foot contour and thereafter keeping undergrowth down and spraying the strip with insecticide. The tsetse concerned is *brevipalpis*. The young man gave us lunch at his camp ten miles away across the lake and was as hospitable as a good Yorkshireman usually is. I developed a very sore throat on this trip, and when we got back to Abercorn and set out to go 25 miles to the Kalambo Fall I was very tired. All the same, I was glad to see this 720 foot fall, the highest in Africa and a sheer drop in a quite stupendous ravine in the sandstone. Marabou storks were nesting in the cliffs (two eggs). The fall is a single stream making a wonderful grey mare's tail. Such a nice dinner with the Gunns and back to bed with a roaring throat.

2 and 3 September 1957

Knocked out with acute tonsilitis. The doctor at the little Government hospital at Abercorn suspected a staphylococcic infection which has been common in the district. Ate nothing on the 2nd but a little on the 3rd, thanks to Mrs Gunn. The doctor gave me three injections in all of penicillin and I certainly felt the better of it but weakly convalescent. Forgot to say that on 27 August when we drove home in the dark, the new moon, Venus and Jupiter made a striking equilateral triangle.

4 September 1957

The doctor let me away this morning and we drove through Tunduma and across the head streams of the Luangwa River into Nyasaland. The headwaters region is being badly beaten up and both gully and sheet erosion are apparent. But the district is exporting maize, so the Provincial Administration is happy. Such export from African chitimene cultivation is just criminal. We came across a quarter mile long game-trap fence running at right angles to the road, as cheeky as that. It contained eleven

A drop-pole trap in the Isoka district found with ten others and one pitfall in a quarter-mile long game trap fence, 4 September 1957.

drop-pole traps and one pitfall fitted with fire-hardened wooden spikes set in the bottom. After taking photographs of this, we set light to the fence and burned the lot. Such traps are strictly illegal but it is highly unlikely that a member of the Game Department has been seen in this district for years. We reached a place called Chisenga at 3 p.m. and set up camp in dwarf *Brachystegia* bush a mile and a half away to await Roelf Attwell's coming. We were now at 5,000 feet under some quite magnificent red sandstone hills which gave escarpments much like those of the Torridonian near my old home in Scotland. Over 200 miles today.

Roelf Attwell, to whom more than anyone else I owe my coming to Northern Rhodesia, arrived at 5.15 p.m. having come from the Luangwa Valley, a long ride and a big change of scene.

5 September 1957

Away before 8 o'clock with Roelf and saying goodbye to young Alan Savory with whom I have travelled so long. He has been a good companion and I have grown fond of him. Roelf Attwell is Provincial Game Officer Eastern Province, a clever biologist, good administrator, and a donnish type – glasses, balding forehead, dry and quite acid in speech, schoolmasterish but with a good sense of humour. He is to be my companion for three weeks now. I met him first at Oxford in 1953 and a little later took him round parts of the Highlands. It was then he proposed this series of visits.

We were headed for the Nyika Plateau, 9,000 feet, not 10,000 as I had thought, and passed through a lot of hilly and wooded country of Nyasaland. Down to a village, Katumbe, where we were able to buy some short but very thick bananas at a farthing each. Soon on our way again uphill, steep roads winding through the *Brachystegia* and then to wider uplands dreadfully ravaged by fire and losing their tree cover. Now we were near the edge of the Nyika and I was much upset by the effects of the fires despite the beauty of the flowers of a geophytic flora. The Northern Rhodesia Government has built a rest home at 7,200 feet on the edge of the western side. N. R. has only 27 square miles of the Nyika; the rest of the 900 of the plateau is Nyasaland. The outlook is magnificent and N. R. has the two largest patches of the former montane forest that must have covered so much of the plateau. The rest-house is rather nice – a long low whitewashed place rather like an Irish house on the west coast. It was cold and windy from the east, so we were glad of the fire and a hot supper after a bath.

6 September 1957

A long day out to see the montane forests. The first patch of 70–80 acres was less than two miles away. It was a contrast to the Mbize forest of last Saturday, in that the canopy is closed and there is not a dense ground flora. On the contrary, it is fairly easy to get about in the forest; the floor is covered with litter and there is plenty of regeneration of tree species. The trees themselves were magnificent. Genera represented were:- *Podocarpus, Olea, Hagenia, Bersama, Entandrophragma, Cassipourea, Macaranga* [Euphorbiaceae], *Neoboutonia, Pygeum, Dombeya, Diospyrus, Cussonia, Myrianthus, Schlefflera, Ocotea, Cola, Chrysophyllum, Ilex, Ficalhoa, Trema, Harungara*; shrubs were *Royena, Garcinia, Ochna, Dracaena* [note in margin: *Macaranga*; tree ferns, *Aphloda, Xymalos,* various *Rubiaciae, Jaundea, Piper, Loranthus, Clausena, Clerodendron, Doyvalis, Maesa, Peddiaea, Tecomaria, Allophyllus, Cordia, Drypetes, Artobotrys.*

We worked through the forest and down across the Chiri Valley and stream and up to the other patch of forest. This is even better than the other in its tallness and close-canopied quality. Attwell has organized splendid firebreak protection round these two relics of some glorious past age. The journey back was hard going – up and down the sides of ravines. We saw blue monkeys at the edge of the forest when crossing one of these ravines. No *Colobus* monkeys here, and no blue monkeys in the Mbize Forest. Fortunately there was a moon to help us home and I was tired out when I got to the rest house.

7 September 1957

By Land Rover across the top of the plateau – a vast expanse of rolling hills like downs, except that they were on granite rock and sour. All these hills are burnt in patches as if they were sheep farms at home, but the plateau is uninhabited. The Colonial Development Corporation (C.D.C.) had the notion that this plateau was going to be a splendid forest venture, but now after making trial plantings of Mexican and Caribbean conifers (which seem to be doing well) are giving up owing to 'political imponderables'. Their worst enemy would be fire sweeping up from the lowlands. We saw one such fire eating into a tiny degraded forest patch in a little combe. The game fauna of the plateau is of the plains type – eland, roan, a few reedbuck and zebra. The zebra seemed to be in unit herds of six or seven and they appeared browner than the zebra of the lowlands. The largest herd of roan was 18. We also saw Stanley's bustard – a cock displaying its great white front of erectile feathers to three hens. Saw the lesser bustard, also the black-bellied korhan. There is an endemic race of francolins here (partridges) and an endemic lark. We went into one or two small patches of forest on the plateau and found the hoofed game were using them pretty well, like cattle getting into the scrub on a downland farm. When we reached the eastern side of the plateau there were scrub areas, say six feet high, in the hollows of the hills. The rainfall here is less than on the west side. Down at 7,000 feet on the east side we came to the patch of 10 acres or so of the forest type which contains the African *Juniperus procera*, less than 100 trees of what must once have been vast areas of such forest. What an inspiring tree it is – a great fine bole, a smallish top, and perhaps 100 feet high. Such forest occur in Kenya but this is the farthest south example in Africa. Fire is the great trouble, the curse of Africa. We saw young junipers doing well, but would they survive in this continent of pyromania? Suffered much from Land Rover neck but rather better in the late afternoon. Home by 7 o'clock.

8 September 1957

To the north of the plateau today and had a look at the Kaulime Pond, which may well have come about by constant wallowing of animals and their taking away mud. We found a young zebra's skull, the result of a lion's kill six weeks ago. Apparently lions cross the plateau occasionally, so do wild dogs and even elephants traverse it rarely. We came on a small village of two or three huts just below the edge. A half naked woman and two children ran across the hillside to our Land Rover. She talked to our Game Guard and through them Attwell talked to her. She said she was happy living up here in the cool and that as she was born here she had no desire to go into the valley.

9 September 1957

We left the great open cool plateau regretfully and came down and down into the Northern Rhodesian plateau country at around 4,000–4,500 feet. It was 181 miles to Lundazi, the Luangwa Valley being on our right below us. I was much impressed by the goodness of the bush in all the middle distance of the journey. By the time we were approaching Lundazi the population was wearing the bush a little thin and there had been more firing. Lundazi I shall always associate henceforth with Errol Button, who was once District Commissioner here and whom I met last year in Lusaka. He has an original mind and also money of his own, which gives him an independence of action few people seem to exercise today. He built the boma at Lundazi and it is the most successful building architecturally I have seen here except the White Fathers' Mission near Kasama. Button also built the rest-house (now the hotel where we stayed) and let himself go, in that he produced a place like a castle in native-made red brick. It is odd and a folly, but lots of fun. And I was most hospitably entertained to dinner by

Mrs Riseley, who invited the present D. C. (Penn) and his wife, and Roelf Attwell and myself. Button also made a dam in the Lundazi River which created a large lake in front of his rest-house, and he also made the lake and two square miles of country round it into a bird sanctuary. So from the castle terrace you look over the water and see bushbuck and waterbuck among the trees on the over–shore. Splendid show.

10 September 1957

A long session with Penn this morning trying to give him an outline of conservation policy. Then away into the Luangwa Valley, down to a nicely hidden private camp of Roelf's on the river bank. We had driven through miles of wonderful *mopani* forest and as I walked the last couple of miles into camp following the Land Rover's wheelmarks, I was able to look around, feel and enjoy. I saw some buffalo and they saw me but did not bother, and I saw lots of waterbuck and impala, and guinea fowls thinking about going up to roost. The camp was under a great Rhodesian ebony tree *Diospyrus* overlooking the 60 yard wide river. The banks were 10–15 feet deep.

11 September 1957

A good day out in this wonderful valley. Up stream till we reached the northern boundary stream of the north Reserve, then across and penetrating into the Reserve. A certain amount of acacia parkland on the riverine strip and then the *mopani*. We saw 30 or 40 elephants during the day and on one occasion we had to get to the top of a high antihill to let a cow elephant, her baby and her two older children pass. She went by at 20 yards, the wind being right for us and they did not see us. There were also zebra, impala, waterbuck, puku, bushbuck, and we saw one rhinoceros at 50 yards. The range is not over-used and looks well. Moreover, fire has not got into it. Tree genera included *Kigelia*, a large *Terminalia, Tamarindus, Trichelia, Sclerocaria* (merula) *Ficus, Zysiphus, Piliostigma, Loncocarpus, Pseudolachnostylos*. We left at 6.15 a.m., back at 11.30 and out again in the afternoon. Watched a family of elephants picking up *Diospyrus* berries near a lagoon.

12 September 1957

Another walk in *mopani* country and then broke camp and came 70 miles further down the river to another point a few miles from where the Mwaleshi comes in from the west. Went out by myself for a couple of hours and had a splendid time upstream. The country is much more open here along the river and because of the dynamic quality of the river there are both dry ox-bow depressions and ox-bow lakes. I saw several groups of Cookson's wildebeeste, buffalo, puku, waterbuck and impala, and for some time I was only 40 yards from two rhinoceros. This country is acacia parkland in the best sense. The river because of its constant change of course does not develop a gallery forest of specific riverine trees. The river is full of sandbanks and low shores below the 10–20 foot bank, and crocodiles lie out on them in the sun. Coming back I heard a throaty snarl in the long grass 100 yards or more away and wondered if a lion had a kill there. At night, from camp we heard the hyaenas kicking up an awful row and in fact they kept me awake in the night. They seemed to be just about where I had heard the snarl, so perhaps they were hanging about the kill, impatiently waiting for the lions to finish.

13 September 1957

A good long day walking from 6.30 a.m. Waded the Luangwa and walked through the north Reserve to the Mwaleshi River where there were distinct signs of over-browsing on the *mopani* and other bushes. We saw elephants beyond counting, lots of buffalo, a rhino, zebra and all the antelope one would expect. This is really terrific game country in which I would like to spend three months.

14 September 1957

A nice walk this morning before breaking camp. Moved through the acacia parkland and into the *mopani* beyond. Saw a colony of bees which made their comb openly under the branch of a baobab, rows and rows of combs black with bees. Old Reuben the Game Guard, who is like our shadow, said 'Come away, as this kind of bee is very fierce.' Then 50 miles or more down river to the Luambe Reserve, a small place of 90 square miles run by the native authority of the district. We camped on the bank of the Luangwa River and watched the puku and impala and waterbuck come out to play on the sand, and with them an elephant who came down as far as camp during the night. The bush is good and mixed here.

15 September 1957

Crossed the Luangwa early in the morning and had a long walk in the Munyamadzi River country west of the Luangwa and north of the Munyamadzi. The country is full of elephants, zebra, wildebeeste, eland, impala, puku and so on, and the rivers have large numbers of crocodiles and hippopotamus: I saw no signs of over-grazing. We saw one enormous crocodile in a tributary of the Munyamadzi. We came down the Munyamadzi (the river of my long foot tour last year) to its confluence with the Luangwa. Both rivers are so dynamic, creating ox-bows and new habitat all the time. Saw a delightful piece of colonization by willow that had not yet been eaten by any of the animals. Found a white-crowned wattled plover's nest a foot or two above the Luangwa River at the confluence. Many elephants and buffalo today. All game in good condition. We also met up with two natives hunting with muzzle loaders. They had missed at an impala but had found a leopard's kill of an impala up a tree and this satisfied them. We also met two primitive natives who had been out to gather baobab fibre for nets. One had features like an Arab's and a sharp warm eye – a legacy of the old slave raiding days. Livingstone came up the Munyamadzi in 1866 or thereabouts. Crocodiles are everywhere but they did not bother us in our two crossings of the river today. Timed a hippo under water – 4 minutes 50 seconds.

16 September 1957

A walk back from the river into a plain, where there were buffalo, many zebra and perhaps 150 Cookson's wildebeeste. The grass *Cetaria* (kasense) had not been burnt but was well trodden down and sheltering the soil from the sun. The soil beneath the grass was not cracked. The grass itself was still sprouting a few leaves of green and was being cropped by the game. I imagine that when the rains begin there must be intense bacterial action and conversion of the dead grass. Broke camp and down the river 60 miles to a pontoon which Eustace Poles put in last year as a means of reaching the southern Luangwa Game Reserve from the east side and from Fort Jameson. We came over and down a dozen miles to Eustace's camp, where Peter Whitehead is now in charge while Eustace is on leave. A lovely grass camp on the edge of the river and under some fine *Acacia albida* trees. A short run out with Roelf and Peter before dark – elephants galore and hippos by the dozen in the river; buffalo grazing on the opposite bank and so on – a splendid concentration of game along the river in the dry season. Several elephants came into camp, but one, a big bull with very straight tusks, is quite fearless in the dark. After I got to bed this night he came alongside my bed under a *Kigelia* tree and seemed very interested in my mosquito net. I was not frightened and he did not bother me. Thereafter he strode away to the river bank, breaking through a grass fence erected there, and enjoyed some fresh green grass which had grown in the protection of the fence.

17 September 1957

This particular elephant has been called now after me. He came near my bed again but was no trouble. He also went to Peter Whitehead's, which is under a grass shelter and he began to shove and poke about with his trunk. Peter shone a bright torch into his eyes and shouted profanities. 'F. D.' retired with dignity but I heard him under the acacias in the early hours of the 18th. A long run down the river today, seeing stacks of game, and elephants particularly. Was impressed with how much of the valley floor is inundated in the rains, and that such ground may grow a mono-crop of some acanthus like thistle. Also, in large areas trees are dying, as if the extent of inundation had recently increased. It appears to me that the concentration of game in the dry season along the river is not nearly such an important environmental factor as the degree and extent of inundation in the wet season. There are so many questions down here that I can't answer. The ox-bow lagoons are delightful – static lakes now for a century or so and attractive to water birds, of which there are many, and to the game.

18 September 1957

Hearing an engine and shouting a mile away we went to see what was happening – 7 a.m. It was 'F. D.' having breakfast from some good grass alongside the track, along which a camp-building gang was coming in a lorry. They were frightened to pass 'F. D.' and were yelling at him from a safe distance. He was indifferent, but when we appeared from the opposite direction he flapped his ears sharply and seemed to say, 'Now, what the hell?' We passed him gently and got the lorry past and left poor 'F. D.' in peace.

Then out 30 miles to the Chikaya Plain, a few miles long and a couple broad. There is a gully within the plain in which there is water throughout the dry season. The plain was full of game – two giraffes, buffalo, eland, zebra, impala, elephant, and showed good signs of usage. The desired grass, nkombuva, was well flattened, whereas an acre of *Hyperrhenia* was still standing. Fortunately the whole plain has escaped fire. A sandy area in the middle shows signs of sheet erosion but they are not serious, and the *Cynodon* is creeping back green and vigorously. It is obvious that the wash-out in the rains changes its course from time to time and the old course heals up satisfactorily and naturally. The behaviour of water in this valley would be a good physiographic research.

19 September 1957

In bed and asleep by 8.30 p.m., but my namesake elephant arrived at 10.30 and kept us awake for over an hour. He came alongside my bed plucking grass and picking up the curled biscuity pods of *Acacia albida*, and through my mosquito net his straight tusks looked very white. Roelf and I perhaps four yards apart left him alone and he did not trouble us. Yesterday we made the men leave the holes "F. D." had made in the fence, and tonight he went through the holes and made no more fuss. What a rumbling his insides make!

Out by 6.30 a.m. making a riverine walk round a great loop of the Luangwa, especially looking at the state of the range. There was close use of the grass by hippos, which are very numerous here, possibly 20 to the mile and for the most part in close schools of 20–30, but the range is not badly over-used. There was heavy use of willow by the elephant, to the extent of killing some of the bushes. Buffalo, waterbuck, elephant, impala, bushbuck and puku in plenty in this big bend. Saw a group of 12 roan across the river, a lot for this valley.

20 September 1957

'F. D.' made his rounds of the camp last night and as usual was quiet and no trouble. We had a walk out in the early morning and saw a lot of all the game species of the valley except rhinoceros. Left the west bank at 11 o'clock, came over the pontoon and came southwards along the east bank and through Nsefu's Reserve. This small Reserve of 40 square miles is full of game and on the Luangwa River. The Chief, Nsefu, has got the idea that to keep his game can be better policy than killing it out. We went down to a grove of *Acacia albida* trees about 500 yards by 250, the largest along the Luangwa. The trees showed a wet season inundation line of 11–12 feet. Some ebony trees *Diospyrus* were among the acacias but on slightly higher ground, meaning a shorter period of inundation. There were further showings that what Eustace calls the Luangwa thistle (*Acanthaceae*) certainly becomes a co-dominant on areas of a foot or two of inundation. Then up to Fort Jameson, best part of another 100 miles. A pleasant township among hills and one of the oldest in the Territory. Some good agricultural ground in the neighbourhood and a failing group of white settlers. The government professes to desire their survival, but it is plain the Government wants them out to make room for Ngomi.

From Fort Jameson, where I stayed until October 1, I went down to the Chilongozi

Ian Grimwood, who played a leading part in Fraser Darling's reconnaissances in Africa in 1956–7, photographed in Kenya in 1986. (Photo: J. M. Boyd)

Reserve of the Luangwa with Jim Robertson-Bullough. Giraffes there. Terrifically hot. Tsetse worse than I have found them anywhere. Also a day out on the Ngoni native lands seeing erosion and bad husbandry and an area of pristine bush which is being bashed to given them more land. Addressed the Provincial Team under the Provincial Commissioner, George Billing. A pleasant time socially with Roelf and Jill Attwell. Then back to Chilanga staying with Fip, and seeing the Member and many others. Fip and I shared a very happy cocktail party on my last night, October 6. Had too much to do and was too lazy to write up this last fortnight properly.

JOURNAL II

Kenya (Kenya Colony)

Introduction

BY 1956, FRASER DARLING had become well known in the realm of international conservation. Eight years had passed since that first personal triumph at the Lake Success conference of the United Nations and International Union for the Conservation of Natural Resources in 1948. Before arriving in Africa he had met, talked to, and travelled with many of the big names in conservation. Word of him had therefore preceded his arrival in Kenya at the beginning of his second visit to Africa in August 1956. East Africa with its magnificent inventory of tropical wildlife, and prestigious suites of National Parks and Game Reserves was becoming the focus of world interest in wildlife biology, which it has been ever since. The first scholars, funded by learned institutions and research foundations in North America and Europe, were already in remote study areas in the bush, keeping company with seasoned expatriate wardens, rangers and professional hunters. Nairobi became the cultural nerve centre of the entire theatre of operations. It was only a matter of time, therefore, before Frank himself was there in company with the top people in wildlife conservation.

Mervyn Cowie, the Director of the Royal National Parks of Kenya, had already met Fraser Darling at a meeting of the Fauna (and Flora) Preservation Society in London. Noel Simon, the Chairman of the Kenya (East African) Wildlife Society, backed Mervyn's proposal to invite Frank to Kenya, and the Society offered financial support. Sir Evelyn Baring, the Governor of Kenya, was consulted in advance. The ecological reconnaissance in Northern Rhodesia was already underway, and though its results were still three years off, there was a clear demand for similar services from Frank in Kenya, concerning the future of the game in Masailand at that time and later in Tsavo. In the former, a formula for the integration of wildlife and livestock management was urgently required. In the latter, there was incipient large scale damage to the vegetation by elephants, exacerbated by drought. In Kenya there were deep divisions of opinion in high places as to what course of action to follow. What better, therefore, than to call in a prophetic figure like Frank to point the way ahead?

Behind these pragmatic reasons for Frank's visits to Kenya there was a yet deeper political ground swell, described to me by Mervyn Cowie in 1986.

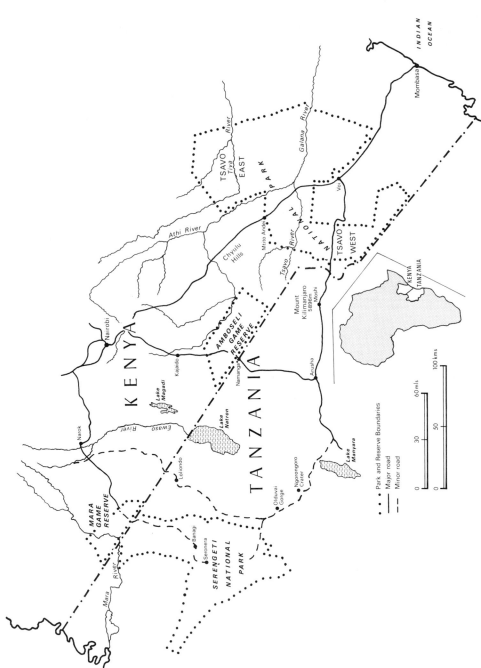

Despite the disruptions of the Mau Mau campaign the colonial government had built a country which was comparatively stable. This was made possible by having a high-calibre man as Governor, and a 'club' administration which resented any operation over which they did not have control. The core question from the outset of Mervyn's campaign for National Parks in Kenya in the late 1930s, was whether they should be independent like the two National Trusts in the UK, or whether they should be a department of government. He believed it right to keep the parks out of government with an independent board of trustees. In late 1945, he succeeded in setting up a semi-independent board half the members of which were nominated by the government, and of which Fraser Darling was later to become an international member. The Nairobi National Park — first in Kenya — was established in late 1946. However, though the government ostensibly accepted the independent board of trustees, the various departments of government did not, and in some cases openly defied the trustees. It was a long and hard struggle to get the system accepted in practice.

The civil administration in Nairobi were most unwilling to release any authority to any body on matters concerning the use of land and natural resources. The setting up of National Parks as a trusteeship independent of government was viewed with the greatest mistrust, and any moves in that direction were bound to meet with resistance. The original setting up of a board of trustees for the National Parks was therefore regarded as something of a triumph for Mervyn and his supporters, but the system had to be tested on the ground. Would the authority of the trustees be equal to the political thrusts of national and local (tribal) factions? Only time would tell, and the first major test came when the Public Works Department decided to take water for Mombasa from Mzima Springs in the Tsavo National Park.

The stink was so great that Mervyn was summoned to see the Governor. On this interview hinged not simply the survival of a few hippos in crystal pools, but the whole future of the National Park organisation and wildlife conservation in Kenya. For that reason it is described at length by Mervyn in his book *Fly, Vulture* (1961). It was a very tense meeting and the two almost parted on bad terms.

In Mervyn's words:

> The Governor was obviously annoyed (by my publicity campaign), and I could see deadlock ahead ... At the same time I could see that he could not make a decision without more information. I started down the corridor when he called me back, I think in the hope that we could find some compromise.
>
> 'Have you seen Mzima, Sir?' asked Mervyn impulsively.
> 'Never! But I should, shouldn't I?'

'How can you judge a place without looking at it?' The Governor nodded a willing acceptance of the point, and Mervyn saw his chance.

'Tomorrow? I will arrange the flight.'

'Not tomorrow, the day after. That's fixed.'

They parted amicably, and with enduring mutual respect. A few minutes after assembling at Mzima two days later the Governor called Mervyn aside.

'I am sorry. I was wrong. We must find a way of saving this place!'

The tide had turned. Sir Evelyn Baring was on Mervyn's side. The authority of the trustees, and the integrity of the National Parks were endorsed by this and many other decisions which Sir Evelyn Baring made as the last but one Governor of Kenya before *uhuru*. Later, as Lord Howick of Glendale, he became Chairman of the Nature Conservancy in Britain, and a leading figure in the conservation scene in the late 1960s and early '70s. He was an avid mountaineer to the end of his life, and I had the pleasure of his company climbing in Glencoe, Wester Ross, and Rum.

The principle of independent trusteeship had prevailed over the power of the government oligarchy, but there were many difficulties and sensibilities ahead. However, Mzima was a test case which was to influence others to come, including the future of the Masai Game Reserves to which the ecological reconnaissance of the Mara by Fraser Darling was linked.

When Frank arrived in Kenya in 1956 the issue of 'independent' National Parks still rumbled on. It was mainly political, but it had a vital ecological dimension. This Mervyn hoped would be provided by Frank – hopes which were not realised for, while Frank was a marvel on ecological matters, he was also a political fumbler.

CHAPTER SIX

Tsavo, Chyulu Hills, and Amboseli

August 1956

THE FIRST VISIT lasted only eight days, but in this time Frank visited the Tsavo National Park with Mervyn Cowie and David Sheldrick, the Warden, penetrating to the Tiva River. This was dry, and he saw the elephants at their self-created water holes in the sand bed, and the way in which a large number of wild creatures, including rhino, depend on these holes for survival in drought conditions. He drove over the Chyulu Hills and saw encouraging signs of forest advance in this dry cindery range. This suggested better portents for water resources in the hot plains below. In Amboseli, with W. (Tuffy) Marshall, he had his first contact with the Masai at the Ol Tukai springs, a contact which was to be greatly developed two years later when he conducted his ecological reconnaissance in the Mara Plains some 320 km to the north-west.

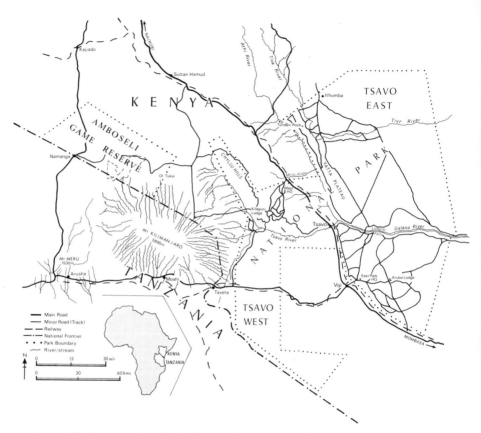

Map 5: The Tsavo National Park, Chyulu Hills, and the Amboseli Game Reserve showing the roads and tracks followed by Fraser Darling and his hosts in 1956. Amboseli is now a National Park.

18 August 1956

Reached Nairobi at 9.32 this morning after a very good flight. Very hot wind at Rome. Hot and still at Cairo: defective oil cooler pump at Cairo so we were delayed for nearly an hour. This was trying because I felt very sleepy, the time being bed time and there are no comfortable chairs at Cairo. No signs of tension at the airport, just the usual shabby, half awake feeling. Slept well on the plane. Had seen Elba, Mount Etna and Mount Stromboli during the day, and coming down to Nairobi this morning we could see Kilimanjaro, snow- and glacier-capped, rising above the haze.

Was met at the airport by Val: I forget her surname but I remembered her well from my first visit as one of Mervyn Cowie's colleagues in the office. She is a big woman, extremely handsome, and hopelessly crippled in the legs by polio. I am wrong to say hopelessly, because there has been that strange polio determination which overcomes hopelessness. She gets around on two strong sticks, drives a car and earns her living by doing a good job well. When I saw Mervyn in February he was driving himself hard with work to overcome the loss he was feeling from the sudden death of his wife a month before, after his return from Britain. I asked Val now how Mervyn was and she said he now appeared to be very happy because he and she had just announced their engagement. I am delighted and remember so well thinking in February what a blessing it would be if Mervyn and she could fall in love. Mervyn himself was brisk and happy and looking much better.

Met a young man called Noel Simon. He is Chairman of the Kenya Wild-Life Society, and in this past week has come on to the Royal National Parks staff. Royal Naval Air Service pilot during the war, now married and with young children. His wife is fourth generation Kenya, the first having been a missionary who came in about the time of Livingstone. I liked Simon immediately and now that I have been through a hefty file of correspondence about the Serengeti, I much admire his diplomatic power and judgement. He came to Kenya a few years ago to farm, getting a farm under a Service scheme. He has done quite well, but has got bitten spiritually in the same way that your husband is doomed. He is putting a manager in the farm he loves as a home, and is going to bring his wife and children to Nairobi. Mervyn and I, exchanging notes, see him as a successor. How wonderful it must be for a man like Mervyn, who is a spearhead, to find someone who can follow him in plenty of time! Boyle had asked me to see Simon and I am so glad to find a man of his calibre.

Mervyn has asked me to his house for tonight. One of these pleasant East African houses with the feeling of quiet comfort and elegance maintained by an adequate staff of servants. The house is on the outer edge of the Nairobi National Park and looks forth over a magnificent vista of plain and hill. He has 320 acres of his own. Mervyn was taking the Austrian Ambassador to South Africa and his wife over the Nairobi Park in the afternoon and asked me along. We drove 50 or 60 miles around the Park roads and saw the quite extraordinary number and variety of game there. Within the three hours we were out, I suppose we saw between 2,000 and 3,000 head – wildebeeste, hartebeeste, zebra, giraffes, gazelles, warthogs, baboons, eland, water-buck, kudu, impala and, of course, lions. Nairobi seemed to be out in force on this Saturday afternoon and one gets the impression that this Park on the outskirts of the capital city must have considerable effect in making ordinary people feel that they have some stake in conserving game. One also feels that sometimes the game must long for nightfall and the disappearance of the motor cars. One lioness lay with an expression of complete boredom with a ring of cars round her. Eventually she got up and walked away, quite indifferent. At another place we saw that a lioness was hunting and when we stopped to watch we could see that there were three lionesses and a lion converging on an impala doe and her fawn. The impala began to jump up and down on all four feet, stiff, and then made a dash, but one lioness was after her at speed and killed the fawn in a moment. The doe went on. The lioness bounded away with the long, loose legs of the fawn swinging from her mouth; another lioness ran after her and then the lion ran in and after a scuffle took the fawn from the lioness. Mervyn says the lion is very often a parasite on the lioness, merely pinching the food she catches. None of them would get very fat on one impala fawn. These big Kenya giraffes are really very impressive in colour and oddity – all that bulk for such a small mouth and pigeon-egg of a brain! The game in the Park is not static. There is a good deal of migration in and out eastwards, and a large area of country to the east of the Park is controlled, in that very little hunting is allowed. Much of it is Masai Reserve and the Masai are not a hunting tribe, just pastoral, and they don't batten on the game as is happening in so many parts for the sale of meat, ivory and rhinoceros horn.

19 August 1956

Mervyn and I left at 9 a.m. with his big Buick shooting brake well loaded, and we drove 200 miles down the Mombasa road to the Tsavo Park of 8,000 square miles. It is a dry area, much of it volcanic, and the rest on a coarse igneous rock. The soil is a bright red. The rainfall is below 12 inches a year, and this year the rain did not fall at all. The trees and ground cover show the signs of the drought and the game has also felt the pinch, particularly the elephants and rhinos. On our way down we had had to stop for vast herds of Masai cattle coming to water at some springs. Herding was being done by boys of about 15 years, long stalks of boys, naked but wearing a cloak of sacking over one shoulder and carrying a spear. I took a photograph of two of them, and then a few more of the dense herds of rather poor cattle waiting for their turn at the water. These cattle water twice a week, and between times trek out to far grazings. We continued through ranges of steep hills though not very high, and it was evident that drought was overtaking the country. Some of the bare trees were in blossom like an apple or pear tree, but there were no leaves. Down now through scrub country to the 2,000 foot level to a vast plain of scrubby forest, with ranges of mountains afar, sticking up to 7,000–8,000 feet.

We entered the Tsavo Royal National Park of 8,000 square miles. It is empty of people because the rainfall is 12 inches or less and there are so few watercourses. We called on one of the Wardens, an English Norman type called Tuffy Marshall. Mervyn has built his Wardens beautiful houses with fine gardens, each with a tennis court. The houses are also placed in magnificent positions. Mervyn says its no good asking civilized people of good background to live in the bush unless you give them the opportunity to recreate their background. Tuffy's wife and two children are down at the coast on holiday, so Tuffy entertained us to a bachelor lunch, simple but served with style.

On again southwards to Tsavo and Voi, only 100 miles from Mombasa. The bush got less in height and more droughty. The last rainy season missed this part altogether. The general appearance is light grey, the grass, the thorn scrub and the haze, but the soil is bright red. It was almost dark when we reached David Sheldrick's house. He is the Warden at the Voi end of the Park, which being nearest Mombasa, is in the heart of the awful poaching area. Rich Indians at Mombasa buy ivory and rhino horn, and poor natives of the Kamba and Wallangulu tribes kill the elephant and rhino with poisoned arrows. The natives are not bad fellows; they are getting a living in the way they know best, but the Indian traders are rich men with absolutely no moral sense in this way. One was caught recently with his car full of ivory, and the policeman took the ignition key to prevent him getting away. A retired legal Civil Servant living in Mombasa, Brian Kelly by name – need one say more? – now acts as counsel for Indians and draws high fees. He is a bully and has a way of finding technical hitches in procedure. This time it was – if the policeman held the car key, the car was in possession of the policeman not of the Indian, and therefore the ivory was not in the Indian's possession. The magistrate dismissed the case. The Administration is not taking the poaching seriously at all. Now Mervyn, who in addition to being Director of the Royal National Parks, a man devoted to his work, is a member of the Legislative Council and is very bold. He has an eye for men, and I am much impressed with the men he has gathered round him. This David Sheldrick is outstanding, young, steady, a born leader. Mervyn has instituted what is called the Fighting Force, a group of 30 natives of the northern province of Kenya – Turkanas and the like. These men are armed and in a smart khaki drill uniform; they are drilled and disciplined by David Sheldrick and from among them he chooses a sergeant and two corporals. David said to me that the disciplined African is a fine fellow, especially if he comes from the desert in the north. I saw something of this private army and it is truly a fine force. They go on foot patrols, split into pairs and reform, and are expert trackers and gleaners of intelligence. Also they are intensely proud of belonging to the Fighting Force. These chaps are doing what the Game Department and the police should do, and what is more, Mervyn and David are going into areas far outside the National Parks and making raids. They are getting the poachers so much on the run that some of the worst of them are coming in and surrendering to David. He then takes them on as trackers and intelligence men and has found them very useful. David keeps a careful dossier of everyone caught and makes up family trees and relationships as far as possible, so that he has good notions where trouble comes from when he finds it. David has two young white men as assistants – lieutenants so to speak – as hard as nails. One of them has won the Military Cross during the Mau-Mau fighting. David and the Fighting Force were able to go straight into action in the Mau-Mau emergency and have been much appreciated. I suspect that the excellent service of the Force has stayed any questioning of the legality of this private army of Mervyn's. Mervyn is most concerned with breaking the trade in illicit ivory and rhino horn, and will deal with constitutional quibbles when he has to. Anyway, they are brave men, because the poachers fight and sometimes with the poisoned arrows. You will not be surprised to hear that David's wife has left him – 'nothing to do down there and David so much in the bush' – but Mervyn says she was always a bonehead and not fit for a man of David's calibre.

21 August 1956

We set off this morning for a remote part of the Tsavo Park. All this southern end is arid, and is in effect bushed desert. The last rains missed it altogether. Waterholes are very scarce indeed. We were heading for Tundani Rock on the Tiva River. We passed through the Gullana area and crossed the Athie River at the Lugard Falls. The water was running here, and there were Egyptian geese and goslings, and waders – all very refreshing after the desert. We then crossed the Yalta escarpment, a volcanic ridge, and continued through waterless country of low bush, much knocked about by the elephants. We saw a good many elephants from time to time, and rhinos as well. Once a mother rhino with a calf charged the Land Rover and I hope I got a photo, but David was stepping on the accelerator. We passed several giraffes in the course of the 100-mile journey through the bush, as well as kongoni (hartebeeste) and impala. I was particularly interested to see the gerenuk several times; a small, extremely slim and lissom antelope with a very long neck. It is able to balance on its back legs and nibble from the upper branches of bushes.

The Tiva River was dry and just a sand bed and as we came to it I saw a lion, a young male, lying under the bank in the shade. He got up in a hurry and climbed up the bank into some low palmetto. We also crossed the river in the Land Rover and found a way up the bank and into the palmetto, which was just below the Tundani Rock. 'Do you think that lion would stop here or keep going?' I asked David. 'I think he is far away by now,' he said; though my own guess was that he would have stopped just in the palmetto, not being used to our kind of disturbance. Anyway, I got out of the Land Rover, and my question had nothing to do with leaving the car. But as I stepped down the lion stepped up about five yards from me and this time he took to his heels in earnest. We saw no more of him, nor did we hear of him in the night. The Tundani Rock is a sugar-loaf of metamorphosed gneiss about fifty feet high and the

Elephants wallowing amid the woodland they have helped to devastate at Buffalo wallows in the Tsavo (East) National Park in 1965, visited by Fraser Darling on 21 August 1956. (Photo: J.M. Boyd)

river bed bends in to the foot of the rock. The rock and river bed structure are such that when the river is dry, water is to be found about four feet below the sand at the foot of the rock. The whole area is so arid that this point becomes the waterhole for large numbers of elephants and rhinoceroses. The elephants dig in the sand, first with their feet and then more carefully with their trunk. When you examine these holes afterwards, you see a smaller hole either side of the funnel, made by the tusks. The rhinoceros takes an elephant hole and pushes his great nose into it, scooping the upper part of the funnel away. He leaves the hole in an untidy state and you can see the central groove in the sand which his horn occupies as he drinks. It obviously takes a long time for elephant and rhinoceros to water at such small areas of seepage, no bigger than their mouths or the tip of the trunk. We slept on top of the rock, but before we had gone to bed at 10 o'clock, a rhinoceros and her calf (toto) were drinking at one hole. David woke us at 1.30 a.m. to say things were in full going order. Apparently the place is used much more at full moon, which it was tonight, perhaps because the animals can see better whether there are any prowlers. A good deal of grunting and crying was going on as we crept down the rock to the edge of the river bank, where we were able to sit down and watch and listen to everything at close quarters. There were ten rhinoceroses in all and about fifty elephants. The noise all came from the rhinoceroses, and all because they were quarrelling about the waterhole. One would be in with his behind in the air; another would come along and want to

The Tiva River party having a lunch break at Lugard's Falls on 21 August 1956. David Sheldrick (right) in discussion with staff.

On 21 August 1956 – 'I was particularly interested to see the gerenuk . . . a small, extremely slim and lissom antelope with a very long neck.' (Photo: J.M. Boyd)

get in; and the one in would be afraid lest he should be caught at a disadvantage. Then there would be a slanging match. At one point there were four rhinos at the hole and none was drinking for fear of being butted. Those with calves were, of course, particularly bothered. Snorts, blasts, cries and roars came up all the time. Mervyn had brought his tape recorder and got a wonderful record. The elephants were quiet and orderly, carefully making their holes and then going to them in order. We saw no attempts to grab waterholes or push in before someone else. One of the elephants was just a little tiny toto and I heard it making little squeaks sometimes. It suckled its mother with its little trunk. We gave up about 3.30 a.m. and went back to bed. We woke at 6 o'clock to find all the animals moving away in rather a hurry. With the coming of the light the wind had done a bit of a change and our scent had carried to the animals. The little toto elephant was still in the river bed, standing alone and seemingly unconcerned. The herd of elephants was now 200 yards away, when one broke away; it was the mother of the toto, terribly upset. Back she came all that possibly dangerous way at the trot and her ears fanned out. The tiny elephant ran towards her as he saw her and she put her trunk round him in a flurry of concern and consternation. Then away to the herd, the little one going as fast as his little legs would carry him. What a state she must have been in when she discovered toto wasn't with the herd! By now the birds had begun calling and as the last of the rhinos went away the birds came over to the waterholes made by the elephants. Flocks of the little doves flew down, the guinea fowls and the francolins scampered over, and all was a new activity. How important the elephant is in this country! He makes it possible for all these birds to drink. When we went down to the river bed to see the holes and read the signs on the sand we saw the doves coming out of the funnels. The action of the elephant in breaking down the thorn bush also keeps open the whole country-side for a variety of grazing and browsing game. Without the elephant this area would just become impenetrable thorn scrub with no grass. I am never tired of watching these great beasts; even when they are doing nothing the movements of ears and trunk are delightful to watch. The elephants here are red in colour as a result of dusting in the bright red earth.

We made our way back in a leisurely sort of way, going by roundabout routes. Always the aridity and the elephants hitting the scrub hard. The two small rivers in the area were used by all the game and the country alongside the rivers, i.e. away from the actual riverine vegetation, was utterly beaten up. The game trails were only a few yards apart. We saw lots more elephants, rhinos, kongoni, impala, and some gerenuk; and I saw my first oryx.

I forgot to say that David Sheldrick found two tiny baby elephants in the bush two years ago. By some accident they were abandoned and were in starving condition. He brought them back in his Land Rover (a few weeks between each incident) and set-to to rear them. In this he has succeeded and they are now two years old, a male and a female. A lion nearly got the female last year and again David dressed the wounds assiduously, getting penicillin and sulphonamide injections for her. The male tended to bully the female, pinching her grub when he put them in a boma and shed at night; so he divided the shed, but the male kicked up an awful shout at being separated. So David left the partition at head height, which enabled the male to put his trunk over and feel the female, and that was all right. They are such perfect little pets. The male learned to undo his shed and get out, but he also went and undid the female's shed for her to get out also.

22 August 1956

Out with David in the morning to see some dams he had made in the Voi River about 15 miles away. The river has dried up for the first time in living memory this year and these dams have held many acres of water which would otherwise have been lost. We saw two separate groups of cows and calves – elephants – come down to one dam, slowly walking out of the bush, but how fast this gait gets them over the ground! They drank from one place but splashed and sprayed in another.

Tuffy Marshall and I left after an early lunch and drove to his house at another part of the Tsavo Park, where we had lunch on Sunday. We called on the way at a prison camp for Mau Mau prisoners of the worst type who cannot be brought back to normal outlook. I saw but a few of the 11,000, but they looked an evil lot. I saw them counted and locked each in his little corrugated iron cell. David and Tuffy were having to have an argument with the Commandant over some of his warders poaching in the Park. Two nights now at Tuffy's.

23 August 1956

Away early deep into the Park, first visiting Mzima Springs where there are some hippos. Watched them gliding along under water and reckon they walk on the bottom with their front feet and hold out their back ones behind the body, close together. The springs are full of barbel, rather like the carp at Fontainebleau, and there are crocodiles. Vegetation of *Raphia* palms, all very tropical, but only a few yards away the dry season bush again. On then a long way to the Chyulu Hills which go up to 7,000 feet from a 2,500 foot floor. These hills are recent volcanoes, and below them are some black lava floes densely covered with bush and full of tsetse fly. The wide valleys carried a good covering of grass, with small mimosa or rather acacia thorns. These bushes carry innumerable galls which do not prevent the development of the long sharp thorns. Each gall is pierced by a tiny hole and within each gall are lots of tiny ants. They must practically depend on this gall for their nests. Up in the hills were patches of true forest which I greatly admired. On the outer rim of each clump of forest were *Erythrinum* trees, which are fire-resistant and lovers of open ground. The true forest develops behind *Erythrinum* and shading it, kills it, but new *Erythrinums* germinate on the periphery, and so the forests are expanding. They are much like the *mushitu* in Northern Rhodesia but cannot be ages old because of the vulcanism. At about 6,500 feet I went down into a forest-filled crater about a quarter of a mile across; very steep and at each step through bracken and lianas my feet went deep into fine powdery volcanic ash. Tuffy and I got as black as your boot, and shoes full of the

The gall mimosa or whistling thorn (*Acacia drepanolobium*) in full bloom. (Photo: J.M. Boyd)

wretched stuff. There was a rhino in the crater while we were there, his steps sounding in the stillness, but we did not see him. The forest got denser as we reached the foot of the crater 200 feet down. In these hills we saw many kongoni and several eland. No elephants; no water for a long way. Home by a forest path, 30 miles or so, a good bath and supper. You hear the hyaenas at night, and in the early morning impala and kongoni are at the bird bath.

24 August 1956

Away by 8 o'clock this morning, going through the Park for many miles and eventually coming on a road which led into Tanganyika. Tuffy pointed out to me a hill where our headquarters was in the first War, then a long line of trenches, then another hill where we suffered a serious reverse and had many men killed, and so on. The last township in Kenya was Taveta, a miserable place in a great valley, where sisal was the main crop. Sisal is a cactus rather like the Mexican magué. Most of the village seemed composed of Indian shops, which are open box-like booths. Often there is a sewing machine on the verandah, and the Indians sit and stand about, usually looking miserable. They are not a cheerful people. Soon in Tanganyika and 40 miles or so on to Moshi, a growing town where there are many Greeks. We lunched magnificently for five shillings. On another 45 miles to Arusha, not a bad town as there are so many Government houses. We were now driving south of Kilimanjaro (nearly 20,000 feet) but too dull to see the mountain. All along this road there are coffee estates, often with the coffee bushes growing in the shade of trees. Shortly after leaving Arusha the country changed to open grass plains, with Masai and cattle. Two Masai youths on the side of the road asked for a lift and Tuffy gave them one. One of them was very lame and quiet, the other spry and talkative. They climbed in, smelling of cattle, and we had some trouble accommodating their long-bladed spears, which were longer than the Land Rover was wide. We carried them about five miles where they wanted to be set down. The spry one came to Tuffy and spoke rapidly. Tuffy does not know Masai very well, but our Game Scout translated into Swahili. The young Masai was asking Tuffy's name because he, the Masai, would like always to be his friend. Then the two Masai proffered their hands to each of us and we said goodbye. These people have not the sense of inferiority which the Bantu have. The Nilo-Hamitic is certainly better stuff. The high resistance of the Masai culture to western civilization is a very interesting study. We got over the plains to two big hills and turned down a road between them into almost desert country. This was the western end of the Amboseli Reserve, which is 100 miles long. The over-grazing is severe. It was dark now and the dust was so deep on the track that we slid about as if in mud. Tuffy and I were red all over, like the elephants of the Tsavo, and I was thankful to have worn my anorak. Then we got on to light grey volcanic dust incredibly fine. The Amboseli Lake is several miles across, very shallow and is now dry, the floor being hard and flat. We crossed it, enjoying the easy going, and then into the dust again till we reached Ol Tukai, the Reserve H.Q.where lives a Warden, now on leave, African clerks and Game Scouts, and there are some simple lodges for visitors. Ours was the one reserved for V.I.P.s and was the mess for the Ealing Film Studios unit when they did the film we saw in Cambridge called 'Where no vultures fly'. From our verandah we look to the great flattened cone of Kilimanjaro.

25 August 1956

We had a long day out in the Reserve. For possibly 10 square miles the country round Ol Tukai is acacia savannah and papyrus swamp. The swamps come from springs, the water originating from the great mountain. The springs mean that there is water in an area where there is otherwise very little, and Ol Tukai springs are the traditional watering places of the Amboseli Masai. They bring thousands of cattle every day. Somehow recently, the Masai have let their cattle graze in the 100 square miles between waterings, and their cattle have much increased because the vets have been needling them with antrycide. The result is disastrous on volcanic soil. I have never seen worse sheet erosion. You see flats of several hundred acres of light grey dust devoid of grass. Such clumps as remain elsewhere are markedly pedestalized. The trees are being heavily knocked about by the elephants and the swamps are opening up through the cattle grazing them. All day long you see dust devils going hundreds of feet into the air and great clouds of dust from the moving herds of Masai cattle. Mervyn said I should find Amboseli to be a Serengeti in miniature. There is a lot of game around Amboseli — elephant, rhino, zebra, wildebeeste, impala, Thompson's gazelle, giraffe and so on, but there is much less than there was five years ago when this extreme erosion began. I thoroughly enjoyed watching the animals at close quarters and hope I have got some good photographs. I also saw a good deal of the Masai and hope I have some good photographs of them watering their cattle. They make mud troughs at the edge of the swamp and teem water into them. The women bring pack-asses to the water and fill 4-gallon petrol tins with the help of long gourds. The cattle stand in close herds awaiting their turn, and even when each herd waters the separate batches take their turn without fuss. The cattle are zebu of all colours, but there are many blacks, and all sizes. The Masai youths in their brown cloaks, holding their spear in the left hand and a long peeled stick in their right, lope around their herds, gently tapping them with their wands. The youths and men wear short cylinders of wood on the lobes of their ears and beaded earrings 4 to 6 inches in diameter. The women shave their heads and wear elaborate gorgets and torques of fine beadwork. The Masai dwellings are of mud and wattle, very low and unventilated and unlighted. Several are placed within a boma of thorns and there are also inner bomas within the acre or so circle, where the calves are kept at night. The children nearly all had running noses and eyes, and you could scarcely see their features for flies, which the children did not even attempt to brush off. There are always flies round the Masai, being cattle people, and they never seem to take any notice of them. I took a lot of photographs of rhino, giraffe, elephant, and so on. Towards evening we went over the lake bed and I photographed Masai cattle crossing to the arid grazings beyond. I also photographed the mirage of water which does not exist. How glad I was to get into a bath and get rid of some of the dust which was through everything.

26 August 1956

Out early this morning before the breezes rose to keep the dust flying. We went on to one of the great eroded flats where there were a herd of 160 wildebeeste and 77 zebra. The calves made their particular little pleasant noises and the zebra made their sound which is something between a donkey and a horse, if you can imagine such a noise. Then both the wildebeeste and zebra began to draw across the flat to a swamp on the other side, a distance of a mile or so. They strung out and did not raise the dust. When they reached the water they took it in turns to drink and came away from the water fairly quickly. When they had all finished they came back over the flat again, this time very happy and full of play. It was one of the loveliest hours I have had in Africa. We left Ol Tukai at 12.30 o'clock and drove through arid and eroded Masai grazing to the east end of the Amboseli Reserve, 50 miles or so, then another 70–80 miles through desert bush to Sultan Hamad where we struck the Mombasa–Nairobi road. Here I said goodbye to Tuffy Marshall, a fellow I liked so much. Taller than I am, nearly as old, very much the English gentleman. My driver then shot me back to Nairobi like a bomb in two and a half hours and another half-hour to Mbagathai, where Mervyn lives. Supper at Val's cottage, just inside the Nairobi Park. She lives there alone, crippled and all, and a week or two ago a lioness rested on the steps of the cottage. She phoned Mervyn and asked his advice as to management of the situation. Eventually the lioness moved away.

27 August 1956

Busy this morning in Nairobi. Out to see Leakey at the Coryndon Museum about the Serengeti affair. Then a broadcast of the usual piffling kind. Then a phone call from Molloy at Arusha, to whom I expressed my dissatisfaction at the lack of co-operation from the Park Authorities and the Tanganyika Government. I also read a speech by Sir Edward Twining to the Masai, in which he told them powerful criticism abroad had caused the Government to appoint a committee of enquiry, but if the Masai would compose themselves in patience, he was quite sure they would find the committee's findings entirely acceptable. I think this is highly improper for a Governor to prejudge the findings of a committee of enquiry.

Mervyn, Val, Noel Simon and I lunched at the Equator, and Mervyn drank my Beloved's health and hoped she would be with us next time. It was such a nice gesture. Mervyn then drove me to the airport and I was delighted to find my plane was a Viscount. How nice it was to fly down to Lusaka that way. Lake Bangweulu looked so much smaller than in April, and below in the dark I could see the grass fires here and there. Ian Grimwood met me at Lusaka and took me out to his place for the night at Chilanga.

*   *   *

In September 1986, in company with Ian Grimwood, my wife and I camped in Amboseli, the Chyulu Hills and Tsavo (West) with the intention of covering some of the same ground as Fraser Darling did in 1956. In all sectors there have been great changes in the vegetation due to fire, the activity of elephants, rising salinity of ground water at Amboseli, and pastoralism in Amboseli and the Chyulus. The acacia forest on the Amboseli flats, and the baobab forest of Tsavo have all been very greatly reduced.

The following is from my own journal of 2nd September 1986 describing the situation at Amboseli almost thirty years to the day since Frank's visit with Tuffy Marshall:

> There have been conflicting views about what has caused the regression of woodland at Amboseli. Some believe that the root cause is salination due to the enhanced watertable, and the seasonal flooding and evaporation from exposed areas of 'pan'. Others blame the elephants. Others blame the Masai with their grazing herds, not so much in the Reserve, but around it causing 'compression' of game, especially elephant within the Reserve. The truth probably lies with them all – the elephants and giraffe are demolishing agents of woods which are predisposed to ill-health and death by salt, and

the situation is not helped by zebra, impala, and Grant's gazelle which inhibit natural regeneration of the *Acacia, Commiphera, Balinites* forest ... The prognosis at Amboseli is towards a treeless plain.
The forests in the Chyulu Hills had also suffered. Though still well wooded, the protection of the high forest by the *Erythrina* trees had not worked as Frank had hoped. Again, from my own Journal of 4th–5th September 1986 in the Chyulu Hills:

We drove uphill through a surviving enclave of cedar forest with the *Juniperus procera* hoary with lichens, denoting much mist. The whole massif seems botanically very interesting, particularly where the areas of scrub and forest have escaped the ravages of the Masai to the west, and the Kamba to the east ... Soon we were among the *Erythrina abyssinica*, which was just coming into bloom. We could envisage the place ablaze with red *Erythrina*, and the many other blossoms which we could not name. It was interesting to see how 'hard' the boundaries of the upper forest were against the open grassland, and of how the transition from the open habitat to thick forest was achieved by a narrow band of semi-fire resistant scrub. This obviously acts as a shield, but gradually succumbs, and thus the forest is continually pushed back. Soon there will be no forest and the whole surrounding country will suffer ... The Tsavo (West) Park now has much less game than when I first knew it 20 years ago. We saw a few elephant, giraffe, antelope, and ostrich, but not enough to travel far to see. The baobab has all but gone ...

CHAPTER SEVEN

The Mara Plains 1: At Oljoro Loromon

September to November 1958

FRASER DARLING arrived in Kenya when the controversy over the independent status of the National Parks was at its height. Mervyn Cowie was working hard to substantiate the independent board of trustees set up in 1945 to run the parks outside the establishment of colonial government, and it was a matter of the greatest importance to gain and hold the support of the Governor.

Nevertheless, there was continuous sparring between the National Parks and the administration, much of which the Governor was above. One of the main issues was the use of Masailand for wildlife conservation. Any suggestion that the Masai be deprived of any land whatsoever for the Parks system was eschewed by the Permanent Secretary and his staff and strongly resisted. The colonial administration had an ingrained desire to give the Masai no cause for complaint, following their displacement from the well-watered farming country in Liakipia, and resettlement in the dry plains of the Rift Valley, Narok, Kajadio, and northern Tanzania. Any suggestion that they should be dispossessed was inadmissible in Government thinking. For fifty years the administration had been bending backwards not to offend the Masai. Yet, some of the finest game lands in Kenya (fine partly because the Masai spurn game meat), were in the Amboseli and Mara plains, swamps, forests, and hills of Masailand. Clearly, the future of wildlife in these areas was intricately caught up with the Masai and their pastoral way of life.

Noel Simon of the Kenya Wildlife Society proposed 'that the northern boundary of the Serengeti National Park be extended to link up with the Mara Triangle and that no time should be lost in coming to an arrangement with the Masai ... for the finest game country in Kenya was almost useless to the Masai owing to tsetse fly ... which would soon be mastered and the Masai will utilise this country for domestic livestock.'

From the beginning Mervyn Cowie knew that the rights of the Masai could not be infringed, but at that time the Mara Triangle was not only tsetse

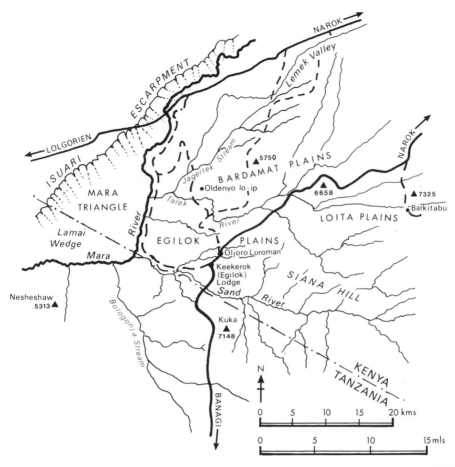

Map 6: Location map of the Mara Plains showing the main roads (solid lines) and main tracks (broken lines). The camp at the Mara River (known as 'Paradise Camp') was situated between the Talek and Sand Rivers.

infested, it was also closed to hunting and an unspoilt example of African wilderness teeming with wildlife. It could be a jewel in the crown of the National Parks but, despite a great deal of international pressure, the boundary of the Serengeti National Park was never extended by Tanzania, nor did it prove possible to set up a contiguous Kenya National Park in the Mara. Cowie took the only course open to him, namely, to persuade the Masai to manage their lands in the Mara and Amboseli for conservation. While the tourist revenue in the National Parks went to central government, so the revenues for the Masai Reserves could go to the Masai local government. Similar schemes were proposed with the Meru north of Mount Kenya, and the Samburu north of the Uaso Nyro River.

How was this to be done? Anything Cowie might do would be construed by the administration and the Masai as the thin end of the wedge of ultimate takeover of these areas by the Parks, with the exclusion of the Masai. Yet something must be done to save the wildlife from destruction.

Professor W.H. Pearsall's impressive ecological survey and report on the Serengeti (1957) had recently demonstrated the great benefit and influence which a figure of such high standing could have in East African conservation. Therefore, at the instigation of the Kenya Wildlife Society (Simon) and with the support of the Royal National Parks of Kenya (Cowie), it was decided – possibly reluctantly by the administration, and passively by the Masai – to invite an outside expert to look at the Mara and make recommendations towards a possible wildlife Reserve there in the context of Masai pastoralism and commercial poaching of game. The choice of expert was Fraser Darling, who was also known to Baring, and acceptable to him. The Parks' finest Warden of the day, David Sheldrick, was seconded as Frank's aide, adviser and guide in the field. Nothing was too much trouble to see the ecological reconnaissance of the Mara Plains off to a good start in September 1958.

25 September 1958

Last words with Mervyn Cowie and Noel Simon and final packing of a new long-base Land Rover and a 5-ton Diesel lorry. David Sheldrick and I away at noon precisely. We turned off the Nakaru road after 45 miles and then on an earth road towards Naroc. From Nairobi, 5,500 feet, the road gradually climbs to over 7,000 feet on the Mau Escarpment. The country to the edge of the escarpment is lush and well wooded, heavily populated by Kikuyu who in general are not taking as good care of their beautiful soil as they should be. Until nearly fifty years ago the Kikuyu dare not have lived in this area because of the Masai coming up from the dry plains of the Rift Valley, but the Pax Britannica let them into this heavily forested country. The primary forest has now gone and there is only secondary stuff for the most part and some planted forest. There are also some European farms with Jersey cattle looking very well. Once down the escarpment we were into much drier volcanic country without trees. The soil is derived from lava and the hills round are old volcanoes. This grassy country of wide horizons was rather like parts of Wyoming. The first few miles were a European-owned ranch and then into Masai grazing lands, not much used because of the shortage of water. The first animals we saw were three giraffes feeding on the gall acacia bushes. Then some Grant's gazelles, five more giraffes and more Grant's gazelles. Now we began to climb out of the plain over some bushed hills showing signs of over-grazing. The bush *leleshwa* was overcoming the grass. The earth road ran pretty straight from the top of the hills to Naroc, crossing one stream where there were Masai cattle and asses watering. On to Narok, where we talked to the District Commissioner on Masai clan grazings and so on, and I left him in no doubt of what I thought about the over-grazing. We were able to fuel up at Naroc, which is the D.C.'s boma, and a few Indian traders' shops a few work for, and that's all. It was interesting to see the unselfconscious mixing of races at the Indian store − the Indian woman herself putting the shekels in the cash box, the few Kikuyu types, a few Masai and ourselves. One young Masai *moran*, long of body and leg and so slender, lounged in his very short toga tied at the right shoulder and open down the right side. He was covered in greasy red ochre which made his body shine. His hair was worn long in ringlets but you couldn't see the hair for the ochre and mutton fat. His spear was in his right hand, his two little cudgels and a purse in his left, and his two scabbarded knives hung over his left bare shoulder.

Incidentally, our outfit is numerous. There is David Sheldrick and myself, a cook and general boy, then a platoon of National Park Scouts, armed and uniformed, and a driver. The Scouts are veterans of the anti-poaching campaign and of the Mau-Mau trouble, very fine chaps.

Ten miles on from Narok through overgrazed Masai country, through almost dense *leleshwa* bush, then into strict Masai Reserve into which one cannot come without permit. A Game Warden, Major Temple-Boreham, and his wife, live in an isolated house here. They were at home and hospitably asked us in to dinner. We made camp half a mile away. All in splendid order and now at 10 o'clock I am about to get into bed. Temple-Boreham a very big man and one of the old school.

The style of this safari is much more comfortable than anything I have known before: large tent and flysheet, (tent necessary because we shall probably have rain), Dunlopillo camp mattress, and sheets, table and comfortable chairs. (120 miles today.)

26 September 1958

An extraordinarily silent night and quite cold by 2 o'clock. Tea arrived soon after 6 o'clock even to the traycloth. Had a longish session with Temple-Boreham, who *does* think there is a fair amount of movement from the Mara into Tanganyika and even to the western Serengeti. He also says he thinks there are 400 lions in the Mara and that their toll of game is 10,000 a year. This would need a good deal of verification.

We set forth over the Loita Plains, a great flat area of volcanic beaten-up Masai grazing. Very dry and much pedestalization of grass clumps. Large amount of almost creeping gall acacia; also a sage-like bush, and some short rhizomatous grass. Such cattle as we saw were in fair order, probably 1,000 or so, and there were flocks of sheep, all close herded. After ten miles saw one Grant's gazelle and then small flocks of them. Coming into rather better and less grazed country of grass and six-foot acacia saw Grant's and Thompson's gazelles mixed, and finally more Thompson's than Grant's. Grant's is more a desert animal, Thompson's more of the plains. Grant's is larger, lighter coloured, and the stripe along the side less well marked than in Thompson's. In this low acacia country there were five giraffe, and only after then, at 20 miles, did we come into wildebeeste and zebra and kongoni, with impala where a more mixed bush appeared. We must have seen almost 100 giraffe today. We began to pass through extensive grass plains with occasional acacia trees of no great height, 10–20 feet, and found kongoni giving way to the allied species, topi. I thought these topi not as dark and chocolate-coloured as those in the Rukwa Valley last year. This is an antelope of dry hard ground where the soil is not acid. Soon we were into a great game country of acacia savannah with herds of wildebeeste, zebra, impala, kongoni, and topi, and innumerable Thompson's gazelles almost constantly in sight. We saw a river line ahead of us and made for it to camp for lunch. There was but a drop of smelly water which, having a supply with us, we did not need to use. We explored the river line after lunch for several miles but found no water. A few buffalo bulls were taking advantage of the shade of the riverine vegetation and we had to be careful when poking about. Those we saw were very large and very fat. All the game is fat, as I would expect from the basic nature of the vegetation. We were now in a country of wide plains and long low hills of quartzite which had they been left alone would have produced a covering of *mateshi*-like bush. Indeed, there were areas of such growth on the hills, with *Euphorbias* becoming common in the hollows and places which would carry the water of the rains. But large areas of the hills had obviously suffered from fire in the past and were exactly like those *mateshi*-covered hills in Northern Rhodesia adjoining Lake Tanganyika which had suffered fire. The upper parts of such burnt hills were practically bare.

We came through to mica schist country with some bush, but still with wide plains. The herds of wildebeeste and zebra and flocks of impala and of gazelles grew larger and there seemed more of them. Giraffe would be in the background against the bush. It was really most impressive. It would seem, however, that the wildebeeste and zebra are gradually working southwestwards towards the Tanganyika border and possibly (we have still to find out) to the western Serengeti Plains. Even as far into the Mara as this, we have found flocks of Masai sheep; they can come into this tsetse fly country without fear of getting trypanosomiasis. The tsetse fly here is *Glossina swynnertonii*, and carries human sleeping sickness of the Gambian type as well as cattle trypanosomiasis. We have set up camp by a stream which still carries some water in the sand and is called Egilok. There is a pleasant riverine vegetation which ameliorates the setting of the camp and gives us plenty of dead wood for fires. I notice that such rock outcrops as I can find up behind the camp are granite in character, pinky grey and coarse. (100 miles altogether today.)

27 September 1958

Out before breakfast this morning to prospect an air strip half a mile away on a plain, which would do very well as a base for our aerial census of the game animals of the Mara. We are going to attempt colour photography of the lot; a bit expensive but it should allow inspection and accurate counting at leisure rather than hurriedly estimating from the air. Forgot to mention we saw a large herd of elephants resting in the shade about lunch time yesterday – perhaps 50. On our going up to the projected air strip we saw six giraffe, many wildebeeste and zebra, impala, and Thompson's and Grant's gazelles. One or two kongoni also, but of particular interest were the topi bulls, which were on their standing places challenging and having mock fights with others, just as in the Rukwa Valley last year. Out again on a 55-mile run, going more or less westwards across the Telek River. Saw 26 giraffe, 17 in one herd, and fair numbers of all the animals so characteristic of this country. The general tendency of the wildebeeste and zebra was to be moving southwestwards. Some were quite definitely migrating. We also saw two lions, males, in the shade of a tree on the plain. We drove up to ten yards and they took absolutely no notice of us. I took a photograph. We also put a hen ostrich off her nest as we drove along – thirteen eggs, all shiny creamy white. The bird went off trailing her wings like a floppy skirt. Walking along the deep sandy bed of the Telek River for a while we found freshwater mussels buried six inches in the damp sand. Their tracks made a groove in the sand five to seven yards long.

Back to lunch at 3 o'clock. Forgot to say I saw three bat-eared foxes yesterday and one large jackal that looked like a coyote.

Out for a walk with David before dark southwards through light bush and anthill which comprises the foothills of the plains, between plains and higher hills. Heard zebras barking and yapping rather like a pack of dogs. Another new sound today was the grunting snarl of the impala buck, which he makes with his head down and his tail up. We saw one bull buffalo and one topi in this poor burnt bush. It is evident the quality of this stuff is not equal to the plains proper.

Have got a few identifications of anthill communities which constitute the bush in these foothills: a glossy-leaved shrub *Euclea schimperi*; *Acacia brevispica*, much eaten by elephant; *Cissus rotundifolia*, a soft-leaved climber; *Asparagus falcatus*, also with a little spine; *Croton dichogamus*, a shrub green-leaved but with white back; another small-leaved shrub *Scutia myrtina*; *Rhus natalensis*, another small-leaved shrub; together with two species of *Aloe*; and *Sansieviera* also present as in Northern Rhodesia. This sort of bush works up into a *mateshi* and is much worked through by elephants, and other things browse lower but elephant keeps it pruned.

28 September 1958 (Sunday)

Mervyn Cowie came in by a Kenya Police plane at 10.30 a.m. on to the air strip which David and his Scouts prepared yesterday. What a wonderful country that an air strip can be finished in a few hours! David went off with the plane, which is to return on Tuesday same time and take Mervyn back.

Mervyn and I went out after lunch to climb one of the quartzite hills on the periphery of the Mara saucer. From the ridge along which we walked through attenuated *mateshi* bush we could see into other valleys. There were perhaps 60 or 70 zebra in here and a few impala, nothing else. It is obvious that these hills have been burnt to death and must once have been covered with a dense 10–15 foot *mateshi* complex of shrubs.

29 September 1958

Away by 8 o'clock in the Land Rover heading for the Mara River, westwards across the plains. Incidentally, we are using a new long wheel-base Land Rover, which is so much better sprung and better seated than the earlier ones. So far I have not had Land Rover neck.

Heading across one plain we saw a single zebra, very lame on his near fore foot. It was obviously broken. So we steered close in and shot it. Within a minute vultures had appeared and soon a score of them were down. Only the smallest kind were brave enough to go to the carcass while we were near, a larger kind was less ready, and the largest species hung about farthest away. Within five minutes two lions appeared half a mile away coming to the carcass. They watch the vultures and seeing them come down in numbers had come along to see what was doing. And 50 yards behind the lions there trotted two jackals, who watch the lions. Soon they were at the zebra and when two more jackals appeared from the other side, the first jackals chased them away. Vultures and jackals would have to wait until the lions had had their fill. When we returned in the evening, two cheetahs were at the spot where the zebra had been but there was no zebra. We saw the lions at a clump of bushes a quarter of a mile away and went over to have a look at them. They were full and distended, and in the bushes we could see the remains of the zebra. The three or four vultures and the two jackals looked disconsolately patient. The lions were indifferent to us, vultures and jackals. Going through the plains we saw several herds of wildebeeste of 200 and up to 500, almost as many zebras, many topi, a few eland and kongoni. Gazelles were numerous and there were a few Grant's near the river. We passed a dozen or more giraffes on the 35 miles to the river and two buffalo bulls appeared near a dry watercourse in which I suppose there were pools. The plains of the last five miles to the river are becoming thick with gall acacia which I think comes as a scrub when the grass by constant firing at the wrong time gets so thin that it will not burn hot enough to kill the gall acacia or whistling thorn.

The river is 30 yards wide, muddy and deep in places. There is a fine fringing forest of riverine species of shrubs and trees. It is obvious this forest is the retreat of the elephants of this country. The forest was braided by their paths but I could not say it looked damaged by their being there.

On the way back we meandered about the plains round the patches of low bush which are the retreat of the lions for cubbing and rest. We saw two magnificent lionesses with seven cubs, and got near enough for photos. They were really a lovely sight, their colour being golden chestnut.

30 September 1958

The Police plane came with David at a quarter to eleven. We were up there to meet it, and the pilot, Mervyn, David and I went for an hour's flight over the Mara saucer, going clockwise along the Sand River which runs through the hills, till we reached the Mara River; then along it for a few miles and up through the plains. There were elephants in the forest by the river, 18 hippos on a bank in the river, 14 buffaloes under a tree some way off, and all the rest of the species where you would expect them. For once I have looked down on a giraffe. The dendritic channelling which causes thin strings of riverine shrubs, gives the impression of hedges unless you are wide awake, and for the most part these seasonal watercourses run east to west. The game trails in the main run to the southwest, although smaller ones cross and recross. This flight confirms my notion that the plains proper were probably always plains but that all the present half and half country and 'almost plain' must have been originally dense short bush. The long years of Masai seasonal pastoralism of sheep and goats have burnt it out. The Siria Escarpment west of the Mara River is an awful sight; what should be densely bushed hills now being burnt raw with occasional fire-resistant trees. To repeat my own expression, 'pyromaniac savannah'.

This evening David and I walked up a valley between two burnt hills through light scrubby bush. We saw a few zebra on the hillsides and a few eland in the valley. A couple of kongoni cows took too great a fright of us, probably because we got too near them before they realized we were there. We have moved camp today from a rather windy spot to a site in a glade of riverine forest. Very nice, with an outlook westward to the Siria Escarpment 40 miles away.

1 October 1958

A great day. We set forth early to the east in the direction of the highest hill surrounding the saucer of the Mara, called Kuka. This meant going up through bush country almost all the time and sometimes we had to pass through areas of dense *mateshi* bush, on the quartzite or mica schist all the time. We stopped at a salt lick of considerable size, with water in the bottom. David says it is one particularly liked by buffalo. I took a soil sample which I hope may be analysed at Maguga. We were seeing zebra all the time, a few kongoni, an occasional topi, and frequent impala. A herd of a dozen eland were near the buffalo wallow, and a little way off we found three buffalo at the edge of dense bush. We also saw occasional single giraffes, and the little dik-dik antelopes which are scarcely larger than a hare. Their habitat is in general the close bush and one rarely sees them more than a few yards from the edge of it. Water holes appeared here and there, and if we had not come upon them the presence of a few waterbuck would have told us water was not far away. The waterbuck here are browner than in Northern Rhodesia. David is very clever with the Land Rover, which he got to the foot of the hill we were going to climb. We had had a little difficulty finding a place to cross the deep channel of the Sand River. It was indeed just a bed of sand now. But we had four askaris aboard who dug a path down, and going along the sandy bed for a while we found a bank the Land Rover could climb. David's veteran askaris were armed and so was David, because the last Mau-Mau gang in Kenya have taken refuge in this country and on this hill in particular. Three months ago the gangleader released a prisoner and told him to go to the District Commissioner in Tanganyika and say that he, the gangleader, was on Kuka, and if the D.C. wanted him he could come and try to find him. The prisoner's wrists were tied with wire and had been for a long time, for the wire had cut into the flesh, which had actually grown round the wire.

We climbed steeply through dense bush in what was more or less a gully; then to more open bush with grass and trees and finally to the ridge which was savannah as a result of constant burning. There were fires going even as we were there. What a wonderful country was before our gaze! To the south and east there seemed an endless array of hills and wide valleys, savannah and forest. Here on the ridge in the immediate foreground were Nandi flame trees in their brilliant scarlet blossom against a very blue sky. There was also a shrub in pinky white blossom that smelled as rich as a gardenia. We walked along the ridge for a mile or two to the highest point, 7,148 feet, from where the hill fell away steeply. The Trigonometrical Survey had set a concrete stake at the summit and on one side of it was pencilled R.M. Mgunda 18/9/58. David thinks this is a bit of bravado by the gang-leader, who was evidently around a fortnight ago. However, we saw nothing of him or his men, and if they saw us they would be only too anxious to keep out of the way.

On the way to and from the summit, where we saw several zebra, we saw 3 and 2 Chandler's or mountain reedbuck and one oribi. From the actual summit we saw a pair of klipspringer on some rocks 600 yards away.

Kuka is actually in Tanganyika but the boundary is a silly one anyway. Coming off the hill we made northwestwards more or less along the border. We were now into excellent bush and the grass looked greener. The ground would still be called hilly rather than plains, and there were no great expanses of grass. Yet here were lots of wildebeeste in country we would not find them 25 miles north and east. We think they are slowly migrating in a southwesterly direction. We have seen the wildebeeste grow fewer on the plains even in the week we have been here. A little later we came upon the main herd of buffalo that live in the Mara; it was quite wonderful to see 600 of these great black creatures moving across a thinly-bushed area. There were lots of small calves among them. The whole bush was dense with game of all the kinds here except elephant. The Mara elephants are all down near the Mara River, in the residual riverine forest I have mentioned. Coming out of this heavily game-populated bush and working northeastwards we were soon conscious of diminution of numbers. Then we struck back east and southeast again, east of Kuka, into pretty heavy bush where we saw only kongoni and one bushbuck. We also saw one male rhinoceros who seemed intensely surprised and a little worried at seeing us. It was just on dark when we reached home after having reached the Sand River again. (90 miles.)

2 October 1958

Out for a round on the plains this morning and to look at the north-east. Soon after leaving camp we came on a pack of wild dogs, eight in all. They were remarkably tame and ultimately allowed us to 25 yards in the Land Rover. I hope I have good photographs. They live on gazelles and impala by hunting as a pack. We crossed the Telek and made for a hill in the middle of the plains called Oldongo Loi-ip. It is not very high but as I had judged it to be bush-covered I wanted to see how it compared with the hills surrounding the saucer. I found the hill to be volcanic lava, dry and devoid of game. Bush was thin except on the cone and there fire at some time had removed all vegetation. The meaning of the Masai name is 'Hill of shade'. The cone was evidently the wet season retreat of a rhino; midden, lying place, little corridors and all complete. Bird song here was noticeable. Northwards again then over laval plains of great extent. These empty dry plains seem to be a bar to the tsetse, and after a few miles we came on great herds of Masai cattle. We spoke with a group of Masai men sitting under a tree; the Headman rose and spoke through one of David's Samburu Scouts, the others curled up on the ground and drew their blankets over their heads. The Headman had beautiful ear ornaments of bead and brass, also a number of rings on his fingers, spear in hand, knife at belt, and plenty of red ochre. He spoke forthrightly and with dignity, but it was also apparent that he was suspicious, giving away no details of fly boundaries and saying that now the Masai cattle go anywhere in Masai country. The fact that they don't made no difference to him. The suspicion evidently is that we want to take the fly area away from them because they don't use it. This is one of the big problems we have to overcome. We were surprised to find large herds of wildebeeste on these laval plains, not showing the migratory tendency in their behaviour. This lot will want watching. The grass on the lava plains was not predominantly red oat, but a more fuzzy type proliferating from the nodes. (70 miles.)

3 October 1958

Decided to give my neck a rest from Land-Rovering. David went off with four of his askaris to prospect the route for us to make our stab into Tanganyika. I have remained in camp, writing and doing some necessary reading. The lions came to our water supply to drink last night and were a bit noisy. The cook is going to sacrifice a goat when he gets home again, for his safe deliverance! Our camp is in beautiful riverine type of trees; a flock of Vervet monkeys lives in these trees and in the early morning I see them come warily to the ground as I lie in bed. There are also bush babies in these trees, small nocturnal furry creatures with large eyes.

David came back just on dusk having failed to find a way through the hills into Tanganyika, so tomorrow we shall go more to the south-west and then strike south-east.

4 October 1958

Away by 9 o'clock in the Land Rover, all set for two nights out and carrying four askaris and Kilonzi as cook. We made southwestwards over almost empty plains till we reached the Mara River near where the Sand River comes in. Then we struck up the side of the Sand River for about six miles, climbed down into the dry sandy bed, continuing up the river for a mile till we could find a place to climb out, the askaris having to do a bit of gardening, so to speak, with mattocks to make the climb out possible. The Land Rover is quite wonderful. We had already begun to see game, including some big mobs of wildebeeste with yearling calves. From the Sand River to the Bologonja River (almost dry) we came on large herds of wildebeeste and particularly of zebra, up to 500 strong. There were also lots of topi, kongoni and impala. We could not find a crossing of the Bologonja for over fifteen miles, the country being fairly bushy with a lot of gall acacia and other *Acacia* and *Commiphora*. We saw a few lions, a couple of cheetah, several groups of giraffes and a few reedbuck. Then we found a crossing and by 2.30 p.m. found a most heavenly spring, enough to make a beautiful clear stream and a stand of riverine forest. We lunched and rested in these ideal surroundings. David knew of this spring and had been trying to get to it through the hills yesterday. The spring is called Bologonja also. Our afternoon drive of 65 miles was through easy acacia bush which hid nothing. The whole countryside, hills and all, had been burnt and the place was depressing. We saw a few giraffe, topi and kongoni to begin with, then it was absolutely empty for perhaps 40 miles, then a few Grant's and Thompson's gazelles, topi and kongoni. A pride of eight lions looked hungry in such a place and followed the Land Rover a little way. In the first ten miles after lunch we did see two rhinoceros cows, each with a calf. We were now well into Tanganyika and came to some rather miserable huts of the Watende from whom we asked the way to Bunagi, which is 13 miles inside the Serengeti National Park. The poor people were on a poor eroded sandy plain. On again till we found a track which soon proved to be a way into the Serengeti. We camped in the darkness, Kilonzi making a fire in a *donga* to be inconspicuous. He turned out a good supper in a short time and we slept under a tarpaulin tied from the Land Rover to the ground, the askaris being in bivvy tents on the other side. The wretched hyaenas were all around. Last night, says our capitas the askari corporal, they woke to find a hyaena over them. Many people have scarred faces from a hyaena coming to them in the night when they are sleeping out. Anyhow, I was awake for a couple of hours in the night and strangely enough there was absolutely no sound of hyaenas. (125 miles today.)

5 October 1958

Away by 7.30 o'clock heading for Bunage where a Game Warden lives, but we knew he was away. Near by, in a fine permanent camp, were the Grzimeks, father and son. This Czech name is pronounced something like 'Grimmick' and they are good enough to answer to that. The old man is Director of the Frankfurt Zoo and quite devoted to the welfare of animals. Two years ago he made a very fine film of African wild life called *No Room for Wild Animals*. I saw it in Denmark last October and was initially put off it a little because it started off with shots of people stampeding the animals and then of someone shooting an elephant, and you saw the mighty rocking and fall of the great creature. Then it went on to show the animals in their own home lives with great beauty, and if I remember rightly, it finished with shots of a poor elephant whose foot had been caught in a snare and was in a very bad way. It left one to answer one's own questions. It is now about to be released in Britain but has been shown privately in London once or twice. Well, Grimmick has made a pot of money out of this film and has ploughed the money back by starting research here which the governments ought to do. Young Grimmick is a post-graduate student and the work they are now doing will do his PhD. I liked the old man a lot and the youngster is a good open enthusiastic fellow with plenty of brains. They have brought a Dornier airplane which has a stalling speed of something like twenty miles an hour and is a high-wing monoplane with a lot of room inside, just the sort of thing to be used in the bad flying conditions here, but our governments will never buy anything of this quality. Britain, in fact, lags far behind in the design of small aircraft. Well, the Grimmicks are doing aerial survey of the game on the Serengeti through a period of two years, linking the movements of the game with the season, state of and kind of grasslands on the great plains. They were entirely candid in showing me their work and I was much impressed by the thoroughness of it and the fullness of mapping. They said they too had lost their game in the past ten days and thought it had moved north. We said we thought ours from the Mara had gone south-west where we knew there had been rain last Sunday. So the Grimmicks suggested we should fly over the ground forthwith. So off we went for an hour and a half, just half an hour too long for my guts, but that didn't matter. It was a splendid 150 mile flight; first over the empty country we had traversed yesterday, then through the hills to our camp to show him our airstrip, then around our plains to the Mara River, seeing very little, and then into the south-west where the grass was green and there were great concentrations of wildebeeste and zebra, and we also saw a lot of buffalo. The wildebeeste seemed to be moving and we saw one or two great herds of perhaps 1,000–2,000 strong moving in a fine convex crescent. The direction seemed to be eastwards, which we scarcely expected. The airplane gave a wonderful view of the country, as there was nothing in the way and it could go slow when necessary. It was very decent of the Grimmicks to give us this flight.

David and I went on into the Serengeti Plains to where we could get a view of the Moru Kopjes, the hills which the Masai were ruining by pastoralism and burning, to which fact Pearsall drew attention in no uncertain terms when he was here two years ago. The Government persuaded the Masai to clear out, but in exchange gave them the Ngorongoro crater. The Grimmicks' work does seem to show that Pearsall was misled by the dry-season boys about the mid-plain movements of game. There is a great rotatory movement there and the new park boundary unfortunately cuts right through it. Pearsall was a clever enough old bird to say that his opinion was formed from hearsay and not from his own observations, but when we talked it over in Athens, Pearsall said he thought game movements were not set patterns, but varied entirely with the incidence of rain, which makes the grass.

We saw several rather mouldy thin-looking lions near Seronera, where the Tanganyika Park has built rondavels for visitors. There is no natural water, only catchment from a granite kopje. I never saw a more depressing sight, and what a contrast to the standard Mervyn Cowie sets and maintains in his Parks in Kenya.

Then headed back to Bunage and through the empty burnt country of yesterday to some granite kopjes where we bivouacked for the night. I forgot to mention an important thing yesterday, that coming from Bologonja Springs through this country, there was a big storm south and east of us, but we skirted it all the time and had no rain. Now we were working back through where the rain had fallen.

It is now *6 October* of course. Anyway, the country had become green in two days and here was the game – large concentrations of wildebeeste and zebra and quite a lot of topi. Obviously they had come in following the rain, and the greening of the country had been very rapid in this climate. Having arrived at Bologonja Springs we stopped for a shave and a glorious bath in the limpid water. I have rarely felt so completely relaxed and at peace with the world as after that. On this little three-day sortie we have had to carry water with us, as we have not been able to camp by water. The journey back to the Sand River and to our own base camp was through country so much greener than two days ago, and it was full of game. Last night we came upon a cheetah resting beside a newly-killed Thompson's gazelle buck. We took a commission, so to speak, of the two haunches and left her the rest. It was a nice bit of fresh meat to have. We got back to base at 3 o'clock, having found particularly easy ways. The plains near camp are much greener, and after 5 o'clock David and I took a walk and found the wildebeeste coming back in goodly numbers, obviously excited by the new grass. The feeling of elation seemed to be communicated to us also. It really is exciting to have animals coming back in definite movement. In short, they are following rain. There was heavy rain here two nights ago – David and I had to get back to camp to escape a heavy shower.

7 October 1958

Away this morning to count wildebeeste on the plains this side of the Telek. The biggest lot were on the big plain near camp, where there were 500 in all. We counted another lot of 375 and another of 290, several between 100 and 200, and the rest 20's to 80's. Our total for the morning was 3,380. We did not count zebra, but we may have seen 750. They have not come in yet from the hills. We saw two lions and a lioness, the lions full and the lioness not very full. Probably she caught whatever it was and the two males ate most of it. There is absolutely no chivalry among lions. Every day we go out we see several jackals, and last night two families of those sweet little bat-eared foxes. On the way home yesterday there were two gangs of striped mongoose, ten or a dozen strong each, running about the bush very busily. The topi are calving now and I see several Thompson's gazelle kids.

As I have been writing at the door of my tent I had not noticed that a numberless caravan of siafu ants had decided to pass through my tent. Suddenly I felt on fire, saw the ground was alive and have had to jump for it and take off my trousers and shoes and socks. Both workers and warriors were busy stinging. Two of the askaris saw me thus employed and had a good laugh. One of them came over to the mouth of the tent and then seemed almost to levitate with a shout. It was now my turn to laugh. On returning to earth he hared off rubbing his legs and came back with a shovelful of ash. The other askari also came for a look. He was wrapped in a sarong, being off duty, but with great suddenness he left the ground also and thereafter helped his friend bring new ash. But I am having to write in our mess tent now until the ants have passed, which may be a couple of hours hence.

The African weather is interesting now the rains are beginning. You get an overcast early morning after sun-up at 6 o'clock; then the sky clears by 7.30 and you have a lovely fresh couple of hours or more. The sunshine continues into the heat of the day, then cumulus clouds appear and get bigger. By 4 o'clock there seems a concentration of grey cloud somewhere and then some distant hill disappears into rain-cloud. By 6 o'clock it may be here and it may or may not rain in the night. You can be pretty sure of the working day being fine, so the rains are pleasant. All these plains are of black cotton soil which gets sticky in rain. We may have trouble getting about after a while.

8 October 1958

Heavy rain and thunder last night, and a hyaena was in David's tent. About midnight I was conscious of something prowling, so turned my electric light on for a few minutes. We have a bulb and flex running through to the Land Rover batteries and it is very convenient. The lions also made a lot of noise as they often do in thundery weather at the beginning of the rains. Poor David tore a hole in his mosquito net when instinctively letting fly at the hyaena. The lorry went away this morning to Narok to fetch aviation petrol for the Grimmicks' census on Monday. David and I have decided to do a sample area between the Telek and the Sand River down to the Mara River. We followed the track of the lorry for twenty miles on our way to the Aitong Plains north of the Telek where we wanted to count wildebeeste. The lorry had had trouble twice owing to the rain last night but it had gone on all right. We saw but few wildebeeste until we reached the lava plains north-east of Oldongo Loi-ip and then came on 3,000 on several plains. We counted the lot and struck across to Aitong picking up odd hundreds of wildebeeste on the way. Some zebra but not many. We then swept westwards making for the Mara River, reached it and were surprised at the dense unburnt grass on the alluvial area near the Mara. Then round to the Jagartek, a tributary of the Telek, and soon after ran into the wonderful sight of half a mile of moving wildebeeste, 430 of them going northeastwards at two to two and a half miles an hour, in a long string led by a cow. It was tremendously exciting, and I felt the same as when I saw caribou migrating in the Arctic. There is such a feeling of purpose as you watch animals deliberately on the move. Our total of wildebeeste or brindled gnu for the day was 4,750, all north of the Telek; and not the same crowd we counted yesterday. Saw a good score of giraffes today, but no buffalo or lion. Soon after setting out this morning we saw a pair of wild dogs with five puppies just toddling. I don't call the wild dog a handsome creature but this party was likable. I got photographs of them. These herds of brindled gnu on the plains look rather like the bison on the prairies of North America. The bison have gone, but need these go too? I do hope they will survive for themselves and the beauty of it. Heavy rain this evening.

9 October 1958

Busy about the camp and writing all morning. Fairly early lunch and then out in the Land Rover towards Sianna Springs. Heavy rain was about and as we got into the hills it was obvious we might get stuck, so we did not go as far as the Springs. Fair number of wildebeeste in the lower valleys of the hills and saw ten giraffe in the bushy country. Lorry returned last night with mail. Some rain later.

10 October 1958

Pleasant overcast morning. Went to Sianna Springs at the head of a valley of the Sianna Hills. Wonderful country of wide valleys, plenty of grassed areas and plenty of bush, through which there are game ways through which it is possible to get a Land Rover with a bit of tacking hither and thither. No game except a few bushbuck, kongoni, reedbuck, and a rhino at the head of the valley, but a fair number of zebra and wildebeeste near the foot. The Sianna Springs well up from the foot of a 20–foot bank of red earth. The main spring came from the bottom of a tiny cistern of hardened earth six inches across and naturally formed. The water was faintly warm and had some taste though brilliantly clear. Heavy rain late this afternoon and into the evening.

11 October 1958

Out over the plains this morning. Wildebeeste seem to be moving more to north-east. Rather more zebra about and we think topi are getting more numerous. Calving of topi going well. Saw twelve young calves with about fifteen cows this morning. Went through a central area of plains where we have not been before. Red oat grass less common and no game at all. Unburnt. Out on my own in front of camp this afternoon to a stretch of acacia bottom land where one would expect to find elephants were they up here, but they aren't yet. All the scrubby bush which lies on the pebbly higher parts of the plains shows plenty of elephant sign, presumably from the wet season. They are all down at the Mara and the Sand River just now.

12 October 1958 (Sunday)

No rain at camp last night. Having got up specially early and made ready for the census flights today and having got the aviation petrol to the airstrip, we heard on David's wireless (with which he can keep in contact with Mervyn at Nairobi and his own headquarters at Tsavo (Voi)) that the flight was off altogether. Grimmick in going to Nairobi yesterday in his Dornier had broken his undercarriage as he left his own airstrip at Bunagi (rough enough as I know too well) and he had to make a guided landing when he reached Nairobi. Fortunately he is unhurt, but his airplane is out of commission for a long time. Mervyn will now have to arrange for a Police plane.

Temple-Boreham of the Kenya Game Department, Senior Warden, arrived at 9 o'clock last night to spend a week or two with us. This is his area and his favourite place, and as he has been here for twelve years he knows the ways very well. This morning therefore we went off together in David's Land Rover, almost over the same ground we covered on October 8 – the Aitong plains. There were obviously more wildebeeste on the ground, possibly another thousand, though we did not count today. They had moved a mile or so north-east. Temple-Boreham brought us across country from the Mara to Jagartek, and again we saw quite another 500 wildebeeste than on this stretch the other day. Coming back by our lovely green plains near camp, which look so much like downland and have the same joyful feeling, there were again more wildebeeste than on October 7, perhaps another thousand or more. Zebra have not yet come back in numbers.

Far across the Aitong plains under the hills, the degree of sheet erosion and overgrazing by Masai cattle and sheep is grievous. They have permanent *munyattas* here. The Aitong Spring has dried up in recent years, doubtless as a result of burning the bush from the hill, and the slight rise where the *munyattas* are and which was once bushed down to the plain, is now bushed but meagrely at its highest place and the quartzite boulders show through the bare ground all the way down to the overgrazed plain. What a problem this is! The ground belongs to the Masai by treaty; if the tsetse fly could be removed, which it could be, the whole of this lovely Mara country would be ruined by the Masai as the Aitong area is now. What is the answer? The Masai do not sell cattle but merely keep them; they live an arrested primitive life and are not evolving as a society. Because they tenaciously remain primitive, money has no great attraction for them – they have nothing to spend it on. The coming of the white man has brought relative peace, though they still make cattle raids on other tribes, so their warrior class is now merely ornamental, and the Veterinary Department inoculates against one thing or another, so that the cattle do not die in the numbers they did. Obviously the British statement of policy should be 'No selling of cattle, no inoculations' but the Administration is too timid and silly.

We saw 31 lions today – 15, 12, 3 and 1. The big pride of lionesses and young ones was really handsome.

13 October 1958

Over the home plains, i.e. Egilok, as far as the Telek River again, counting all the wildebeeste. Heavy thunderstorm last night, but as our camp is inside the bit of riverine forest the soil is highly absorbent and we are in no trouble. The black cotton soil of the plains gets gummy and greasy with the rain and our river crossings are getting a bit more difficult. We had a long day at the counting and over the same ground as October 7, and found 5,090, so they are definitely coming in. Rather more zebra also than on October 7. At one minute before noon we came on a Thompson's female gazelle giving birth. The kid was fully born at noon; it struggled to a sitting posture at 15 minutes and appeared to reach successfully for a suck of milk. It stood on all four legs shakily at 17 minutes but fell over again. Several more stands and attempts at walking, movement almost unco-ordinated, and at 25 minutes stood and sucked a good meal. The two walked away at 12.30 p.m. Not bad going. The jackals were much in evidence today, looking hither and thither about the plains. They were on the look out for anything that might be going, but in particular these Tommie gazelle kids being born. Two of them come; the mother gazelle chases one and the other snaffles the kid. The gazelle mothers do the same as the red deer hinds, i.e., leave the kid while it is very young curled up in a bit of a hollow in the ground. We passed one such kid and I took a photograph of it. We saw a cheetah chasing a flock of Tommies, but as it saw us it desisted and stalked off across the plain. Showers all around us today but we came home dry. There was a lion and lioness over a zebra kill near camp, with four attendant jackals optimistically and deferentially waiting for any crumbs from the rich man's table. The topis are more numerous than earlier and they are in full flood of calving. The calf is kongoni coloured, i.e., sandy golden, not chocolate like the topi.

14 October 1958

Temple-Boreham and I out early this morning for two and a half hours and back to breakfast at 9.30 o'clock. We counted a plain to the south of us which should have been included in our earlier count of 7 October. We picked up 705 wildebeeste in two large herds. Absolutely glorious country and these strings of animals over it in the morning sunlight are an exhilarating sight. Have taken a few photographs but despair of getting the full sense of wide country and many animals. We also went into a shallow valley and plain where T.B. said a rhino lived. The rhino was on view out on the plain. Lots of bat-eared foxes about and occasional packs of striped mongooses. The stone curlews are on the plains now, reminding me more than ever of the Downs. There are many crowned coursers, blacksmith and crown plovers. And larks. Saw a brilliant flame-crested hoopoe yesterday. Writing most of the day because David is going to Nairobi tomorrow, 7 hours' journey, to fix up about maps, airplane and so on.

Temple-Boreham, David and I out again late afternoon, going westwards more or less where we went this morning, through lightly bushed country with clumps, much over-burnt, but now green and lovely. We met an old lion who hadn't much to say to us. Then half a mile farther on there was a young wildebeeste alone and going round in circles. It allowed us close to it and was obviously ill. T.B. shot it and almost immediately there were vultures overhead, and while we did a postmortem they came down. Then we saw the lion walking up, and as we finished and stepped back into the Land Rover he sank gratefully to his free meal. Once more he had kept an eye on the vultures, who had given him his cue. Two miles farther on we were going to climb a bushed hill scarred by burning, but we saw two lionesses farther up. T.B. said he thought this was Sally, so he called her and the two lionesses came down to us and sat only five yards away. In fact, Sally came so near that T.B. could have touched her. The story of Sally is this. She is now between 11 and 12 years old. When she was two, T.B. found her very ill with a hole in her side and apparently dying. He brought her water in a tin basin and then fresh meat on which he sprinkled sulphamilamide, camped beside her for a week or two and did what he could. She recovered, and every week or so during her convalescence T.B. came down to see her and give her some leonine dainty. The lioness came to know him so well that she would come whenever he called, if he was in her neighbourhood. She bred six months afterwards and produced cubs and still remained tame. She has now had four litters of cubs and all of them have learned to accept T.B.'s presence. The large family of twenty or so keeps very much together and Sally is the matriarch who looks after grandchildren with as much care as her own cubs. This lioness has a rupture in the region of the navel, to do with the bad wound she had, so is easily recognized from afar. I had the impression of complete trust of T.B. by this lioness. Some time ago T.B. was in camp when it was raining hard. Sally appeared and lay under the fly sheet at the entrance to his tent. The rest of the family were around but would not come so near. It is really one of the nicest animal stories I have heard, in that Sally has not forgotten her friend, though he may now go two or three months without seeing her.

15 October 1958

Temple-Boreham and I out for the day, working our way gently down to the Sand River where it joins the Mara, passing through slightly hilly, lightly-bushed country which should have been dense bush but it had obviously a long history of burning. We found a young cow rhinoceros that had just taken her bath in black mud. Later in the day we saw a mother and calf: how ridiculously small and far forward is the eye of the rhinoceros! And later in the afternoon we encountered a bull rhino who saw us off at a smart pace. Rhinos are more numerous than I thought. The Sand River is now running and we walked down to its junction with the Mara, over rocks of mica schist very fine in texture. There was a score of hippos in the pool and one or two largish crocs. The water of the Mara River is coffee-coloured and opaque. T.B. and I had lunch by the Mara, enjoying its tropicality. We had two Game Scouts with us, one being a Ndorobo, an almost gipsy tribe who live by hunting. This man had been a poacher but T.B. made him a Game Scout and he is delighted. The Wa Ndorobo are very poor and primitive and therefore know all there is to know about the bush. They are splendid trackers, very keen in the nose, and they are confirmed and expert honey hunters, following the birds known as honey guides. Well, when we had finished our lunch our pair had disappeared but we heard knocking half a mile down river. Eventually they returned with several pounds of honey comb. They devoured the comb containing the grubs themselves, and gave us the honey. I am having it at breakfast.

We were among game, including lions and a cheetah all the time, and when around 6 o'clock in the evening sun we sat on top of a little hill Temple-Boreham calls Roan Hill and looked over a valley to the Sand River and another wooded hill and lightly-bushed areas, we had a view long to remember. Down near the line of riverine vegetation was a herd of 50 elephants playing about; a herd of 93 buffalo slowly crossed a plain to the water; eight roan antelope were just below us; there were two or three giraffe and eland in the landscape; a herd of impala; several knots of zebra; one or two old bull wildebeeste; several small groups of kongoni; a few topi and some Thompson's gazelles; warthogs here and there; two reedbuck and one lion a good way off. Where else could you see as much as this at one time? It was almost dark as we came over the plains and the wildebeeste in their hundreds were in silhouette, so deeply moving. It has been a great day, one of the days in the life of a naturalist. Dinner with T.B. at his camp and very nice too. Yesterday we picked up the long leg-bones of a giraffe which had come to grief some time ago. This bone gives a good thickness of material of beautiful texture which can be worked. One of the askaris has made a rod 14 inches long, three-quarters of an inch thick at one end and half inch at the other, to serve as a handle for a wildebeeste tail as a fly switch, but I like the rod itself and its feel so much that I shall keep it as it is, as something to twiddle in conversation, like the amber beads on a short string twiddled by the men in Athens when talking in the cafés. The bone looks like ivory.

16 October 1958

A long day out with Temple-Boreham to the Bardamit Plains, going by way of the foot of the Sianna Hills and the point called Barkitabu. Only a few wildebeeste till we reached the plains we counted on October 8, and now we counted 4,656 and did not do the part from Aitong Spring to the Mara and Jagartek. T.B. thought there would be very few down there, but we couldn't have managed it today and we came home in the dark as it was. In the acacia at the top of the Bardamit Plains we saw a single

Temple-Boreham's camp at Oljoro Loroman in 1965, where the party dined 15 October 1958.
(Photo: J.M. Boyd)

bull elephant – no great ivory, but he was enormous. The wildebeeste seemed to have
moved westwards between one and two miles and they were in pretty big lots up to
600–700. Counting through field glasses and with the heat haze is very tiring, and
both T.B. and I had had enough by the end of the day, our eyes watering if we
squeezed them.

17 October 1958

Temple-Boreham and I out along the little hills towards the Sand River and the Mara.
We walked up the Sand River for a mile or two in bare feet and then struck across
the hills. Very little doing, but we saw a cheetah and two or three lions taking their
ease. An old buffalo bull was lying quiet and we passed him at 120 yards, he not seeing
us at all. But we had to cross his wind to get back to the Land Rover at 200 yards
range. He rose suddenly and was off from his position facing away from us and he
never so much as looked back to see who we were. Men, and that was enough. We
found a Thompson's gazelle newly dead and unmarked, so we postmortemed it and
reckoned it was poisoning, because the whole gut was inflamed and the spleen
engorged. And coming back over the plains we found a late wildebeeste calf alone and
scarcely able to walk. T.B. shot it, but we could find nothing wrong. It was in very
poor condition, and as the calves are still suckling, we think possibly its mother had
been taken by a lion. Our Ndorobo showed us a plant which is poisonous – a tree
which I must now get identified. We noticed at one place that two plants of a

geophyte, which I have called rabbit-brush for want of a better name, had been dug up by an animal I guessed to be a porcupine, and our Ndorobo when asked said it was he who had arrows instead of hairs, so that was good enough. This plant appears to have a deep-seated swollen root the size of a melon, which Mr Porcupine had been digging down to and eating, but spitting out the fibrous structure. Our Ndorobo told us his tribe also dig these and chew them for the moisture contained. The texture of the root is that of a coarse old turnip. T.B. away in the afternoon 30 or 40 miles northward. Our blower this morning had a message for him saying there was a lion attacking Masai cattle. As he says, he will go and show the flag and leave a couple of Scouts, and he should be back tomorrow evening or on Sunday. The slim bones of the Thompson's gazelle have been saved for me. They are exquisite in fineness and cleanness.

18 October 1958

A good walk westwards in the morning to the valley where we saw so much from Roan Hill three days ago. There was a cow rhinoceros and her calf there today. A honey guide led us to a bees' nest in a dead acacia in an ant-heap, distance 350 yards, in about 5 hops and almost direct. We met three Masai going to Narok [blank in MS] miles, with one sheep, to sell it. One asked to look at the rhino through David's binoculars and they all giggled like schoolgirls. I have noticed this tendency to giggle before. Out in Land Rover late in the afternoon back along the way to Uaso Nyire, and found 200 wildebeeste where there was but a handful two days ago. Explored a pleasant valley leading into the Sianna and saw one rhino bull. We found a place among some bushes where a couple of Masai had had a fire and cooked a sheep some time today, for the fire was still alight. The bones were picked clean. David says that occasionally Masai will have an orgy of meat-eating. Two or three men will brew an infusion of a particular species of acacia which gives them a tremendous appetite for meat. They will gorge themselves, and then sleep and wake up with a further craving for meat which they indulge again and so on, till the bullock or sheep is finished. They say that after such an orgy they are left with a wonderful sense of wellbeing.

19 October 1958 (Sunday)

A quiet day spent reading the Game Policy Committee's Report, the Serengeti Report and so on. Am gradually getting myself tuned in to this new set of politics in Kenya. Just as involved and unpleasant as in Northern Rhodesia, but at least the Governor here, Sir Evelyn Baring, is a straight and fine man, who feels for the game. On my early morning walk I found the 500–700 Egilok River plain group of wildebeeste closely massed and running southwards; something must have frightened them though I could not discover what. In the evening they were well spread out and feeding quietly in the bottom plain. David and I went a run northwards in the late evening. Our downs were practically empty of game and we found them on the lower plains nearer the Talek River. We have now had a week of fine weather and the grass on the plains has lost its freshness. The lower plains are in slightly better shape obviously, than the higher downland type. Temple-Boreham went away on Friday afternoon and has not yet returned. The weather is warmer, and to sit by the fire in the light of the half moon is such delight.

20 October 1958

The airplane arrived at noon with Pilot Supt. Bearcroft of the Kenya Police, and Verdcourt the botanist of the British East African Commission who works at the Corydon Museum, Nairobi. Temple-Boreham also came back at nine o'clock this morning. We had a merry lunch, the five of us, for Bearcroft, nicknamed Punch, is a humorous ex-RAF type with a turn for descriptive though often improper language. David and T.B. flew with Punch till sundown, doing the Sand River and Bologonja, the wedge between the new Serengeti Park and the Mara Reserve, which is full of game and ought to be put in the Serengeti Park because it is heavily poached; then to the Bardamit plains where there was a heavy concentration of 6,000 wildebeeste now, 1,000 having gone over from the plains this side of the Telek. Verdcourt and I were out botanizing, and he has been a tremendous help to me in identifying the grasses, herbs and shrubs of the different habitat types. I noticed how the rain we had has brought the small fine grasses into flower, such as *Sporobolus, Hapachne, Arogrostis, Aristida* and *Microchloa,* but the taller heavier grasses have only sprouted vegetatively. This is rather a nice bit of evolutionary behaviour, in that if these finer grasses were to flower in the big rains as the coarser grasses do, they would be overwhelmed, whereas by flowering now they have plenty of room for pollination and growth. We did the low plain, riverine forest, and area of inundation.

21 October 1958

Away early, going with Verdcourt over the higher plains, the lion-bush complex and to Oldonja Lo-ip to see if the change to volcanic rock made much difference. We did find a few more species. Punch was flying David and T.B. all day.

Seeing the tiny little grasses coming to fruition in the little rains and the taller grasses producing only vegetatively, set me thinking on the whole matter of conversion of living grass into decaying organic matter which is lived in and by, by other and tiny organisms. If the game population here, large as it is, was once as large again, then twice as much grass was converted into manure for reintegration with the soil and that much less remained as dried dead grass to be burned to ash. A full conversion cycle means the optimum population of animals. Less than the optimum means a bottle-neck, and annual fire has to take the place of more complete grazing, which got more organic matter into the soil.

The fliers came home with good results, which David has converted into little dots of red (wildebeeste), green (zebra) and black (buffalo) on to our sketch maps. It all looks most impressive. We had a splendid evening of good companionship round the big fire.

22 October 1958

Up early and wishing Francesca many happy returns on her third birthday. Flying with Punch Bearcroft just after 7 o'clock, taking advantage of the early morning calm air. We went down the Sand River to the Mara and then into the area I call the Wa-Tende Wedge [Lamai Wedge], i.e. the acute triangle of country, about 100 square miles, between the Mara River and the International Boundary which also happens to be the boundary of the Mara National Game Reserve in Kenya. The latest Tanganyika White Paper on the Serengeti problem gives a large northern area to the Park, coming up to the Kenya border, but by stopping the westward boundary at the Mara instead of continuing the line across to the border, they leave this wedge of country into which the Mara Plains game goes from time to time. The Wa-Tende from the top of the Isuria Escarpment descend into the wedge and slaughter the game by means of poisoned arrows and nooses of stranded wire. The Wa-Tende are doing this poaching in gangs for commercial ends. As a tribe they are not admirable, being low and useless for the most part. Flying over the Wedge this morning I could see a lot of game in lovely country and there were many dense patches of bush much less punished by the elephants than in the Mara Plains. Then we flew over the Mara National Reserve where there were more thousands of wildebeeste and zebra. A central series of salt pools and marsh was an interesting feature. Again there were dense patches of bush. Aground again by 8 o'clock, having flown over 100 miles. Punch was so good as to congratulate me on my power to conquer two rounds of nausea: said he'd never seen it done before by anybody having reached the green-faced stage. He and Verdcourt flew off on their return to Nairobi, and we found the Land Rover packed ready for our sortie into the Mara Reserve. We drove over the plains to Mara Bridge and bushwhacked into the Reserve, a journey of four and a half hours. By the time we reached a place at the far end of the Reserve where there was a temporary pool of water and suitable for camp, we had seen four rhino. There seem to be 20–30 in this triangle. The wildebeeste were all round us gnu-ing as we sat at our late lunch. In fact, as one lay in bed at night, or any time, this sweet sound of the wildebeeste could be heard and the laughing of the zebras. We walked to the salt marsh after our late lunch; this is warmer country than on the plains proper and the dense patches of bush I mentioned are higher than on the plains. This Mara triangle is obviously well watered by a higher rainfall than on the plains; the triangle is really a gentle slope from the foot of the Isuria Escarpment to the Mara River. The scarp is volcanic lava, amorphous basalt, and the foot of the talus from the scarp is lovely green ground much used by the zebra. Once a year the Masai from the top of the scarp drive their flocks down the face and through the triangle to the salt, then back again. There is no human habitation in the triangle. Temple-Boreham has hammered the poachers good and hard so that there is now no poaching within the Reserve. When flying over this country early this morning I saw, nearly three miles from the International Boundary and in the wedge, a lorry body high in an acacia tree. I knew it had been got there and hauled up by blocks and tackle by a Boer called du Pre who has a mine (gold) twenty miles away at the tip of the escarpment in Tanganyika. Somehow, as I looked down, I felt a dreadful antipathy to that man. He is an honorary Game Warden of Tanganyika and he says there is no poaching in the area. Indeed, the Game Department and National Park and Police of Tanganyika organized a poaching drive a few months ago through the whole border area from the Bologonja to the scarp and they found nothing. Nor would they anyway, travelling as a crocodile of seven Land Rovers and a lorry. They never poked through a patch of bush. Catching African poaching gangs means getting out on your flat feet, creeping up the *dongas* to the hide-outs and so on. Mr du Pre had unfortunate family matters which called him farther north at the time of the operation, but T.B. says his informers say du Pre was down at his lorry body in the tree three days before and that shots were heard. I got the feeling as I looked down on that tree that du Pre was in the racket somehow. A Dutchman rarely does anything for nothing, and I didn't see what that lorry body in the tree gave him because it was not a point where a lot of game could be expected to come beneath.

23 October 1958

Temple-Boreham's suggestion was that we should make a sortie into the wedge in Tanganyika with as big a load of askaris as we could manage, so away we went with eight aboard. Our path lay along the foot of the escarpment until we reached the Tanganyika border. Just over it we saw three Wa-Tende women at a *shamba*. They wear a think blanket from waist to knee, and brass bangles below the shoulders and above the biceps, and another set below the biceps and above the elbow. They are shaped in such a way that the muscles of the arms must be constricted, though doubtless they achieve accommodation to the bangles which cannot be removed. While T.B.'s corporal talked with them a tallish man came up, apparently from nowhere, wearing a little red felt hat with no brim and a cotton blanket tied on top of the right shoulder and drooping below the knees. He had a trace of moustache and beard. Looking down his naked right side I could see he was powerful, though in no way thick. He shook hands with us and talked a little with T.B. He was one of T.B.'s sources of information and he told us a gang was working in the wedge at that moment. Without further ado he got in the back of the Land Rover to join us for a couple of days. We descended from the foot of the scarp into the wedge and before two miles saw two men less than half a mile away in the thin bush. They set fire to the grass as we approached and ran for it. We shot through the fire and chased them. We caught one who had begun walking towards us and who was the most injured man in the world at being picked up, especially as he had no impedimenta. Then after the other, who was haring away with bow and quiver and spear. Of course, the Land Rover went faster and our askaris dropped overboard to pick him up. T.B. opened the quiver and there were six poisoned arrows; so he joined his friend in the back and they were handcuffed together. Then we returned to a patch of close bush near where we picked up the first man. The askaris worked through it and produced a bag which, when emptied, showed some dried game meat and a bang pipe. The Wa-Tende are confirmed smokers of bang (hashish) and after a round of it tend to get into murderous condition, running amok. The evil-looking little man to whom it belonged immediately said it belonged to his companion, which brought forth expostulation from the thick little man with the bow and arrows. These two were coming back to fetch the women to carry loads of meat from some hide-out in the bush. We moved farther down into the wedge and came to du Pre's lorry body high in the *Acacia albida*. It certainly wasn't there for viewing game from, but I noticed it was beautifully placed for signalling to the top of the escarpment several miles away. This man is probably feeding his mine labour on game meat which he either poaches himself or gets the Wa-Tende to poach for him, and he, having got himself into the position of honorary Game Warden, effectively precludes anyone else being that in that place, but as T.B. as a senior Game Warden in Kenya is also an honorary Game Warden in Tanganyika he can come around anywhere with powers of arrest: du Pre and the Wa-Tende are probably scratching each other's backs. Well, coming to a *donga* and thick bush within a mile of du Pre's tree, we poked into it and found a hide-out of the poachers. It was a horrible sight in that within the shade of the bush there were the remains of over thirty zebra, wildebeeste and topi. A grass hut and drying poles half filled the area but the remains were all over the place. A hyaena had crashed away as we reached the place, and it was the bones pulled outside by the hyaenas which had first attracted us to look inside. This place must have been left two or three weeks ago. We poked about for another mile or two and one of our prisoners took us to another old hide-out, still within a mile of du Pre's tree, where we found remains of another dozen animals. Off again, and the man in the red hat directed us to a point between two little hills of dense bush and a quarter of a mile above a water hole. There we found a quarter-mile bush fence with stranded wire nooses all set. T.B. got quite excited and said their hide would be in thick bush near the water. We deployed our forces and crept up on the bush but there was neither hide-out nor poachers. So we came back to the little wooded hills and came smack into a hide full of fresh meat, but the poachers had flown. Examining the meat surfaces of the dead animals we could see the cuts were two or three hours old, and we deduced that we had been spotted earlier in the day when we had come under the hill on the other side. T.B. put a guard on the hide, lifted all the wire nooses and prepared for an ambush on the path up the escarpment. During the night one man came back near the hide but would not come in. The askaris were restrained from shooting him because it would have taken a bit of explaining why Kenya askaris were raiding in Tanganyika where control is very lax.

The sight of that hide with the guts and opened bellies lying around, the grinning zebra heads and sad-looking wildebeeste heads, the legs all a-tumble to the sky, and the stench, was unforgettable. Looking around I was struck by the complete lack of system in the butchering, meat and filth and the men's lying places all together. Examining the portions of carcasses still carrying hide, we found the deep cuts of the wire nooses as the beasts had struggled. Later in the plain we found a dead zebra which had been caught by the hind foot in a noose. The wire had cut through to the bone. This hide was within half a mile of the Kenya border and Mara Reserve. There was little more we could do this time, but towards the end of November we are going to have another sortie and get a force of about forty askaris. It will be done quietly and on foot.

24 October 1958

Went to the hide again and dragged out the carcasses for the vultures to clear up. Then making for the Mara River on the Kenya–Tanganyika boundary, through small hills of quartzite covered with gall acacia. I don't much like country of perpetual gall acacia or whistling thorn. It is faunally poor and you have to weave this way and that to get through. The spines are appallingly sharp and strong, well over an inch long. A day or two ago, watching a rhinoceros closely as it was eating, I noticed it was going for young gall acacias perhaps 6 inches high with a spread of a foot. He would so mouth the thing as to get hold of the stem at ground level and pull up the whole plant. An adequate population of these animals might well keep these bushes within bounds. When grown they go up to ten feet with a long straight stem and a bush at the top, looking rather like young standard apple trees.

We started this sortie with plenty of meat, for Lynn Temple-Boreham found a topi bull with a broken foreleg, which he put out of its misery. Then near the Mara we scared four lions off a bull buffalo they had killed. The men cut themselves large hunks of meat from the leg, rump and along the spine, and skinned part of the ribs to use the skin as wrapping for the meat which was going to be carried in our Land Rover. There was still plenty left for the lions, for the buffalo weighs over a ton. The men are in high fettle anyway, having had lots of meat, and the poaching arrests have set them on their toes. The jailer for our prisoners is a Ndorobo who was an inveterate poacher, and who, on his last release, asked Lynn T.B. (the man who had caught him) for a job. So he took him on, gave him the dignity of a Game Scout's uniform, drilled him and has made him into a loyal retainer who is quite happy catching other poachers. These people rarely bear malice. Our men are really so little different from the poachers, being quite savage, but nevertheless well drilled and disciplined. They have also been allowed as much of the poached meat as they wished and they have done themselves well.

We came back to our base camp at Oljoro Loromon and Temple-Boreham went up the escarpment to a meeting with the District Commissioner tomorrow. I took a photo of our two disconsolate prisoners and felt a little sorry for them because I have a feeling for all prisoners and captives. When someone is utterly in your power you feel the need for restraint in behaviour. While they are on the run one could happily shoot them; I don't think I should have felt restraint to Japanese or to Mau-Mau terrorists. Both T.B. and David had a lot of experience of both.

Back home in the almost dark, the wildebeeste being back on our plains in large numbers. Seeing their silhouettes against the darkening sky, I felt these poor things were at rest and had successfully run the gauntlet of those dreadful wire nooses in the Wa-Tende wedge. We must try hard to get that area into the Serengeti Park.

25 October 1958

A quiet day of writing with a run over the plains morning and evening. The fresh green grass is now all grazed away and we need more rain if the herds are to stay with us round camp. During our evening run we saw some topi panic (which they rarely do) and went up to a patch of lion bush to see what was happening. A male lion had just caught a topi calf and was carrying it away. A lioness was near but took no notice, remaining seated. A hyaena and a jackal were already in evidence, waiting for the crumbs. Then we saw another lion apporaching, an old one thin and lame and gone in the feet like an old dog. We got close to him and found he had no teeth. He looked bad-tempered and when he got near the lioness sprang peevishly at her. She answered by giving him a good clout over the head and then sat down again. The old lion then stood a few yards away and roared. Then he walked through a patch of bush to the lion eating the topi calf and rushed at him. We expected the younger lion to set about him, but no, he ran into the bush carrying his bloody meal. From the row, we gathered the old lion had got the remains of the calf. How did he maintain his authority when he had nothing with which to reinforce it? That lion must scrounge all his keep now.

26 October 1958 (Sunday)

David away to Nairobi this morning on important business, coming back tomorrow. I went for a seven mile dander towards the Siannas, through a big patch of lion bush and out on to the plains and round back to camp. I had set off alone, intending only a mile or two, but I soon found a couple of askaris were following me to look after me. So I took the farther walk and thoroughly enjoyed the chance of it. More writing. Half way today to my return home. Another little walk as the sun went down.

27 October 1958

Away with two askaris by 8 o'clock this morning hoping to climb Sianna, or perhaps it should be called Mwusito, 7,362 feet, and to continue round the ridge of the whole valley, returning to camp at Oljoro Loromon at 5,300 feet. We started with Lathithemine 6,518 feet (no, Latmethine) and crossed the valley to Mwusito. There is still quite a bit of scrub on this high hill, but there are also bare patches on the ridge at the top. The view is absolutely splendid, better I thought than from Kuka. There are magnificent grassy valleys below. The air was quite glorious and fresh. It had taken over three hours of pretty hard going to get here, and with the half hour's rest I knew I would never get round the whole valley, so we descended a very steep side of quartzite boulders and crossed the valley to a spring where we washed our feet and dabbled our hands and wrists. The askaris drank but I made a rule not to drink when I am walking. Better to wait till the evening. I had certainly bitten off a bit more than I could chew and I was glad to get back at 3.15 p.m. to a lot of tea preceded by a bush cocktail of lime juice, water, salt and Eno's Fruit Salt. A very fine drink. There was a lot of grass up the valley but little game except at the foot. Saw no Chandler's reedbuck or klipspringers at the top. There is a rhino far up the valley, and going up the high hill I saw a leopard go out of a bush in front of us. He was a very big one, and as he went away from us, the sun shining into his coat, he looked dark and tawny and the spots were scarcely visible. Have had a marvellous bath and change and must now start sorting out the associations of plants made with Verdcourt last week.

28 October 1958

A quiet day spent talking and writing. So much seems to crystallize from talk about the subject. David came back at 1.30 this morning when I got up to meet him, though I had slept in my chair by the fire until 11 p.m. And after I had read the mail I lay awake a long time, so I did not feel like going far today. Neither, I think, did David.

29 October 1958

A long day northwards to country outside our area but on its border where the Veterinary Department is bush clearing to make a tsetse break between the Mara and the country north again. They are also ranching Masai cattle (steers) on the cleared land. Caterpillar bulldozers and weighted chains dragged between two of them are being used for dragging up the bush. When they had a lot of prisoners they sowed stargrass seed broadcast on the cleared and burnt land, but now they do nothing. As we came to the cleared land we were confronted with a good growth of a pernicious imported weed of arable land, Mexican marigold. The area looked devastated, and was. Where the clearing had been done two or three years ago the shrub *Euclea* was coming back strongly. We called at the headquarters of the ranch – in what I imagine was an inopportune moment, for as it was explained to us by the three officers who were in conclave, the money advanced for the clearing was to be repaid from the proceeds of the ranch, and a wave of contagious pleuropneumonia struck the cattle some months ago and is still going, so they are pretty well broke. However, they were very kind to us and talked about the scheme and gave us a drink of tea. I asked where the river came into their reckoning and whether they had cleared the riverine forest. The leader (Lewis) said, yes, they had cleared the river for a mile or two but they hoped not to have to fell the big trees; so they had grubbed all the bush and burned it and had then found the heat had killed the trees. It had not occurred to them that trees within a forest were fire-tender and not like the fire-resistant savannah trees. We went down to the river afterwards and it was enough to make anyone cry. Scrub cattle were wandering about among dead skeletons of trees and the river banks were gullied and eroded by wear and wind. It is time in all the African territories that major physical changes in land use should be considered by a land-use ecologist as well as by departments who don't know the consequences of what they dream up as practical schemes. We drove on into a wide valley where the whole floor was forested with cover rather between riverine forest and lion bush in character. It had water and was obviously well patronized by elephants and buffalo. There were two or three small areas of grass, perhaps 100 acres each, and on these were flocks of Masai sheep. We then turned into the Lemok Valley where the bush was going back and the grass areas were showing overgrazing by Masai stock. How lovely it is to get back into the Mara Plains where only the game graze.

30 October 1958

Over towards Kuka today and west of it in Tanganyika. It is undulating bushy country quite difficult to get around in with a Land Rover. We went a good way seeing little game but giraffe, and then seemed to run into a lot of wildebeeste, zebra, topi, kongoni, buffalo and eland. We must have seen 100 giraffe today. One herd of eland was of over 50 animals. Grass was greener and more plentiful than on the plains. Little pans of water all over the place, and we also found springs in the range of hills to be seen southwards from Roan Hill. Donald Ker, the well-known white hunter, is down in the Mara with an American party and had camped in the forest a quarter of a mile away from us. David and I were asked to dinner. The American was a Mrs Brokaw with her two sons in their twenties. Very nice party indeed. They are shooting nothing, merely seeing the game and taking photographs. Donald Ker has almost given up taking shooting safaris and goes only with those content with photography. He is President of the East African Professional Hunters' Association and one of the greatest forces for wild-life conservation in the Territory. In this he has quite a bit of trouble with the less responsible elements of his Association. He is a little dark man, restrained, polished and civilized. I liked him immediately. The Brokaw family are interested in the conservation movement in America. The two sons are in a family law business.

31 October 1958

Eastwards today, over 50 miles to the Loita Plains and a hot spring on the southern edge of them appropriately called Majia Moto. The Loita Plains, vast expanses of flat country with occasional volcanic cones, are outside the fly area and are therefore subjected to the full force of Masai grazing, cattle and sheep. The condition is frightful, a semi-desert of *Justicia, Sida* and *Pennisetum,* with dwarfed gall acacia. There are many patches of bare blown yellow earth. Majia Moto is a tragedy. The hot water comes from the earth as a clear bubbling stream, fresh to drink. Its gradual descent into the plain is marked by yellow-barked acacias and what has been pleasant bush. David and Donald Ker remember the area when it was pleasant, but Masai pressure has increased and the acacias and bush are going. Two or three Indian dukas have set up near the spring in the midst of the desolation. The Masai come there the more, if not to buy then at least to finger the blankets and argue prices. The country improved all the way of our return and we were thankful to get back to this paradise, the future of which concerns us so deeply. If it is made safe for Masai cattle, these plains will soon look like the Loita. Round for another chat with Donald Ker and the Brokaws in the evening.

1 November 1958

David and I busy with a design for a wild-life service for Kenya in place of the present decrepit Game Department. Round to the other camp between tea and dinner for an hour with Donald Ker and the Brokaws. Two young male lions near camp.

CHAPTER EIGHT

The Mara Plains 2: At the Mara River (Paradise Camp)

November and December 1958

O F ALL THE PEOPLE Fraser Darling met and respected in his African days there were a few with whom he struck a deeper chord of friendship and trust. David Sheldrick was one of these. Frank acknowledged him as his 'friend and teacher'. Until he died, he looked upon these ten weeks in the Mara with David as some of the most emotive of his life. Under the great physical effort of the field survey, there was the grief over Averil's death, and the anxiety about the children and the work back home. 'When Averil died', writes his friend Noel Simon, 'Frank wanted to get away for a few weeks. This was exactly the opportunity he was looking for, and accepted at once. To accompany him, there was no finer game warden in East Africa than David Sheldrick.' In the sublimation of that sorrow, and the relief of that stress, David played a great part. His uprightness of bearing and spirit, and his personable warmth was a balm to Frank, who was already deeply in love with Africa.

In 1965, I was myself the guest of David and Daphne Sheldrick in the Tsavo (East) National Park, and flew with David in search of elephant, rhino, and poachers, just as Frank had done a few years before. I had the opportunity then of hearing from him at first hand his impressions of Fraser Darling in the Mara. He said that his thinking on wildlife management had been greatly influenced by Frank, who, more than any other, had taught him to look for the underlying interdependent forces in nature, and to mistrust superficial observations. For example, the emphasis which Frank laid upon indicator species of pasture grasses and shrubs turned David's interest and attention from game animals *per se* to the detail of their vital food supply. He regretted that he did not have a scientific training, saying that he would then have felt better qualified in the company of scientists, and better able to tackle the complicated ecological problems posed by his job. David underestimated the value of his own intuitive appreciation of these problems. A scientific training might have been an advantage in discussions with Frank and in solving the elephant problem of Tsavo, but academic life might well have blunted his instinctive understanding of wildlife which so impressed Frank. Sadly, David Sheldrick died suddenly of a heart attack on 13th June 1977 aged 58 years, his work for the wildlife of Kenya still in full cry. Today, that work is continued by Daphne and the trustees of the David Sheldrick Wildlife Trust, Box 15555, Nairobi.

2 November 1958 (Sunday)

Moved camp today down to the Mara River. The altitude at Oljoro Loromon was 5,300 feet and at the river it is 5,000. I am sorry in many ways to leave this lovely place on the wide plains. There are plains nearer the river as well, but there is more bush and more gall acacia. Along the river itself there is the beautiful riverine forest, on the edge of which we have camped. In front of us westwards is a plain of perhaps 500 acres, completely surrounded by forest. The plain is what, in the old royal forests of England, would have been called a lawn. Towards dusk buffalo, waterbuck and bushbuck and a few impala were grazing. David and I walked along the river in the beautiful forest after our late lunch. The elephants are too thick on here and are damaging the big trees by ring-barking them. A reduction of 500 would be welcome, but it is a major operation to do this. Otherwise the forest is pristine in that man has never touched it and though fire is nibbling at its edge, no fire has gone inside it. We saw two troupes of Sykes' monkeys, smallish very dark people with lighter faces. Baboons of course everywhere in the forest. They are the darker forest species. There were Egyptian geese on the deep river banks and we saw a furry-necked stork on her nest, which was placed far out on a branch of a large tree. We saw cinnamon bee-eaters, but not the carmine, which to me are so much part of the African scene.

3 November 1958

A lot of lion roaring in the night and this morning two large male lions were sitting on a plain 250 yards from camp. They had killed a yearling buffalo bull in the night. I thought I heard the bellowing. We shall take a liberal commission of the meat for it is young enough to be tender before it goes too far. The elephants squealed a bit in the night but did not come close to camp.

We went exploring today, trying to find a way through to where the Mara crosses the International Boundary. This meant crossing the Jagartek and the Talek Rivers somehow. We got into a lot of rough country, first in basaltic boulders and then in mica schist. We came to what we thought was the Talek and Jagartek joined in one, twenty feet below the plain and the water running very slowly indeed over shimmering mica schist of very fine grain and lamination. The ripple of mica in the rock made it appear like the side of a mackerel newly caught from the sea. We found what could be a crossing with a bit of gardening, and in the long pool above were seven hippos. They appeared to have a side pool into which some of them went, and on going to examine it we found it to be in the nature of a fault in the rock, making an absolutely straight canal a few yards across and running for more than half a mile from where we had come on the hippos. Tropical riverine vegetation lined the sides of the canal and almost met overhead. There were places where the hippos or elephants slide down to the water and these were sunsoaked, hot and still. All very beautiful, especially when the brilliant malachite kingfishers alight on a stone or fly rapidly down the river, and the carmine and cinnamon bee-eaters settle on the branches over the water. By the end of the day we could not be said to have found our crossings for a quick journey down the Mara, for we got mixed up in a complicated *donga* system, but it was all most enjoyable. There was a lot of game everywhere but precious little for them to eat. We met one bull rhinoceros who chased us properly, but he had been in trouble, having lost one eye and with a wound in a hind leg. These animals do fight so senselessly. He had the usual 'sweat sore' at the base of his throat. What and why are they? Nearer the Mara River we have found five dead buffalo 6–8 months old calves being consumed by vultures. We are afraid rinderpest has struck the herd, for we found the young wildebeeste with it over the Tanganyika border a few weeks ago. Temple-Boreham said he had heard it was moving north. Long years ago, 1895–6, rinderpest came to Africa from Europe and was extremely serious among African cattle and the game, especially buffalo and wildebeeste. Now they have acquired some immunity and comparatively few die. We also found a big bull giraffe that must have come in from Tanganyika. He had had a noose on his off fore-leg and it had made a terrible wound. The leg was swollen from shoulder to foot but he had at least got rid of the wire and the wound looked clean. He was in quite good order, so we let him be.

4 November 1958

Another day of exploration of the country bordering the left bank of the Mara River. We put in three hours' work on the Talek crossing, having got five askaris in the back of the Land Rover. We also had to do some landscape gardening to one or two other *dongas* before we could get across. We lunched late by the Mara River, David giving the askaris some tea and sugar to give themselves a hot sweet brew. They had brought half-dried pieces of buffalo meat for themselves (from the lion kill) so they were very happy and full of fun. They amused themselves teasing the foolish and inquisitive baboons on the other side of the river by making sudden movements this way and that. The baboons get into a state of half-fearful, half-thrilling excitement; I, going down to the river to rinse a cup, pretended to throw it at one and it nearly fell out of the tree; the askaris laughed so much that the baboons got cross. They can't laugh and when they see a man laughing and showing his teeth by so doing, they think he is showing anger of some kind. The country on the over-side of the Talek was unburnt this year and was empty of game of all kinds. These undulating low hills, sometimes basalt and sometimes quartzite, carrying gall acacia and *Acacia hockii* do not seem to have much attraction. One of the askaris saw a young acacia which seemed to him just right for the bark to be used for an infusion for drinking with his buffalo meat, so hacked it off with his simi. He was a Turkana, but evidently several of these tribes other than the Masai have this trick of increasing their capacity to eat meat. We did not reach the Sand River, and it was well into the dark before we got back to camp, very thirsty. All the way back our askaris, two Samburu and two Turkanas, sang wild Turkana songs which were really rather pleasant. One would sing a story and the rest would sing a continuous accompanying chorus. The solo singer was in harmony but not in unison, and the chorus was full, bold, rhythmic and in unison. If you get yourself a bunch of wild savage Africans like we have with us, discipline them to a state of pride in being what they are, they are quite magnificent fellows and a joy to have as a gang for work, tracking and so on. Their singing tonight took away all tiredness and tedium of the cross-country journey.

5 November 1958

At midnight I woke to feel myself being crawled over by myriad feet. And so I was. A train of siafu had, most unusually, climbed up between the mosquito net and my mattress and got into the bed in thousands. Until I moved they did not bite and sting but once I had put a light on they were well into me. My hair was full and one of the soldier caste of them bit between my big and second toe. I went out to the fire and slept in two chairs till 3.30 o'clock, by which time the ants had left my tent and I was able to settle down again.

A six-mile walk in the forest along the river this morning, quite hard work for all one's body twisting, ducking and stretching through the honeycomb paths made by the elephants and over fallen boughs. There is no thick field layer, but an ample shrub layer and of young trees, so that one cannot see farther than a few yards ahead. The big trees, such as *Warburgia* and *Ficus*, are so very magnificent; the fig trees have enormous round branches that spread a long way horizontally and almost each one is a joy of sound because the bulbuls are there, eating of the large fruit clusters which are like great bunches of grapes. Sometimes there is a flash of cerise and blue as a bird called a trogon flashes among the lower branches of the trees. The brilliant louries run and hop in the high branches like squirrels more than like birds. Pied and malachite kingfishers are over the river. We saw one very large crocodile this morning and got to 25-yards range. We still don't know what we did that made him suddenly take to the water. We came to a small lawn of a few acres with one or two acacias on the side of the river, a heavenly place for a camp if one could have carried in one's gear. The river bank in one place here was low enough for the hippos to lie out and for the elephants to come down and disport themselves. We crept close to the ten hippos lying there and watched smallish birds pecking at their wounds – all adult hippos seem to have wounds or scores of some sort. We did notice that the two young hippopotamus had no wounds. Eventually a flurry of wind took our scent to them and they got up hurriedly and made for the water, except one old bull who took his time and looked as if he was saying, 'These women do make such a fuss.' We got the impression that the elephants were not harming the forest proper. Their damage is on the edges and outside by ring-barking, stripping, and knocking trees down. The forest proper must be used mainly as cover. Occasionally we hear them trumpeting and squealing, but they are very shy here and we seldom see them.

The askaris clearing some long grass in front of the camp today. Finding a rotten log they heave it to the 30-foot river bank and heave it over. A wonderful and much appreciated splash. David interprets for me: they found it so heavy they are sure it will sink, but no, it floats. Ah! how strong the water is to hold up that log! Then one asks the others if they remember when one of their fellows, a Somali called Saladin, walked into the lioness who had just had cubs. The others remember so well they have to stop work for a good laugh. Then one imitates the proud Somali walking backwards praying to Allah, 'Allalalalala inshallah' and so on. It takes the men a few minutes to overcome their laughter. And so life goes on in this camp. Another thing delighting them is the fishing. They go down with fish hooks baited with buffalo meat on string and little rods cut from the forest. They catch catfish to 4 lbs, barbels and barbets to 1 and 2 lbs. Each catch is a moment of excitement for all. Kilonzi before dinner is all smiles, showing us his catch, and the askaris' toto emerges from his dormat existence and shares in the triumph of catching a large fish.

The special fun today among the askaris was getting a bees' nest from the top of a high tree on the edge of the forest. You never heard such excitement and fun. The tree itself was hard to get up but they made hooked sticks and climbed those hungover branches. Dried elephant droppings set smouldering on the end of a stick were used for subduing the bees. By darkness they had reached the honey, and in dropping the large combs out of the tree one dropped plump on to one askari's head. You never saw such a mess, but everybody was wildly happy, even the man with honey in the hair. They returned to camp with a bucketful of honey and said there was still more in the tree.

6 November 1958

Heard by wireless this morning that Noel Simon would be coming down by air from Nairobi. We went up to the Tsetse Ranch airfield at Aitong-Mara in cloudy weather with some rain. It would appear the short rains have started rather suddenly. When the little Cessner came down we found it piloted by a thin slip of a girl called June Wright. She is one of the crack bush pilots of Kenya. She also has the name of being mischievous, as on the occasion of a yacht race on Lake Naivasha, when she dived her little airplane just behind the leading yachts and took the wind out of their sails. Noel was accompanied by Sid Downey, Donald Ker's partner. He was such a nice quiet man who, like his partner, has given up shooting altogether. Noel and I were busy talking over affairs of the future of the Game Department in Kenya with David. Took the party back to the Cessner 182 and they took off at 5 p.m. We reached home at dark. Got some mail.

7 November 1958

Lovely clear morning. We got ready for our stab down the left bank of the Mara River into Tanganyika, packing a tent fly and a few days' rations, and taking four askaris and Kilonzi as cook. The route across the drainage to the Mara River which we have been pioneering in the last few days now stood us in good stead for the first half of the

Sheldrick's uniformed askaris making a passage for the Land Rover through a *donga* on the way to the crossing of the Sand River on 7 November 1958.

journey. We had one more crossing of a *donga* to shape up before reaching the Sand River, which took us best part of a couple of hours hacking. After that we went through a good deal of unburnt grass and undulating bush country, having some expanses of close bush to circumvent and several more *dongas* to cross. The askaris had had enough, I think, by the time we reached the Bologonja River, here at its lower reaches a deep *donga* of fairly stagnant pools so different from that lovely limpid spring and stream I mentioned on our stab to the Serengeti. We had eventually to come almost to the junction of the Bologonja with the Mara before we could cross, but having done so we came on to part of the Mara which is all rapids and quite delightful. So we called it a day — 45 miles of hard going and great heat. We were dusty and dirty and hot, because after the last *donga* before the Bologonja we had got into newly burnt grass that had not had a shower on it yet. The flying ash went straight for my throat, as always my weakest part. Now, camping by the rapids of the Mara River, we were able to have a proper bath and hair wash, sitting in the river on submerged granite rocks, untroubled by crocodiles, hippopotamuses or mud. It was absolutely glorious to feel so clean and relaxed. At lunch time we had also stopped by the Mara River a little above where a granite outcrop almost spanned the river. A hippo cow was there half out of the water, and her very young calf, possibly two days, was in a tiny pool on the granite outcrop. The two were quite delightful to watch; the mother so attentive and the calf so playful. A playful little hippo is quite something. As darkness fell at camp, we could see a rhino and her young calf not far away, and late in the evening they came nearer and we could hear the strange little mewing note of the rhino calf.

8 November 1958

Lovely morning, and David and I away for one and a half hour's walk before breakfast, leaving before 7 o'clock. We went a couple of miles or more down the Mara River and found it continuing as a lovely river of rapids and rocks, lined by Phoenix palms, a few *Podocarpus* trees, *Syzigium* trees, and a little bush. We found signs of poachers having camped by the river, some of their light noose rods being propped against a tree. Going into a dense patch of bush which might have been used as a poachers' hide, I nearly trod on a 6–7 foot mamba, one of the most dangerous of snakes. I recoiled in utter revulsion and fright, for I am not familiar with snakes, nor do I like them really, but on this occasion instead of rearing up, it slid away between David and me and quickly disappeared. After breakfast we set off in the Land Rover to go where we could get. Dense bush was a bit of a trial but we got through one mile of it on an elephant track with the askaris using their pangas and simis to cut the overhanging fronds of bush. Again two or three *dongas* to cross without much trouble, but we got to one deep one with dense surrounding bush and we had to turn up into the higher ground among a lot of granite boulders and outcrops of that rock. Everywhere had been burnt and it wasn't very pleasant in the fierce heat. Eventually we reached the top of a hill where the herbage had been too short to burn and we settled under a *Launia* tree for lunch. The country is practically empty of game – 3 oribi, 1 reedbuck, 4 topi, a few kongoni and a herd of zebra – all we saw all day. Three miles east of us was a granite kopje called Nesheshaw, 6,313 feet. David and I walked over to it with one askari. The walk was not really enjoyable because it was all through black newly-burnt ground, with dead trees still smouldering. When these are lying down and gradually smoulder away, they leave a white ash picture of their branched shape on the ground. There were springs, rather sluggish, here and there at the foot of the kopje, but before the era of such dreadful burning this must have been a well watered place. There were a few buffalo around in the wet wallows in dense patches of bush. We found portions of a giraffe's skeleton near the kopje, and as there were three shin bones, there must have been two giraffe, and you would not get two giraffe dying together unless they had been killed. A gang of poachers must have come on a herd and shot two with poisoned arrows. Back to camp then, as black and dusty as sweeps and with a beautiful thirst. Another ineffably glorious bath in the river and a relaxed dinner beautifully turned out by the good Kilonzi. This rapid and rocky section of the river attracts the common sandpiper, a bird so familiar to me in the Highlands, and so well loved. A pair of Egyptian geese near camp also. Elephants trumpeting in the night but we do not see them – extraordinarily shy. The rhino and her calf came to 15 paces from the tent last night. Lovely to lie and hear the running water.

9 November 1958 (Sunday)

The way David has pioneered into this lower Bologonja-Mara country or, as I suppose it should be called, the Ungruimi, is now so good that we got back to camp in three hours. No game until we get well into this side of the Sand River, and then precious little. The plains are parched and the wildebeeste have none of that prodigal sparkle of movement they had when the grass was green. Excitement back at camp, told to us in dramatic style by Frederick the cook and the corporal of askaris, was that a pack of wild dogs had rushed through camp this morning early and that a lion and lioness had come into camp at the same time. Dogs and lioness looked at each other hard but nothing happened. Why should the wild dogs dash across the plain towards camp and the lions come into camp? Animals do things we can't explain. The men have done a lot of work at the new airstrip. Some anthills to be planed off now. Busy writing the rest of the day.

10 November 1958

Lorry to Narok this morning to fetch petrol, so we got a mail away. Another brilliant morning: when out early I looked at the upper branches of a species of fig tree that had fallen while the tree was in full leaf. This happened on Saturday, so everything was very fresh. The delicate grey of the smooth bark was patterned with circles of lichen growth, making a wonderful mixture of greys and greens. These circles of lichen on acacia branches as well as on other trees must have had something to do with the development of a leopard's spots. When he is lying along an acacia branch with the sun dappling everything, he is almost invisible. Coming back to camp I saw three elephants on the other side of the river, perhaps 80 yards away, feeding on the *Panacium* and *Pennisetum* grass. Writing all day except for going out on the great lawn in the evening to see two pairs of lions. I imagine both lionesses were in season, and when they come to have their cubs they will doubtless join up and keep house together for the benefit of the cubs. Sid Downey opened my eyes to an interesting train of thought the other day. He said when numbers of lions got so low that two lionesses could not cub together they were in grave danger of extinction in that place. If the mother has to go off and hunt and leave the cubs alone they are prey to hyaenas, vultures and even jackals. Many carnivores have their young below ground, but when you get the size of a lion it is a different matter altogether making a hole in this tough soil. Sociality is evidently very important in the survival of lions.

11 November 1958

David and I set off for the country on top of the Isura Escarpment. It was not what I expected. Instead of being bare it was well clumped with bush, mostly on anthills, and small patches of thick woodland. The grassland in between did not look much overgrazed. Of course there was too much burning. After the flat and waterless top of this volcanic escarpment we came into an immense landscape of hills and valleys of parkland. Acacias were few and the trees and patches of woodland were nearly all broad-leaved and very green. The country up here was not as parched as down in the plains. The grass still carried a bit of greenness. The landscape was really glorious. There was one police post, Olgurion, and what appeared to be a disused mine in a hillside, but otherwise little sign of habitations. The Masai own some of this country and we saw one herd of cattle and a flock of sheep. Several miles away to the south-west we could make out through binoculars the round huts of the Watende people on a bare ridge. We saw some sign of elephant up here, and as there is nothing for them to disturb, possibly they are tending to come up the escarpment again. A few spots of rain this evening. Saw a dozen or so of lions near camp on the way home and took a few pictures. One young lion was up a tree but the light was not right for a satisfactory photograph. Took one all the same.

12 November 1958

Overcast this morning. David and I out to see if there were any lion kills about. We saw one male lion who had evidently been fighting. He looked all in, and had a scratch over the eye. Suppose he would be all right after a rest. A little farther on we saw an old bull buffalo who was crashing into things. He was completely blind. So David fetched his Rigby .416 and shot him through the head. The poor old buffalo went down and never moved, though we noticed his heart went on beating for some minutes. He had wounds underneath him which looked as if the hyaenas had been grabbing at him. Both eyes had gone and pus was coming from one socket. Ticks of bright black, green and yellow geometrical pattern were numerous in his axes, and the hair of his shoulders and forelegs carried a great many botfly eggs. His ears were almost gone and the frontal boss of his horns was wearing away and almost gone. At least he is saved having a lion grab him by the snout to suffocate him while another jumps on his back.

Just after dark Lynn Temple-Boreham arrived, bringing Gerry Swynnerton and Hugh Lamprey, the Game Warden of Tanganyika and his biologist. Also with them was George Rushby, almost the last of the old-time elephant hunters who crossed and re-crossed Africa in their quest for ivory. Eventually he changed and has actually served as Deputy Warden to Gerry on his leaves. He certainly has a lot of sense. Dinner was late and being later to bed than usual and having a mail to read, I was very late getting to sleep.

13 November 1958

The six of us, and askaris, away in two Land Rovers this morning for a long day of wandering about the plains. They were very dry indeed. Possibly 2,000–3,000 wildebeeste and but few zebras were towards the south side, verging on the hills, but small groups were to be found anywhere. We saw one bull herd of wildebeeste, perhaps 150, in which there were several bull calves of this year, which means that the mothers must have been weaning them and chucking them out of their own groups. The dry weather and shortage of grass must have cut down the milk. The rain came today and the air got pretty cold. We saw several lions including the two lionesses and their seven cubs which Mervyn and I saw on September 29, and they were not more than 400 yards from where we saw them that day. Heavy rain all evening but fine again and a starry night by 10 o'clock.

* * *

In Nairobi, in 1986, Hugh Lamprey described to me his meetings with Fraser Darling. At the time Hugh was a young biologist whose boss, Gerry Swynnerton, the Chief Game Warden of Tanzania (then Tanganyika), thought it wise to have Hugh discuss his work with Frank, the eminent ecologist, while he was within reach.

Hugh regarded this as a great privilege, and jumped at the opportunity. Before travelling to the Mara to have audience with Frank, therefore, he put together two year's work on seasonal variation in densities of animals. When they met, however, the event was an anticlimax. Frank seemed bewildered by the graphs and figures, and gave the briefest of responses. It took him about three minutes to say: 'Yes, I'm sure that's okay', and nothing more. Hugh then realised, what I also discovered in my relations with Frank, that he was not a numerate person. He disliked statistics, and found them a conundrum. Though his preoccupation with the 'broad ecological view' (on which he made his name) may have at times been used to cover for mathematical weakness, Frank's ecological discourse rang true for his students, without statistical back up. Lamprey, the young professional biologist with an appetite for research, was disappointed by Fraser Darling; Swynnerton, the senior game manager and naturalist, was very impressed.

On a more positive note, Frank was impressed with what Hugh Lamprey said, both then and when they met in 1961 in Ngorongoro, about the ecological separation of species of ungulates in the graze–browse spectrum of herbivores (pages 222 and 292).

14 November 1958

A longish day, crossing the Mara River below camp with two Land Rovers, through the Mara triangle and into the Lamai wedge. There was more game down there than in the triangle because the grass was green. We must have seen 3,000 wildebeeste and as many zebra, and four rhino. We showed Gerry Swynnerton the hides of the poaching gangs, the bush fences where the nooses are strung, and du Pre's lorry body in the tree. Gerry is also suspicious of Master du Pre. Gerry says the Lamai wedge was included in the new schedule of the Serengeti National Park but was cut out because the Watende objected, saying they had *shambas* down there, which is scarcely accurate. The damn fool District Commissioner also objected, saying it was his intended area for Watende expansion. Both the agricultural and veterinary johnnies have given written opinions that the area is unsuitable for cultivation or livestock. But the area was cut out of the proposed extension to the Park all the same. Gerry says that though we may fail to get it into the Park, there is very good reason to believe it may be made into a Game Reserve which would set it on a much firmer footing for protection and would scotch Master du Pre. It was just on dark when we got back, and Tuffy Marshall and young David Lovatt-Smith had arrived. Mervyn had arranged for them to have a week down here with David and me. Good to see Tuffy again after nearly two and a half years. He is much thinner, but just his own drawling humorous self. Being such a crowd of us, sundowner time dragged on in the interminable African tradition and it was 9 o'clock before we sat down to dinner.

15 November 1958

A long day up the plains and over the Bardamit and Aitong sections. There were more wildebeeste there than I had expected to see, for there seemed no grass except a fuzz of dried *Pennisetum*. A bunch of 350 were all bulls, with only two bull calves among them, and farther down there was still a preponderance of bulls but there were some cows and calves among them. Back by 4 o'clock and some good talk in the evening.

16 November 1958 (Sunday)

Gerry Swynnerton and George Rushby away today, heading for Bunagi via the Sand River and the Bologonja. We convoyed them as far as the Egilok stream. Gerry broke the main leaf of his trailer spring going up the plains, so we had to do some improvising to get him roadworthy for a long rough journey. Quite a few wildebeeste to the south of the plains, possibly 2,000; we came down the southern side via Roan Hill and into the unburnt stuff near the Mara. Then home by the route across the Talek we pioneered a week or two ago.

17 November 1958

All of us to the Lamai wedge, i.e. David, Tuffy, Lynn, young David and myself. Across the Mara River and down through the triangle. We came upon great herds of wildebeeste and zebra which seemed to be on the move because they were close-packed. Have mislaid my fine-point pen. There are many less *dongas* or *karongas* in the triangle and the Lamai wedge than on the other side of the river. The main obstacles are littered boulders which tend to cover the ridges of undulations. The boulders, of course, are basalt or lava. We soon came upon the extremely casual game fences in which nooses are set, finding a lot more than the last day we were here. We also found a new hide very near to the habitual tracks of du Pre's lorry. They had evidently pulled out with the meat yesterday, for the bones and skins of zebra were quite fresh. It is so upsetting to be just too late. We are wondering now whether the mine is a blind and that du Pre is really running a big *biltong* racket. The next thing we want to know is whether the wheels are going round at du Pre's mine. If not, on what would this Dutchman be wasting his time? At another place we followed Land Rover tracks up to some boulders where they stopped, and there we found a fresh carrying pole covered with congealed blood. Two men had obviously brought meat to be loaded in the car. There was also a handful of palm leaves which must have been used as a shoulder rest under the pole. The nearest palm was down by the river four miles away.

18 November 1958

A quiet day spent talking for the most part, and getting things arranged round camp. Coming home last night we saw two impala bucks fighting; one eventually caught the other on the skew and pushed him backwards several yards. The fight broke off and the loser had a broken front leg. He disappeared into gall acacia where we could not follow quickly enough to shoot him. By watching this encounter we missed Lynn and Tuffy in the other Land Rover, and as dusk was approaching we had to press on hard and took another way than they did, which involved us in some volcanic hills. It was dark when we found the opening in the forest that would take us to the river crossing. And when we reached it we found we were ahead of our friends. They came into camp at 8.45, having waited for us at the hole into the forest and having lit a fire there to guide us.

19 November 1958

Out in the morning outside the forest on our side of the river. It was obvious several hundreds of zebra had come in during the night, almost certainly by crossing the river. We searched for the game crossing and found it a few miles north of camp, north of a straggling marsh obviously much used for watering. The game crossing was at a point where there was no forest either side of the river. The animals will not face a narrow way to the river for fear of predators springing upon them.

20 November 1958

Out this morning heading for the tsetse-clearing scheme northwards beyond Aitong. Just outside the great lawn we saw two lionesses on a bull topi. So we stopped, pushed the lionesses off and took our commission as we are now short of meat. We got the back and fillet steaks and one hind leg. The askaris took liver and heart, some ribs and a foreleg. We have not really robbed the lionesses, who were already full, but the vultures which were perched on the acacias waiting for the lionesses to have done.

Slipped back to camp with the meat and then went on. Back 5 p.m. to a wonderful meal of tendered topi steaks, soup, fruit and cheese. Life is pretty good. Sixty elephants on the great lawn this evening, with giraffe, waterbuck, reedbuck and buffalo. David and I walked up the lawn and found there was a herd of buffalo cows and calves on a deep bed of *Pennisetum* grass on the other side of the river. The wind being right we sat on the bank of the river and let them graze towards us. Sometimes they watched us curiously but untroubled. Then the fireflies came out and gleamed to and fro across the dark shapes of the buffalo. The good moon shone and threw shadows. A gust of wind came at our backs and blew across the river and in a matter of two or three seconds it reached the buffalo, who then retired without panic into the forest. The elephants had also retired into the forest and were making some very loud roars and growls.

Yesterday afternoon we crossed the river and drove northwards outside the forest. There were depressions that side which seemed like oxbows of the river but were not deep enough. They carried water and were undoubtedly significant to the game. Near the crossing, in the forest, we saw for a moment or two, two giant forest hogs, my first glimpse of these animals. They are black and big as a donkey, and they inhabit dense broad-leaved forest of this kind.

21 November 1958

I stayed behind to do some writing, and the chaps went eastwards to the Jagartek stream to see if the zebra which came to our side of the river two days ago have now gone up there. They found the Sianna valleys green and full of wildebeeste and zebra, and saw two cheetah and a leopard.

While writing I watched some baboons on the other side of the river come to the bushes over which was growing the purple convolvulus. They were sitting in the bushes pulling the convolvulus from the branches, stripping the vines of their leaves by stripping them through their finger and thumb and then eating the leaves. A few days ago we saw a rather amusing incident with a male baboon. A lion was lying unconcernedly under a tree and the baboon was in the tree wishing to come down and get away. He was too frightened to do this: he would come down the thinnish trunk and jump to the ground and go a couple of steps, but then he dare go no further and leapt back up the tree in fright. Then he would try again, with the same result. He dare not make the dash of 30 yards to the next tree. His fore end was trying to show his courage and aggressive qualities but his hind end was suffering nervous diarrhoea. So many of us are really in this dilemma. Our arrival let him away from his potential but not active danger.

22 November 1958

David, Tuffy, David Lovatt-Smith and I away to the Lamai wedge today, to make as deep a penetration into that country as we could. We were well down into Tanganyika before we saw much game and then it was very wild. We saw a string of Watende women carrying loads of firewood near the foot of the escarpment: they took to their

'A lion was lying unconcernedly under a tree . . . ,' Mara River, 21 November 1958.

heels when they saw us, dropping the loads of firewood. This day we followed several shallow *dongas* and found so many poaching hides and bush fences that we lost count. Killing game is just an industry in the Lamai. I am to see the Governor of Tanganyika at Dar-es-Salaam on 3 December and wonder how far my tale will influence him. We saw a lot of lions today, a pride of five hanging around du Pre's lorry body in the acacia. Presumably they get some feeding there when du Pre is hunting. The country is so dry just now that the game killing is off for a few days. It is obvious that these men work intensively when the game is moving to and fro. We do not think there was anybody in the area today. We struck the few remaining roan antelope, eighteen in the Mara Triangle but too near the border, and five well down in Tanganyika. The country down there, undulating, bush and grass, and the shapely volcanic hills, should be a paradise for game, and its beauty almost hurts when one is aware of what is going on. These charnel houses of the poachers' hides make one loth to go into a patch of bush over a *donga*.

23 November 1958 (Sunday)

A quiet day writing at my report, after Tuffy and young David left at 9.30 this morning. David and I went into the forest down the river on our side in the late afternoon. The river meanders tremendously and gives a good block of forest. It gets very thick, with bush near the river, and one travels bent double along buffalo tracks. Old bull buffaloes are obviously using the banks for grazing here and there, and this bush for cover. Sometimes we would hear a snort and a rush in the dense bush and would wonder if the creature was coming our way. Back in the forest proper David almost trod on a 3-foot mamba. We did not kill it but watched it glide away, such a lovely almost translucent green, and greeny-yellow below. Such a pity they are so deadly.

I had been some way into the same patch of forest by myself in the morning, but along a dry shallow oxbow which led into the forest almost like a ride. The great trees, and the masses of the fig trees in particular, made an almost romantic landscape in the style of Fragonard. Last night, as we came to the hole in the forest leading to the river crossing, we saw a lion with a newly-killed zebra, and only twenty yards away was a lioness with another newly-killed zebra. Rather unusual two together like that. During the morning we had come upon another young buffalo bull dead from rinderpest. We found our way to him because of the wheeling vultures.

24 November 1958

A run north along the river and then into the plains where there is a good deal of young acacia. Some hundreds of zebra, wildebeeste and topi on the river flats and near the open marsh, which saves the game having to go through bush to the river. There were also a few hundreds in the young acacia area, but no farther up in the plains which are now so dry. Saw a pride of eleven lions, but they had no kill around from which we would be glad to take a commission. Food is now getting a bit short and for the last two or three days we have cut down to two meals a day. This will do us more good than harm for the next few days. Got on with some of my report.

25 November 1958

Lorry to Narok today to fetch petrol and food. No mail. David and I went off for most of the day exploring the river flats. We found several lagoons or oxbows of varying depths. Some had surfaces clear of plants, others were at the lily-pool stage, and some, being almost free of water, grew an abundance of sedge or scirpus growth. The lily-pool stage was richest in birds – jacanas, gallinules, moorhens, black ibis, egrets and Egyptian geese. Green sandpipers and greenshanks were at the muddy ends of these lily lagoons. Two lions and a lioness on the great lawn as we came in. One of the lions was newly scarred from fighting. I think the males must keep their numbers in check somewhat from fighting. Wrote a bit of the report in the evening.

26 November 1958

Writing in the morning. Lovely weather. Just now it tends to cloud up in the evening. Out for a run on the lower plains in the evening. Came upon the pride of fifteen lions where we once met them before. We went up to 10 yards and still no sign of fear or moving. Judging from Temple-Boreham's experiences with Sally, her pride occupies a territory of about 10 square miles, but that happens to be one through which game is constantly passing, so in a poorer area the territory of a pride might be twice that size. Looked at a bunch of vultures in which there were three species, including the griffon. Yesterday we stopped by a flock of vultures demolishing a young wildebeeste

Elephant and waterbuck share a water hole in the Tsavo (West) National Park.
(Photo: J. M. Boyd)

Masai Cattle watering at Amboseli on 25 August 1956 – the tribesmen make mud troughs at the edge of the swamp (see p. 165).

Kilimanjaro seen above the acacia woodland and eroded flats with giraffe at Amboseli, seen by Fraser Darling from his verandah on 24 August 1956 (see p. 163). (Photo: J. M. Boyd)

The Mara River in November 1958 showing the high riverine forest and the heavy siltation of the swollen stream (see p. 199).

Ngorongoro Crater in Tanzania from its southern rim showing the Lerai Forest, the crater lake and the northern escarpment 15 km away. Visited by Fraser Darling on 21 January 1961 (see p. 220). (Photo: J. M. Boyd)

On 29 September 1958 – 'seven cubs . . . a lovely sight . . . golden chestnut'. These (two litters) are similar, near Oljoro Loroman (see p. 175). (Photo: J. M. Boyd)

On 13 October 1958 on the Egilok Plains 'we saw a cheetah chasing a flock of Tommies, but as it saw us it desisted.' This cheetah in the same area has killed a Tommy fawn (see p. 184). (Photo: J. M. Boyd)

Wildebeeste and zebra on the Egilok Plains looking to the Sianna Hills in September, with patches of bush surviving in the middle distance (see p. 173). (Photo: J. M. Boyd)

Counting game at sunset on the Egilok Plains as Fraser Darling did with Temple-Boreham and Sheldrick on 14 October 1958 (see p. 184). (Photo: J. M. Boyd)

The water hyacinth in Lake Aliab near Yirol, Bahr-el-Ghazal, on 22 February 1961 (see p. 265).

The market at Nyala in Dar-Fur on 5 March 1961 is immense and is 'a haze of indigo' (see p. 279).

A street in Zalingei on 6 March 1961 – '. . . this lovely place is just a huge grove of *Acacia albida*, thousands of them' (see p. 281).

Terraced landscapes on Jebel Mara on 8 March 1961 – 'Presumably these terraces were made in some pluvial period more than 2,000 years ago' (see p. 282).

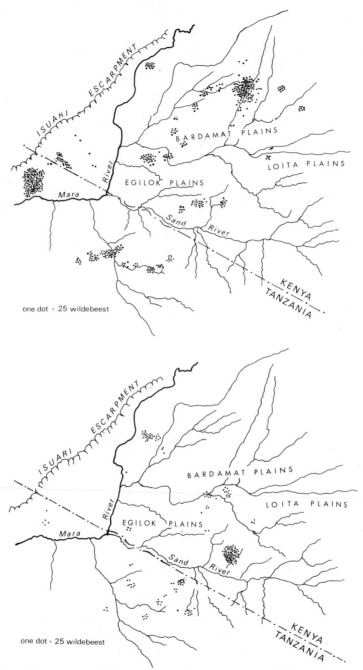

Map 7a (top): Distribution of wildebeest over the Mara Plains, during the first aerial census on 20–21 October 1958.

Map 7b (bottom): Distribution of wildebeest over the Mara Plains, Kenya, during the second aerial census on 27 November 1958.
(Both maps from Fraser Darling, 'Ecological Reconnaissance of the Mara Plains in Kenya Colony,' *Wildlife Monographs* No. 6, August 1960.)

fairly newly dead. One vulture had its head into the neck a long way, so we were able to hold it and feel the neck writhing about. How strong it was! When we let the creature go it continued its exploration for a while and then came out, as it were, for air. It was surprised and puzzled, and doubtless thought David's restraining hands had been other vultures pressing in. Once the vulture was out we were interested to find how much of the neck meat had been picked off and guzzled without the skin being disturbed.

27 November 1958

After expecting 'Punch' Bearcroft and his Kenya Police airplane at 7 a.m., he eventually turned up at 10.45 a.m. and having left his sandwich breakfast in his Land Rover at Nyeri, he was hungry. David and he went off for the game count at 11.45 o'clock. A good deal of mail came, including a lot of annoying requests from UK needing urgent replies which they won't get.

David and Punch away again in the afternoon, and I for a walk by myself which I much wanted and enjoyed. This may be my last day here by the Mara River and in the forest. The weather was overcast and very still in the later afternoon. I looked at the silent river, at the grace of the Phoenix palms and the strength of the great fig trees. The cinnamon bee-eaters perched on twigs beside me, so brilliant; reedbuck and waterbuck bothered very little about me, and I saw no other game. Only the baboons sheered off and shouted. The poor things are too clever and crafty, to the point of seeing trouble where there isn't any. If animals could have duodenal ulcers, baboons would. After all, they are susceptible to all the infectious diseases of man and they make so much fear and trouble for themselves.

28 November 1958

My last morning in the Mara. Punch Bearcroft flew away by 8.30 o'clock and we left at 9.30. Before then I had a last walk into the forest over the great lawn. In this time I lost my much loved Conway Stewart fountain pen of large proportions. And a hornet hovered over my head and stung me on top of the left ear. Yet how lovely were these last few minutes.

We came out by way of the Lemek Valley, through the great eroded plains north of Esuvatai, to Narok, where I realized I had lost my pen. Khushi Mahomed at the dukar welcomed us with a flourish: we had come to pay our bill, but before that an iced orange drink and biscuits were put before us. Then on, after saying goodbye to that cold fish of a District Commissioner, Galton-Fenzi, towards the Mau Forest. We climbed and climbed over a roughish road through a most depressing country which should have been inspiring – vast burnt-out areas of the Mau Forest belonging to the Masai. The trunks of the dead junipers – *Juniperus procera* – locally called cedars, stuck into the sky black and maimed, like a forest of telegraph poles, like some hurtsome painting of Nash or Klee that one used to see in the bad days of the 1930s. This continued far up towards the summit of the Mau, 10,002 feet, but eventually we got clear of the fire's worst devastation. Where the forest was suffering there was that dense high herbaceous growth of compositae which I saw, did not understand, but which disturbed me in the Mbizi Forest in Southern Tanganyika last year. This dense herbaceous growth in a forest is all wrong. As soon as we took the left-hand road over the Mau summit with its fine high junipers we got into a country of wide grass and sedge moors like Galloway, pastured by Masai cattle but infested with Kikuyu squatters. The Kikuyu are forest destroyers and wasters. But again, the Administration does little about keeping these forest rats within bounds. The Forest Department, which has little use for the juniper except as a capital resource to be used up, allows the trees to be cut and replaced by quick-growing cypresses, and then lets Kikuyu in as squatters to cultivate the newly-bared forest soil between the planted rows of cypresses. The country down as far as Elburgia was lousy with Kikuyu and cypresses. But I am getting on too fast. As soon as we began to turn downhill on the Mau we passed through miles and miles of most glorious parkland, stretches of good grass among stands of juniper and olive. Undergrowth in the forest was insignificant except for regeneration of the great trees. The vast undulating sweeps of grassland and forest made one imagine a world that has gone, of rolling landscapes and ideal forests of Fragonard, of half a century ago when all this country was empty, and one could have ridden on horseback through it all instead of in a Land Rover. Had I lived fifty years before I did and come here, I would have taken up a great area for myself had I come here. There is water, grass, forest, and at that time there must have been game as well. Now eland and zebra have gone, and the most one might see is a bushbuck. One we saw was quite black, as the forest type is.

Galton-Fenzi had asked us to give a message to a man called Moen, a Norwegian who was managing a sawmill on the Mau for a Masai Co-operative. We stopped to set up camp for the night in the forest-parkland, then went down into the forest to find Moen. His house was new, made of juniper, and set against the forest looking forth over the parkland. He was a tall good-looking Norski who came to Kenya in 1940. Took us inside to drink tea and we met his wife and one little girl. House inside panelled in juniper and *Podocarpus*. Quite lovely and smelling so sweet. I would love a room panelled with juniper. It is the wood used so often for pencil slats. Was glad to hear Moen say they loved living where they did. Back up to camp four miles away, to a great juniper fire; a cold wind blowing all the time.

29 November 1958

A cold night, during which I got up three times to make up the fire: it seemed warmer hopping about outside than it was in bed.

Down through the planted forests to Elburgia and through much more forest and farming country to Molo, which is in the White Highlands. We climbed from Molo nine miles to Noel Simon's farm at 9,000 feet. The country up there is almost exactly like the Berkshire Downs. Was much struck with the delightful home Noel and his wife Bets have built from scratch in twelve years. Good *Cupressus macrocarpa* hedges round the garden close-clipped like yews, a fountain and lily pool, flowers in profusion and lovely roses: lawns thick and dense. Fine sunny weather. Noel and Bets have retained some area of broad-leaved forest which are a great help to the garden in providing background masses. The house is Tudorish, coming out that way largely because they used so much home-felled juniper and olive, and home-baked bricks ... [the concluding page of the Kenya Journal is missing].

*   *   *

Although the Governor, Sir Evelyn Baring, was in favour of Fraser Darling's reconnaissance, the Government were not, and the seven-month delay in the arrival of his report in Nairobi did damage to the cause of Masai wildlife Reserves. There was no point in pressing the National Park principle into Masailand, since it would be politically unacceptable to both Masai and Government. It was important, therefore, that Frank's report did not unduly criticise the current system of landuse. To push too hard for measures requiring significant withdrawal of the freedom of the tribesmen to graze and water their livestock, and burn their grasslands in season as they wished, would be highly unpopular. The ecological facts, as Frank saw them, bespoke catastrophe if some action was not taken quickly; for example, to save the riverine woodland and 'lion' bush, or reduce sheet erosion around manyattas (villages), and watering-places. Frank had been fearless in his utterance of this truth similarly expressed in Scotland, Alaska, and Northern Rhodesia; was this a time to be less candid? He did not think so.

It was of the greatest importance in the promotion of the Masai Reserves that Frank's report be non-political. Its perceived role was to provide the ecological rationale for the political stratagem. In the event and to Sir Evelyn Baring's great disappointment, the report failed to separate the ecology from the politics. It was not accepted by him on political grounds. Mervyn Cowie tried in vain to persuade Frank to amend it, and has written in a letter to me: 'this valuable survey of the Mara was therefore wasted and in no way could it serve to strengthen the standing of the trustees.' – meaning also that it failed to serve the cause of wildlife conservation in the Mara.

In September 1958, when he went to the Mara briefed by Cowie and Simon, Frank could not accept the postulate of an African Wildlife Park in tribal management, in which the revenues would go to the local (tribal) authority instead of central government. His perception of the 'National' Park transcended anything that might be thought up on a 'regional' or 'district' basis in Africa. He saw the financial, technical, scientific, and cultural aspects of such Parks set at the national and international scale. Three months later at the end of the Mara reconnaissance, however, he had been 'half won over'. In May 1960, after his report, he conceded that the tribally-managed Park was the right way ahead, which he deemed 'prophetic of a future elsewhere in Africa'.

Nonetheless, there were lingering doubts about the changes on the lifestyle of the Masai which had troublesome portents. In August 1960 he wrote:

... The Masai as a tribe are outstanding for the fact that up till now they have lived alongside game, tolerating it and hunting only occasional lions ... The present attitude of the Masai towards game is not so tolerant, because their herds are increasing as a result of prophylactic needling which veterinary science and a benign government have made available. There is now definite competition, not so much for food as for water. The Masai still do not kill game, but insofar as there is competition there is no attrition, and game animals are on the losing side ... This is not the point at which to argue the questionable wisdom of a governmental policy that brings about an increase in the cattle population (with its inevitable problems of soil erosion and habitat deterioration) among a people not yet developed to the state of parting with many of their cattle: merely is the statement made that, with a shrewd insight that there is every likelihood of not losing a large proportion of their herds to disease, and that the tsetse fly is no longer the barrier to large virgin areas that it was, the Masai are developing an attitude to game much less tolerant than it was even a few years ago.

In 1961, the Egilok, the deep African wilderness lying to the west of the Great Rift Valley, which had attracted big game hunters for over half a century, became the Masai Mara Game Reserve, 1,120 sq. km. in all. 320 sq. km. south of the Talek River is managed on the lines of a game sanctuary and viewing area, without Masai settlement and domesticated livestock. The remaining 800 sq. km. is settled and grazed by the Masai, but no shooting is permitted. Today, the Masai Mara Reserve and the Amboseli National Park are outliving their founders. The visits which I have made in the 1980s to Kenya indicate a widespread change in the distribution of wildlife throughout the country as a whole. In the face of a rapidly increasing human population, wildlife is disappearing from the developing countryside, and becoming concentrated within the National Parks and Reserves like those in Masailand. The 'human population' factor was not one which dominated Fraser Darling's thinking in the late 1950s. He was preoccupied with the effects of grazing, burning, and poaching within the wildlife habitat. These pressures still exist, but they are now compounded by increased land hunger in the country as a whole, and the rapid growth of the tourist industry. In the Mara and Amboseli, the 'compression' of the game, and pressure of visitors pose great problems. Fraser Darling would be impressed that the Masai County Councils have managed their Reserves as well as they have done over forty years, but he would have some strong advice to give in the management of tourists and in curbing the activities of drivers and rangers who break the law for unscrupulous clients.

CHAPTER NINE

Ngorongoro Crater and the Serengeti Plains

January 1961

T HE EAST AFRICAN JOURNALS conclude with a brief visit of eight days
to the Crater Highlands and the plains of northern Tanzania. It was now
four years since Averil's death, and three since the expedition in the Mara. A
few months before had seen the publication of *Wildlife in an African Territory*,
and 'An Ecological Reconnaissance in the Mara Plains of Kenya' in *Wildlife
Monographs*. Fraser Darling had become an authority in Africa as great as he
had been in times previous in the United Kingdom and the United States.

The achievements in wildlife conservation in East Africa with the establishment of the great Game Parks and Reserves in Uganda, Kenya, and Tanzania,
were bringing world renown to these countries which were fast approaching
independence. The scale and splendour of the wildlife scene epitomised by the
annual migration of the plains game of the Serengeti had gripped world
interest, and had been moved centre stage by Grzimek's best-seller *Serengeti
Shall Not Die* (1960). The lives of those concerned with wildlife in Africa were
deeply touched by the conservation message borne upon a wave of exciting
discoveries, spectacular films, and thrilling books. Louis and Mary Leakey's
discoveries at Olduvai Gorge, which brought man closer to his origins, were
particularly poignant. There was the feeling that, cradled within the grandest
panoply of life on earth, there were, awaiting discovery, the secret beginnings
of man himself.

Frank's life had stabilised at home. On 2nd July 1960 he married Christina
Brochie who, until then, was house mother to the three children of his second
marriage – Richard, James, and Francesca. She was now in charge of the
household in the fine Georgian house at Shefford Woodlands in Berkshire
which, through Averil's trustees, now belonged to the children and was
managed by Frank and Christina. His links with Edinburgh University and the
Nature Conservancy were, for the time being, at an end, and he had been
appointed a Vice-President and Director of Research of the Conservation
Foundation in New York. He was therefore able to pursue his career with an

unfettered composure the like of which he had seldom enjoyed in his life. This happy state is reflected in the sure-footedness of his Journals in Tanzania and Sudan, when he again seemed to be far more the master of his own feelings than in earlier travels in Africa.

In 1959, financed by the Ford Foundation, Frank made a flying tour of fourteen African countries to arrange a scientific, economic and administrative conference as part of government policy. The conference, at Arusha in September 1961, was jointly funded by the Co-operative and Technical Commission for Africa and the International Union for the Conservation of Nature and Natural Resources, and was attended by leading figures from the new African states, scientists and administrators from Africa, Europe, and America, and representatives of UNESCO, FAO, and ECA. The conference is remembered as a milestone in conservation in Africa marked by *The Arusha Manifesto* — a liberal statement on conservation as it would apply in Tanganyika signed by Julius Nyerere, then Prime Minister and later first President of Tanzania. FAO formally announced its support of wildlife conservation in Africa. Frank was in his element.

The push was on in earnest to save Africa's wildlife. The scope for research and management was enormous, and was matched by a brilliant intellectual focus on primary needs of wildlife conservation. Opinions varied about what should have priority. Among a few others, Lee and Martha Talbot had already led the way, as a man and wife team, in their classic study of the wildebeest in western Masailand. Funds were being sought for imaginative research on key species, not only for their own sake, but as the pathway to deeper understanding of the ecosystems of which they were a dominant part. Support was obtained for a succession of research scholars from Europe, America, and Africa working later in specially formed institutes like the Nuffield Unit of Tropical Animal Ecology (NUTAE) in Uganda, the Serengeti Research Institute in Tanzania, and the Tsavo Research Project in Kenya. The needs for training led to the formation of the East African Wildlife Management College in Tanzania. A great deal of this broadly based effort throughout the 1960s stemmed from the Arusha conference, for the success of which Frank took much credit.

20 January 1961

Reached Nairobi at 11.30 a.m. and was met by Phil and Alma Glover and Ian Grimwood. Phil and I immediately got things straight: he has given up the Ngorongoro Conservation Unit entirely. He had found himself in an impossible position and backed down. Alma said she was glad I had written firmly in November because it had given Phil the shock treatment he needed to prevent his dissipating his energies on too many things. I am to lunch with Russell on Saturday next week. Whisked off by Ian to his house for lunch: an amazing place nine miles out of Nairobi near the Ngong Hills, built some years ago for a film company, just like a castle, courtyard and all. There is also a dam 100 yards away which attracts a lot of waterfowl. Ian keeps a couple of good ponies and is very happy to have cleared out of the suburban Government house that was provided for the Game Warden in Kenya.

Lee and Mardy Talbot came to lunch, so we got down to talking shop almost at once. It seems that the Mara wildebeest are decreasing alarmingly, a rinderpest-like disease attacking the eight-month old calves and killing about 40 per cent of them. He has also found that the occupational risks of maternity in wildebeest are considerable, defending their young and so on, whereas the male wildebeest seems to go on and on into old age.

21 January 1961

Away early this morning to fly down to the Ngorongoro Crater with Phil Glover and Noel Simon. We flew over a parched Masailand, for the short rains failed altogether. Then into the forested highlands over Embugai Crater. The lake held thousands of flamingoes along its edge. Then round the Crater highlands to see cattle grazing at the forest edge and new Masai bomas up there too. There was even some cultivation and Waarusha huts newly built. And very soon we found fires going in the high forests. The undertaking of the Masai when the Serengeti Park was bisected at 1 July 1959 and the great crater went to the Masai was that no cattle would be allowed in the highlands, there were to be no habitations up there and there was to be no fire in the Conservation Area. These undertakings are being disregarded and the relatively junior officers of the Conservation Authority are being given no administrative support or authority. They are extremely frustrated young men. The letter of the Nihill Report of 1957 is being fulfilled but not the intention. The Crater highlands are degrading rapidly. Lastly, at least 34 rhinoceros have been speared and killed by the Masai since 1 July 1959. The whole idea of the Conservation Area is being slow-timed, yet the Colonial Development and Welfare Fund gave £182,000 of British taxpayers' money and the Nuffield Foundation gave £20,000. It is being frittered away. These chaps asked me point blank whether I was going to be one more visitor who would tut-tut in indignation and go away and do nothing. But I anticipate. We flew low into the crater but saw no Land Rover awaiting us on the dried mud of the lake floor. The pilot felt he could not trust it without confirmation so we climbed out of the crater again, flipped over the edge and over Tony Mence's house. He is Game Warden in this part of Tanganyika now. I had a week with him and Gerry Swynnerton and Desmond Vesey-Fitzgerald in the Rukwa Valley in 1957. He waved to us and the pilot then took us to Oldeani, an airstrip 20–30 miles away. There is nothing else at Oldeani. We got out and sat under an old acacia tree for a session of talk – Noel, Phil and myself. We had been there, ignoring the future, for less than an hour when the airplane came back to tell us that as he flew over the crater he had seen the Land Rover and had landed on the hard lake floor to speak to Tony Mence and Hugh Lamprey. Hugh is biologist in the Tanganyika Game Department and I have known him for some years. We last met in the Mara, when he and Gerry Swynnerton came down for three days to talk. So we took off again in the Piper Tri-pacer and landed in the crater a quarter of an hour later. First things first – some tea and food, made ready at the edge of the acacia (fever tree) grove of Lerai. The chaps talked and I got some indication of the depth of their feelings. They are being let down, yet will be held responsible for the failure of the Conservation Authority (which has no authority) to which the rest of the world is looking with trust. I have been hot under the collar for over a year about this log-rolling policy, but am always asked, particularly by Barton Worthington, to hold my horses. Barton's latest excuse is that we should say nothing till after the forthcoming CCTA/IUCN Conference at Arusha in September. Huxley met these chaps and was upset by what he saw, but was then muzzled by Barton Worthington when he got back to England. I myself feel so cross that I decided then and there to keep quiet no longer. By nightfall, when I had been 75 miles around the crater and counted six fires in the highlands, and heard that no administrative action had been taken on any of the infringements of the Conservation Area agreement, I had decided to write an absolutely dead-pan plain account of facts as a letter to *The Times* of London – as a private individual of course and from my home address.

The condition of the crater is excessively dry, but that does not matter really. It will come back after the rains. I think the resilience and stock-carrying capacity of the crater floor to be very great indeed. Its present state should not frighten one into thinking the failure of the Conservation Authority has anything to do with it. One's eyes should be raised to the hills and be truly shocked by what can be seen there. We saw many wildebeest, zebra and Thompson's gazelles today. Near the swamp where the soil is still damp I saw the tremendous feeding power of this crater. The wildebeest have begun to calve, so most of the Masai cattle are out of the crater itself because of the malignant catarrhal affection cattle can get from the wildebeest afterbirth. But there was one large herd of cattle at the watering place where the spring bubbles up clear out of the gravel. The Masai herders regarded us impassively but not incuriously.

We climbed out of the crater just before dark, and Noel, Phil and I set off for Lake Manyara Hotel for the night – 40 miles southward. Tiring journey in the dark and we were almost too tired to eat at half past nine when we got there.

22 January 1961

Away early this morning to Tony Mence's house, from where Tony, Hugh Lamprey, John Newbould (a young Oxford botanist), Phil, Noel and I set out in John Newbould's new Land Rover to cross the Serengeti. I wanted to reach the Moru kopjes which the Masai evacuated and which are now in the Serengeti Park. I wanted to see how the grazing was looking, after the cattle being out of it for over a year. The journey was a long one over dried up plains absolutely devoid of game. The shifting sand dune gradually crossing the plains is a sign for the future for those who care to read. We lunched among the few fever trees near the dune, after having called at the Leakeys' diggings in Olduvai Gorge. Mrs Leakey was on the job, tireless in an arid and uncomfortable situation. We saw where the jaw was found, and some new sites which have yielded some highly important material this season, secret as yet because of National Geographical support. The wealth of bones in the gorge is staggering. Fine tooth-comb is the right expression for the Leakey's sifting of the material. They have found a lot of shrew bones, the jaws so delicate yet perfectly preserved. As Hugh and Tony and I looked over these tiny relics of a microfauna we thought of Gerry Swynnerton who had given so much attention to the existing shrews and mice in East Africa. He would have been thrilled with this stuff. On again over empty desiccated plain towards the Moru kopjes: then we started seeing groups of Thomson's gazelles with a few Grant's among them. Soon we were meeting large flocks, then a herd of 150 topi and by the time we reached the kopjes there were the wildebeest and zebra in long stringing herds. The kopjes are of granite with parklike expanses of grassland between them. There is also good watering. My opinion is that even after so short an exclusion of cattle there has been appreciable come back in the grazing. Phil Glover agreed with me. Of course, granite base holds its soil and moisture better than the volcanic country we had come over. The kopjes are a lovely bit of country in which I take a small personal interest because in 1956–57 it was I who asked specially that if the crater was to be excised from the Park, then the Moru kopjes should be cleared absolutely of cattle and sheep. The idea was taken up and incorporated in the agreement. The Masai made one grazing raid into the kopjes a year ago, but John Owen is a dynamo and he had them out. Forgot to mention I had an interview with John Owen the first night in Nairobi. He is going flat out to get what he can into National Park and he is ready to take money from anywhere to set up a scientific station for the Serengeti. The German Government has already given a considerable quantity of equipment and this has been done through Bernard Grzymek. Uganda has now its National Parks Scientific Field Station with the link with Cambridge. Kenya has nothing and when I saw Mervyn later in the week he said how much he wanted scientific help.

The Moru kopjes, Serengeti National Park, 'are a lovely bit of country in which I take a small personal interest.' Visited on 27 January 1961 and shown here recovered from Masai pastoralism in 1965. (Photo: J.M. Boyd)

It had now got too late for all of us to get back to the crater and three of us to Manyara that night. So Phil, Noel and I stayed at Seronera and had dinner with Gordon Harvey, whom I had not seen since the Luangwa Conference. The following morning we went to one of the quartzite hills where all 'lion-bush' vegetation has long since gone and only an *Acacia* savannah shrub remains. We tried to persuade Gordon to run a fire-break right round the hill to keep the fire out as an experimental hill. Then back to the house to sit down and compose the letter to *The Times*. The others read through and approved so far. John Newbould came back with our gear from Lake Manyara and June Wright appeared with her airplane to pick up Phil, Noel and me. What a fine lass she is, small and slight, very modest and reckoned one of the best bush pilots in East Africa. Devoted to airplanes and mere men don't get a chance. Had a very bumpy ride back to Nairobi in the heat of the day, and believe it or not, I never felt the slightest bit sick in one and a half hours. Perhaps I was too ashamed to look a fool sitting beside June Wright. Spent the whole afternoon talking with Noel Simon. He has now got the Mau Forest Reserve gazetted. Wonderful – 105,000 acres of high forest containing a lot of elephants and bongo. Tells me there is heavy mortality among bongo this year, which the Wandorobo (a tribe of good naturalists) attribute to poisoning from a plant which flowers once in seven years and is then poisonous. Noel has gathered the plant and had it analysed and it does contain a poison, though the bongo eat it heavily in the other years. I immediately saw the possibility of this phenomenon being a population control factor on the bongo, which has no other enemy except leopard on younger age groups. Noel is also making headway with the idea of getting the Masai to conserve game throughout their country and to appoint a European warden. Ted Crosskill, Minister for Forests, Game and Tourism, has taken up the idea also and is rather proud of it. He has even gone down to Tanganyika with a view of getting the Masai there to join in. Had dinner with Ted Crosskill that night at Ian Grimwood's and he enlarged on the scheme, but he will probably be out of office in six or eight weeks when this damn-fool election has gone through. The primary election has come out strong for stability, but the secondary may undo all the good. Should think this election system was dreamt up by a leftish don at Oxford after a goodish dinner at high table. Nairobi is as gay and unconcerned as ever, but underneath all that the Kenya European is getting ready to look after his own. It is almost certain some Africans will make a mistake, when they find that their ticket for a European's car or a European's wife is not really valid and that this is not what *uhuru* means. Then there is going to be a sharp response and people here will not expect the British Government to do much for *them*. Exports from Kenya will be heavier than ever this year but the money will not come back into Kenya, as it is being salted away. The banks are calling in overdrafts, though not mortgages as yet. They are probably realistic enough to know they wouldn't get them. The British Government has ruined Kenya's economy and should make reparation. I said this to H.E. The Governor, Sir Patrick Renison, later in the week, and he said he had just come back with a fairly generous gesture from London.

23 January 1961

A round of interviews among Government people – Gerry Swynnerton's brother, who has handled the Swynnerton Plan for redistribution and consolidation of holdings within the Kikuyu Reserve, and who is one of the steadiest and most farsighted people in British Africa. Also the Acting Director, Veterinary Department, and Leslie Brown (a good ornithologist) who is Director of Agriculture. Saw Webster again, who is permanent secretary of Crosskill's ministry. There is no doubt whatsoever that the game situation in Kenya has taken a considerable turn for the better administratively. Ian Grimwood has changed the Game Department picture entirely, the notion of game utilization has taken hold, and the tourist value has also got home. The Masai situation has also changed and they have agreed to 200 square miles of the Mara being an absolute sanctuary and free from any grazing of their cattle and sheep. A further 500 square miles are to be barred to any shooting, but sheep grazing will continue. The Game Department, by co-operation of Lee Talbot, are arranging for a cropping of adult bull wildebeest on the Loita Plains. I hear rumours that this wild man, Royal Little, has dreamt up some notion of leasing the Mara as a Reserve. It is sincerely to be hoped that someone will stop him before he starts talking to the Masai chief and elders, or he will spoil a favourable situation which has taken a long time to contrive. The very word 'lease' is anathema to the Masai. Little is at present out of sight and ken in Tanganyika, and the general feeling is of relief that he is out of the way, but concern as to what gaffes he may be making. His waterhole complex is very trying because there are not many places where these things can be made without detriment to the future of the grazing.

24 January 1961

Forgot to say on the first day that when the plane stopped at Entebbe, I had previously arranged to meet William Logan, Director of Forestry, Uganda, who put up to us a study of the elephant. I sent it back to him to re-cast because he was mixing up research and economics to such an extent as to make nonsense as a study. He now showed me his new effort, which is much better and could well be the means of getting done the large three-year ecological study of the elephant which is so much to be desired. I think we might well ask Rockefeller Foundation to be interested in this. Uganda would provide house, all transport, and African personnel: the help needed from United States would be the salary and passage of a senior scientific worker for three years. I would like to see Buechner doing this job, and when I mentioned his name to Logan (more or less to get his immediate reaction) he was enthusiastic. I also note that the retiring Minister for Agriculture, Killick, who had been rather shirty to Buechner, has now cleared Buechner's application for leave to do the kob study which has been put up to National Science Foundation. The elephant job is more important.

Have also talked to many people about immobilization of animals by shooting a syringeful of some drug into them. I am more than ever convinced that experimentation should be in the hands of qualified people only. There are too many variables to call the technique anything else but extremely chancy. Many animals are being killed. Yet it is one of those new toys that any pseudo-scientific pip-squeak thinks he can handle. Each species almost needs its own drug, the weight of the animal must be accurately judged (animals in the Ngorongoro Crater are 15–20 per cent. heavier than elsewhere), the syringe does queer things in flight, and the animal itself reacts differently at different stages of the grazing season. Furthermore, rhinoceros seem to be unduly susceptible to sepsis. Catching rhinos by roping means the death of 3 in 10 or 12 if the catchers are experts. Immobilization cannot equal this figure yet.

27 January 1961

Off early by Land Rover to Narok, the administrative boma for a large part of Masailand. Phil Glover and I were to see the District Commissioner, Denton. Once we got into the Rift Valley, the drying up of the country looks pretty bad, and the last fifty miles to Narok worse still. Yesterday I had seen Phil Glover's beautifully prepared composite air photographic map of the Mau massif from the 1948 aerial survey. The fire tongues were so obvious, reaching far up narrow valleys and getting into the interior of the Mau. Now he has got the 1960 aerial photographs and is in process of making a new composite, which will be very revealing. We found Denton a good young man, perhaps a trifle pompous and a little bit patronizing towards research and all that. He was apt to think Phil's warnings about the Mau too alarmist

and he asked me about the urgency. I gave the Mau Forest ten years if no preventive
and rehabilitative measures were taken and that it was urgent to exercise discipline
right now. I told him the price paid for political appeasement in Africa was from the
capital of the land. We pushed on to Uasi Nyero to see Temple-Boreham but he was
down in the Mara. Came back and had tea in Naivasha where there were some oil
paintings round the walls for sale. A woman called Vera somebody or other can
certainly use a palette knife and I would gladly have bought one or two of her
landscapes of Masailand at 300 shillings, the price asked. But another young gentleman, whose landscapes were utterly wooden thought himself worth 500 shillings.
Back to Nairobi and gave dinner to the Talbots at the New Stanley. On leaving Africa
in May they are going back to the United States via Asia to do a second look-see for
Hal Coolidge. They are going on totally inadequate funds and are bothered. They also
showed me a letter from Boyle of the Fauna Society more or less telling them to get
the report of their Asian trip to him within six weeks of their return. So I wrote to
Boyle, telling him if he was going to call the tune it would be a good thing to pay
the piper. In other words, the Fauna Society should cough up £600 to enable Lee and
Mardy to do the trip properly.

28 January 1961

A terrible day of interviews and trying to catch up, which I never quite did. Some of
my personal shopping had to go, also going along to Wild Life Society for my home
mail, so shall go a long time before I hear anything from home. Eventually had to leave
the Talbots point blank. The Glovers took me to the airport and I dined them there.
Am leaving Kenya in good state as far as our organizations are concerned, and I think
I have done a very necessary piece of work. Joined the BAOC Comet at 10 p.m.
and arrived Khartoum 2 hours 35 minutes later, 1,220 miles away.

*   *   *

Co-incidental with the Arusha conference in 1961, the situation in the Tsavo
(East) National Park had become critical. A severe drought had forced the
elephant to move in search of water. Many rhinos, which depend on elephant
holes in dry river beds for dry-season water (see p 159), were dying, and the
tree cover of the Park was being devastated by the displaced elephants. Frank
had seen some of this for himself when he visited the Tiva River with Mervyn
Cowie and David Sheldrick in 1956.

Cowie had already sought the advice of Julian Huxley about what should
be done in Tsavo, and had been told that 10,000 elephants should be killed
to relieve the pressure and stabilise the habitat. This would have been a
gigantic and violent exercise, which would be widely misunderstood and
deplored, as well as being of doubtful benefit in the long term. Huxley was
a theoretician not a practical man, and, with the support of Phil Glover, Cowie
invited Fraser Darling to advise the trustees on what course of action to

follow. Both regarded Frank as a pragmatist capable of matching the theoretical and practical elements of the situation.

Frank's advice was non-committal yet sensible – no action should be taken to kill elephants until research was done into the structure and distribution of the herds and their likely long term effect on the habitat. This suited Cowie. It bought time, and allowed the crisis of 1961 to pass without slaughter. It also cleared the way for a substantial programme of research in Tsavo. Cowie's appeal for funds for scientific research was disappointing, blunted by Huxley's summary pronouncements on the need for immediate heavy culling. However, the Ford Foundation paid for two biologists on a programme drawn up by James Glover of the East African Agricultural and Forestry Research Organisation to provide the basis for a management plan for Tsavo. In the late 1960s, Richard Laws, formerly Director of the Nuffield Unit of Tropical Animal Ecology in Uganda, Murray Watson, formerly of the Serengeti Research Institute in Tanzania, and Ian Parker started a new programme of elephant research which sadly was never finished.

JOURNAL III

Sudan

Introduction

DURING HIS FLYING TOUR of Africa in 1959, Fraser Darling visited
Sudan. In Khartoum he met Santino Teng-Deng, Minister of Animal
Production, and members of the Game Department. It was put to Frank that,
against the developing economic importance of stocks of wildlife in the Sudan,
there was a dearth of expertise. The scientific training in wildlife management
and advice on game policy in line with conservation principles were not
available in Africa. The proposal was made (by whom is not clear, possibly
Frank) that the Conservation Foundation should bear the cost of a senior
consultant (himself) to visit Sudan and that the Sudanese government should
act as host during an extended tour in early 1961. George Merck, Secretary
of the New York Zoological Society, was a late addition to the party, affirming
the close working relationships between the Foundation and the Society in
African conservation. Fairfield Osborn was a moving force behind this liaison.
He had been Frank's champion since the Lake Success conference of 1948, and
it was to him that he felt accountable. When he set out in Africa in 1956, he
had addressed his Journals to his wife, Averil. Now they were addressed to
Osborn:

2 February 1961

Dear Fair,

George and I are all packed ready for disappearance into the far
south with 2 two-ton lorries piled high with gear for our six weeks'
safari . . . More [of my Journal] soon, but not frightfully soon. Love
to you all.

Yours ever,
Frank

The diplomacy was sensitive. If care were not taken, the origins of this survey
might be traceable to Frank. The conditions of wildlife and landuse in Sudan
gave cause for concern, and he had a burning desire to see these for himself,
preferably as a guest of the Government. Having had a personal discussion
with the Minister in the aftermath of the reconnaissances in Northern Rhodesia
and Kenya, he hoped that an invitation would be forthcoming for a similar
exercise in Sudan. When no invitation came after nine months of waiting, on
22 November 1960 Frank wrote a rather despairing letter to Sayed Medani,

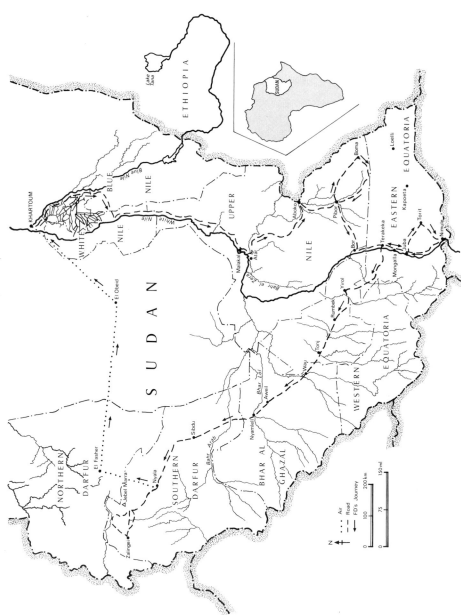

Map 8: Location map of Southern Sudan showing the route (broken and dotted lines with arrows) followed by Fraser Darling and George Merck in the ecological reconnaissances of 1961, compiled by JMB from Fraser Darling's Journal.

Director of Game and Fisheries, asking – '. . . Could you please give me definite news as to whether your Government wishes me to come or not?' By 15 December he had an affirmative reply. All was well and the tour could go ahead. With a note of triumph Osborn prefaced Frank's report thus: 'This survey . . . was carried out as a result of a spontaneous invitation from the Government of the Sudan – a circumstance much at variance with the usual "western world activities in Africa", which are imposed on African people rather than being requested by them.'

Edited excerpts from Frank's report to the Conservation Foundation, Zoological Society of New York and the Government of the Sudan, entitled *Towards a Game policy for the Republic of the Sudan*, are interspaced below with sections of his Journal tracing his tortuous journey in wild and remote country in southern and western Sudan.

CHAPTER TEN

Upper Nile

January and February 1961

THE EXPEDITION motored south from Khartoum to Malakal, passing from desert country through a large area of cotton growing into acacia bush. The dwellings changed from the adobe houses of the Arab culture to the round thatched huts of the Nilotic peoples. Malakal is intensely interesting as a considerable town where different cultures meet. A barge pushed by a wood-burning tug took us sixty miles up the White Nile to the Zeraf Reserve (450 square miles of *toich* or seasonally inundated country) to see Mrs Gray's (Nile) lechwe. As we disembarked in the grey light of the morning, the local Shilluk were swimming herds of cattle across the river from Atar, the men astride rafts of bound stems of *Aeschenomyne*, or ambatch. We saw fully a thousand lechwe in the course of a morning's walk.

On this three-day voyage up the Nile there was a wealth of bird life: pelicans, ibises, herons, egrets, geese, ducks, fish eagles and other birds of prey, and innumerable waders. Whatever be the fate of Africa's mammals, we feel that the birds have sure refuge in this great country of rivers and lakes extending from the Sudan to the Zambesi.

From Malakal we drove the Sobat River and by a detour to Waat in the country of the Nuers. Our admiration of the Nuer thatching was unreserved. Then we went to Akobo on the Ethiopian border, and to Pibor across vast grass plains of *Hyparrhenia* grass and *Balanites* bush where there were great numbers of francolins. The country was overburnt and we saw no game nor many vultures indicating its presence.

Our route then lay south-east to Boma, through an area of softer grasses, on the line of migration of the white-eared cob which passes to and from Ethiopia and northern Kenya each year. We saw many hundred cob in a reserve. Our camp was below Boma Mountain. Rainfall was well distributed and the mountain was intensely green with new grass showing little or no signs of grazing. The hollows still held original high montane forest with blue monkeys, but the ridges and the sweeps of land where fires could run had lost the forest trees, and had degraded to tamarind, acacia and *Terminalia*. The mountain should be totally covered with forest. The plains held but a few head of Lelwell hartebeest, zebra, tiang, reedbuck, and giraffe.

20 January 1961

Arrived Khartoum about 2.30 a.m. and met Medani's second-in-command who had come to the airport for a talk while the aircraft refuelled. It seems that all is in order for our reconnaissance and that in addition to the main job in Equatoria we are to have the safari to the Dar-Fur. This should be quite wonderful, for that massif has fascinated me since I was at school. Game is by no means scarce and we should see some fine herds of addax. Got back into the airplane feeling rather thankful, because it has not been too easy making contact by correspondence.

29 January 1961

Was met by Khalifa, the Fisheries man, and taken to the Grand Hotel. It is a large place on two floors not frightfully grand, but grander than anywhere else in Khartoum. Have stayed here before. Cuisine adequate but far from inspired. Wondered if George Merck would show up for breakfast. He did not, and enquiring at the desk found he had not arrived. Spent the morning away with Khalifa and the Director, and in the afternoon a call came to my room. It was George, whose plane had come in five hours late. He had had a good sleep and was now ready for tea. So was I.

29 January–2 February 1961

A rather trying but nevertheless interesting period – trying to me because it involved a lot of waiting, waiting, and not being in charge of one's own goings on. Our friends had nothing organized at all for our journeys and set about arranging it all by telephone while we sat and sat and sat. Admittedly their speed and efficiency was remarkable on that instrument and many difficulties were just brushed aside. The Director of the Ministry of Animal Production and Resources is a veterinary man called Dr Ibrahim Khalid. He is a big northern, nearish white sort of Sudanese, speaks English well and I like him. We have had many talks, more on attitudes than on details, and I find him with a large view opposed to vulgarity in exploitation of tourist and safari potential, and with an idealism in behaviour towards peoples, habitats and animals. Nevertheless, he is a practical man and knows ideals are never attainable, but his patience in action does not let him get bitter, nor does he shamelessly trim his sails to meet every little breeze of popular demand or governmental opportunism. We have been asked to see several films these days intended to show the safari and tourist potential. One of them, in colour, described the safari of Tito and his entourage in the southern Sudan. It set my teeth on edge. Everybody moving in a crowd, the Europeans fat and vulgar, shooting at elephants and buffalo at much too long range, with the local

Game Scout shooting also to give the *coup de grace*. There were native dances with the people wearing oddments of European dress, and the Yugoslavs joined in with dreadful joviality. The English commentator said the natives were so pleased to have their visitors join it. I wonder. At another native dance where the dancers dressed naturally, one was happier, but the thing was ruined by the wretched Yugoslavs capering about in among the dancers taking photographs at all angles. When the Yugoslavs shot anything they looked high-chested and arrogant, and the picture of their shooting a young lioness little more than a cub, and helped by the use of fire to get her out of the long grass, was quite horrible. Yet this film is being used for advertisement purposes. Ibrahim Khalid is disgusted with it and he says the Yugoslavs were awful, disregarding all canons of behaviour in the field. Yet, as he asks, what can you do when this was a State courtesy visit and future relations of two countries may depend on keeping one's temper? He and I were agreed that at least such a vulgar film with a luscious-mouthed English commentary need not be used for advertising Sudan safaris. In fact, safaris here are run with very close supervision and not in the style of the *nouveau riche* Chicago gentleman in Northern Rhodesia who did not want to move far from the Land Rover. 'You line 'em up and I'll knock 'em down. That's what I'm paying for.' Anybody who has stalked deer in open hill country knows there is very little sport at all in shooting African animals. It is little more than poking a gun in their ribs and pretending to the danger in the smoking-room. Anyway, Ibrahim Khalid has a fine contempt for all this sort of thing. He does not want to see his country following western vulgarity under the impression that it is western civilization. Incidentally, talking of vulgarity and power figures, I came on an interesting case in Kenya and Tanganyika last week. A person called Burns, who is president of the Shikar Club in the United States came to Kenya last year, took out his licence for the lot and asked for one for his son aged fourteen. This is not allowed in either territory. Burns gave the Kenya Game Department $5,000 but did not get the additional licence. Off he goes, gets his animals and then when he gets back to the States has to do what the rest of them do, let his hair down in melodramatic style in the sporting papers. His article was headed 'The Biggest Poacher of them All' and described how his boy shot this, that and the other in East Africa. The East African Game Departments take these papers just to see how fact and fancy tally. Burns now applies for another licence for this year's safari, but Grimwood points out that the behaviour described in the article precludes the granting of another licence. Burns is rather nettled but swallows hard and says, 'You don't mean to say you believe everything you read in an American magazine!' Grimwood says, 'Nevertheless' and Burns says, 'Of course the boy didn't do the shooting'. But he doesn't get the licence. Bruce Kinloch in Tanganyika goes further. Having read the article he gets the white hunter concerned on the mat and finds the boy shot fourteen animals. Kinloch takes away the white hunter's licence for three years and when Burns applies for his licence Kinloch writes him a letter (I saw a copy) which was a masterpiece of politeness and firmness, quoting facts and figures and suggesting the best thing Burns can do is to make a clean breast of everything. What I particularly like about this story is the retribution awaiting the vulgarity of writing up in sporting magazines. Substantial presents to Game Departments will not buy British officials. Good or bad, they are not corrupt. The wealth and strength of the Sudan is in herself, in the variety of her own cultures, in 4,000 miles of the Nile River and the great quantity of alluvial soil which will develop as water is brought to it. We have had talks with Shawki and Hamad of the Forest Department and my opinion is that in few countries would you find a Director and Assistant Director of equal calibre to these men. Medani of the Game Department has not equal scientific knowledge and he is curiously slapdash but I think he gets there in the end. George is to advise on the Zoo organization and management. We went to a party one night given by the delegate to the United Nations for David Owen whom we know in New York. I am to help Forestry and Game and Fisheries to get technical help through David Owen for conservation education and wildlife management. Our assignment here is not going to work out quite as we had thought.

Another morning we went on a trip in a launch to the junction of the White and Blue Nile and up to Gordon's Tree along banks which were bird sanctuary. The birds are legion. Large flocks of Egyptian geese, marabou storks, ibises of one sort and another, pelicans, ducks such as pintail and teal, and lots of European waders such as sanderling, stints, curlew and godwits.

In the bank one day to get some money changed with some trouble. A horde of camel men from the province of Dar-Fur came in to bank some thousands of pounds gained from camels taken from Dar-Fur northward through the desert to Egypt. They were very lean, brown men; small and small-boned, quick and volatile. They greeted each other with beautiful courtesy, laying their right hands on their friend's left shoulder and slightly bowing. Their unbleached cotton turbans were very large and wide, their robes of cotton with good work round the neck and yoke. Their robes were the colour of the desert with the dust collected. Prices in Khartoum are high and everything is dearer than in Kenya. Have had a bush jacket and trousers made from Sudan grown and woven unbleached cotton that looks like coarse linen but is much kinder to wear. Costing me £5 as against £3.10s for khaki drill in Kenya. Shall use this suit for changing into in the evenings on our safari, and I can use it in the summer at home. It is extraordinarily kind against the skin. Finally, our programme and our amended programme have been made up and we are to proceed by road from Khartoum to Malakal, 525 miles, in two 2-ton Kommer lorries, starting at 5.45 a.m. on Friday, 3 February.

3 February 1961

George and I are down to time and we wait. One of the drivers does not arrive and he has the key. Eventually he is fetched and the trouble found to be that the driver lives eight miles out and as this is a sabbath there was no occasional passing transport to give him a lift. We get away by 7.30 sitting hunched up in the cabs which the blasted British manufacturers put on their lorries. The road is merely car tracks across the desert, a series of tracks 200 yards wide. We pass through miles and miles of damn all, as you might say, sandy gravel with an occasional shrub. We stop at 9.30 at Gateina for breakfast. This is a small Arab town with school and mosque and market, which latter was in progress. We stopped at the rest-house and our men began to pull the loads to pieces. Our major-domo is a Game Inspector called Bushir el Hagani. He is a little lean brown man with a bald head, who spent six years in the British Army and who speaks English after a fashion. It soon became obvious that whoever packed the loads was not unpacking them and things got a bit frantic. The breakfast stop took two hours although the breakfast itself was lovely and nicely served by our handsome waiter Ahmed. If he were to come to England he would cut out our young men completely with the lasses. Sorry, he is called Mahomed. I told Bushir to cut out lunch entirely or I cannot see us reaching Malakal tomorrow night. We drove on and on, eventually reaching irrigated cotton soil and crops. We began to leave behind the flat-roofed adobe houses of Arab culture and get into round thatched huts with pointed roof. In other words, we are getting down into Africa. The dress remains Arab, but I fancy the Prophet has not much say in the villages we are now passing through. George and Bushir are in the front lorry with a driver; I am behind with Ahmed who must be a camel-racing jockey or something, because he likes to be up close in the dust. I have a job to keep him a quarter of a mile behind out of the dust trail. Once we got stuck in deep sand and had to use metal treads to get ourselves out. Some of this flat cotton country is slow because there are constant culverts to cross over the irrigation channels. The people seem to live in the fields a lot of the time, their goats, sheep, cattle, camels and donkeys hanging around. Eventually to a one-horse little town called Rabbuk, where the super-polite Bushir asks if we would care for him to find an empty house for us to lodge. I say no, what is wrong with the desert further on? So we go on a further ten miles and stop as it goes dark at 6 o'clock. I also say, no need for tents, for I can see time being wasted, but we do want mosquito nets. It is obvious our outfit is not accustomed to camping travel and it will take some time to work them up to a quick routine. George, having had a cold through sleeping with the damn air conditioner running in Khartoum, decides to have a whisky for his health. Bushir thinks it an excellent idea and has three himself. Don't much like this when I have myself set the standard for abstinence in a Moslem country. The whisky should be kept for emergency. Noticed four men making up my bed. Enough to send anyone crackers. There is a canal near us though we are camped on desert, so mosquitoes are there all right. My tea well flavoured with paraffin (kerosene).

4 February 1961

A cold night, and George found it very cold. Told him this was a result of drinking whisky. I woke at 4 a.m. and heard everyone snoring. At 4.30 I got up and got folk moving, for it will take us all our time to be away by 6.30 o'clock. Tea and biscuits only and through a misunderstanding with Bushir breakfast was made up of sand-wiches to eat later with Thermos tea. George's face fell. Got away five minutes late and had a slow trying drive to Renk, a most ugly nondescript settlement of petty officialdom. We are now in black Africa with Islam as the dominant colonial power as it were, but we must not put it like that. The road from Renk, six hours' hard driving to Malakal, is at least made, in that we pass through acacia scrub cleared and stumped. On and on into Shilluk country. Clothes are either shed altogether or a saffron toga is worn by the men. Farther back the women wore a drab skirt with an invariable blue veil. Now the veil has been shed. The Shilluk men have one or more lines of pustulated scars across their forehead and the women wear a nose ring in the right nostril. Some of the men bleach and straighten their hair and wear it like a fur cap. A long broad-bladed spear is carried, and a knobkerry. The Shilluk are not a beautiful race of people. They are essentially Nilotics and do not show Hamitic influence. We drove on and on through an enormous grass plain that must once have been forest. We are never really far from the Nile, which makes life possible. The cattle are predominantly grey and better than those farther north. We passed for some miles through groups of Shilluks in full paint, spears and shields. Women also dressed up and waving to us excitedly. They are gathering for what must be a gigantic dance. Ahmed gets sleepy, so I drive for a couple of hours. The driving position is even more cramped, I find, but I managed to keep up the speed. And so we arrive at Malakal dead on time at 5 p.m., ready to go aboard a steam barge of two decks got specially for us. George fading from hunger, but like us all he had to wait till 9.15 p.m. for his dinner. All the gear came aboard and our cook prepares the food on a charcoal brazier. Nevertheless we had a bath and a shave and felt pretty good after a hard and uncomfortable day. The water aboard is pumped up out of the Nile. The lavatories empty into the Nile. As we look around at the scene we see numerous people carrying out their thorough ablutions in the Nile. Cattle come down to drink and lounge about in the Nile. The Nile is the river of life. Last week I had the first upset of the guts I have ever had in Africa, spasmodic abdominal pain lasting for a full week. Can only hope we don't get another nice little lot of trouble from contributing to the theory of the Nile as the river of life. Making tea from a stinking elephant wallow on Lake Mweru Wa Ntipa years ago was nothing to this.

5 February 1961

Supposed to be away by 7.30 a.m. but we did get moving up river by 11 o'clock. I had a good night's sleep, but coffee at breakfast was a mistake. No more of it for me. Better by lunch time, which meal I almost skipped. One way of getting the belly down. We proceed up river for a dozen or fifteen miles watching lots of birds and naked men fishing. The Shilluk men, in common with other Nilotic tribes, have extremely long private members. Then, ha ha, we stop at a place where there are acres of firewood in neat cords, and we spend five hours refuelling. I sit and write, but George finds it hard to relax and curses the fact that we shall continue up river in the dark. This is Africa. The Shilluks carry the wood aboard on their backs and the black Arab sits on the bank and plays dominoes. If there is any daylight on this cruise we are supposed to see a Reserve for the Nile lechwe, called the Mrs Gray.

This morning at Malakal, a most interesting and colourful town, there was high jinks for the opening of an American laboratory to study kalahazar. A programme of seven speeches began at 10 a.m. as we could see from our ringside seat on the steamer. The first man in a brown suit was still talking into the microphone in front of the marquees when we left at 11 o'clock, so God help the assembled company we could see there trying to look interested. Our siren blew long and loud at intervals before we steamed up the river, which must have been hard on the gentleman in the brown suit. Some sort of a Shilluk dance was going to be staged for the assembled company and we saw canoe loads of highly decorated black gentlemen coming over the river in dug-out canoes. George and I, after our talks last week, have the gravest doubts of the good the crowds of half-baked ICA technicians are going to do in the Sudan. The United States is being had for the mug around the world, but with its strong evangelistic spirit it is cheerfully having itself for the mug. The Shilluks are still loading wood and singing monotonously. The cords of wood we are getting are those farthest away from the bank. The five hours will soon be gone.

6 February 1961

Managed to get dinner by 7.30 last night and into bed by 9 o'clock. We anchored by Atar school after a lovely night run. This is border country between Shilluks and Nuers. The latter have incised lines across their foreheads. We rose for 5.30 a.m. and were away by 6 o'clock over the marsh and flood plain to look at Mrs Gray lechwe. Before we actually left the boat we were treated to Shilluk tribesmen bringing their cattle over the Nile to graze. The Shilluks make beautiful little high-prowed rafts of ambatch (*Aeschenomyne*), a pith-filled leguminous plant, and one or two men sit on these and lead the cattle which swim after them. One raft brought a new-born calf over, but it is remarkable how young the calves will swim across at the side of their mothers. The rear of the herd is accompanied by dug-out canoes with men and gear aboard. Shilluks and Nuers are kind to their cattle and in this long grass of the flood plain will build grass wind-breaks for them and burn smudge fires at night to keep mosquitoes from the animals. They themselves dust themselvs with ashes which makes them look ghastly. The cattle are tethered at night and I was interested to see and photograph how a rope tether was made with a swivel. In one corner of a wind-break I found very young calves and tiny puppies playing happily together. We saw fully 1,000 Nile lechwe in the course of our walk, but they did not let us much nearer than a quarter of a mile. The females are darker than the red lechwe, but not very much. The males, with lovely horns, grow black with age but have a white patch on top of the neck and chine. Very handsome marking. The Reserve for the Nile lechwe extends to 450 square miles and there must be some thousands within it. George suggested that a 50-foot lookout tower would be a wonderful thing for people to see the animals, which of course it would be, with all this long grass, but I doubt whether many tourists come here. Sorry to find the animals so shy, but there are too many Shilluks and cattle about where we were to expect the lechwe to be quiet. We steamed up the Nile for another hour and a half through flood plain country and *Vossia*-lined banks. At one point on the left of us, slightly rising ground had allowed a large grove of *Acacia albida*

to develop, with a few red-barked and yellow-barked *Acacias* among them. I thought
the bush there looked good and was told it was forest reserve. The Sudanese seem
to have a pretty good control of their southern tribes in these matters. As our part
of the ship is really a barge and the engine part is a separate pushing tug on the
starboard quarter, we experience motion without any sense of engine. Quite delightful. Last night we heard beautiful bugle notes, which we understand are cows' horns
blown by Shilluk headmen to invite the neighbourhood to a dance. The birds on the
lagoons this morning reminded me of the Kafue Flats, so many of the same species,
and always spectacular. Went ashore at Atar school on the way back this afternoon.
A Shilluk master in white shirt, khaki shorts and solar topee showed us round. Built
by the British 10 years ago, all rather gaunt and ugly I thought. The garden was full
of various fruit trees – lemon, orange, grapefruit, mango, guava – irrigated by
pumping from the Nile. Now we are coasting rapidly down stream past small floating
rafts of this new weed which has come into Nile and Congo – the water hyacinth. Very
few hippos on the river. This part is pretty poor in game I feel. The name of the lechwe
Reserve is Zeraf, and is really an island of flood plain called *Toich* in Arabic, meaning
temporarily flooded land as against that which is permanently flooded.

   The Veterinary Officer, Malakal, came aboard as we reached Malakal at 6 p.m. and
asked us to a dinner party at the rest house. All the chaps from the celebrations for
the kalahazar laboratory were there – Minister of Health, Minister for Economic
Affairs, Chief of Police, and a large tribe of Americans and their wives. Quite a do,
but rather slow. No drink of course. We have had a nice intelligent young Game
Ranger with us these two days – Mahdoub Abdulla Bedawi.

7 February 1961

Our 10 o'clock start occurred at 11.30 a.m. and another quarter-hour stop in Malakal
town for some reason or other. Noticed a nomad Bagharra woman loading an ass with
stores. Arab-negroid features, reddish-brown complexion and the lips and lower part
of the face blue with kohl. She had a girl child on her hip of about a year old, most
immaculately turned out with fancy headdress and cloak. The child had fine features
and was most charmingly expressive, as if it might have been five years old. Malakal
is a remarkably pan-African sort of town. You will see veiled Arab women, Arab
children of the official class in little cotton European suits and dresses, naked Shilluks
with bleached-to-ginger hair carrying long spear and stick, and portly official types in
shorts and open-necked shirt wearing topees and riding slowly on bicycles. The solar
topee is obviously a thing of rank and prestige. Later in the day I saw a Nuer wearing
one. It was the only article of clothing he wore. In such circumstances the topee
became slightly shocking. We travelled over grass plain to the River Sobat where we
had to cross on a ferry hauled along a wire. There was a prison on the bank and we
talked to the Dinka governor who took us round the garden and orchard. Charming
man with good command of English. I like the Dinkas and shall look forward to getting
into their country. My friend Joe Broadhurst, who worked among them 35 years ago,

used to tell me how mild and wise they were. And that is how they strike me – very big men, dignified, gentle, smiling. When one of them speaks English, you feel you can communicate straight away.

We passed through a large expanse of red acacia scrub with occasional green trees of what I think is *Balanites*. Then a *Balanites* savannah, but I must confirm *Balanites*. We are in the acacia tall-grass country, *Panicum turgidum* and the grass is 7–8 feet high. We are passing through vast expanses of grass plain in the country of the Nuers. They are tall and thin and not attractive. The men go naked and in the remoter villages the girls are also naked. The married women wear a skin apron triangulated fore and aft, but many of the women now wear a toga and even a toga over a cheap cotton print dress. A string of beads worn round the waist is common both in men and women. One Shilluk chief we saw the other day had so many waist bracelets that they hung down almost as an apron. The drivers said at 1 o'clock that we should halt for two hours as it was too hot for the cars. So we stopped at a Nuer village with two or three shady tamarind trees near a lagoon. The village manhood and boyhood were lying beneath the trees: Bushir, the fussy little official of small authority who is with us, strutted up and waved them all away. This was too much for me, shifting people from the enjoyment of their own village shade, so I stopped him, and we occupied one part of the shade and the Nuers the other. The Headman came to shake hands and all was well. British colonialism (sic) has gone and Arab colonialism has taken its place. But the yattering small nations and the United States say it is alright! Mohamed brought us tea and a light lunch and we sat down to watch a Nuer idyll. Boys bathed and swam in the long lagoon. Herds of cattle came by, and tiny boys playing with little calves. Some of the idle warrior-class braves came down to bathe, washing off a clouded-leopard effect on their shoulders and upper arms brought about by ash and ochre and touches of oil here and there. Very artistic. These Nilotics are definitely hair-conscious. One brave came along with this hair hard encased in clay. He leaned backwards, broke the clay cast and then shook his head in the breeze. The white and ephemeral cloud of dust, the gleaming body so black, and the sun and the lagoon made a fine picture. He then ran into the water and thoroughly washed his hair until it appeared copper-coloured. Two belles came down to the lagoon for a little chaff with the fellows, and one or two married women came down for water. I suppose we saw a typical Nuer day, and I felt that as long as the United States and Russia bid against each other in pouring money into Africa, the Nuer may continue to live the good life.

On and on again until we reached a remote village at 6 p.m. We had done the 110 miles we were supposed to do and I expected soon to stop, but the wretched little Bushir had learned at the prison camp that the road to Akobo was impassable and so we were to make a detour involving sixty miles extra. He had said nothing to us about this and now we went on and on in the dark over grass plains where fires were in view. So we had to go on to Waat, arriving 8 p.m., which is no time to make camp and start cooking. I gave Bushir a pretty serious lecture and added a few more details, such as getting washing basins ready and smartening up timing on making and breaking camp and so on. The little twirp is obviously itching to come all the way with us and I know

The Nuer lagoon – 'boys bathed and swam in the long lagoon. Herds of cattle came by . . .' on 7 February 1961.

he will try to put off the man at Bor who is to take his place. But Dr Ibrahim told me the other night we could certainly have the new man and that there was nothing important enough to override his coming with us. This morning Bushir helped himself to a large bottle of beer though we refused to have any, and this evening I refused to allow the whisky bottle to be brought out. I fancy George thinks me a bit of a martinet. The Sudan is doing us so well and we have a good bunch of chaps, that with an officious and crafty little twirp like Bushir it is necessary for me to take charge and boss him around. We must make the safari a success. Waat is a police post and we camped in the middle of the parade ground. There are a few prisoners here in a compound or in chains. Cattle stealing, clan wars, and homicides mostly. The police keep a few Barb ponies for cross-country work; hard, sleek little horses with hammer heads, not Arab types at all. Bushir has left our mosquito nets in Malakal, so we enjoyed the night sky the better.

The Nuer are very fine thatchers, as you might expect in a country of long grass. I have never seen better. There is another tribe from Pibor eastwards to the Abyssinian border and beyond called Murle – pronounced Morley – who scarcely thatch at all but merely make shelters. These people are much smaller people than the main Nilotic tribes and are very primitive. The men wear nothing at all and being so beautifully made and so black they look really magnificent. The women wear an apron. They scar their bodies in fine patterns, every individual different. They are hunters and food-gatherers and are practically nomadic. One sees a family walking along, the women

A Nuer village in Upper Nile (Jonglei) near Waat on 7 February 1961. 'The Nuer are very fine thatchers . . . I have never seen better.'

usually carrying their simple gear on their heads and the men often carrying a small child but always with his spear in his right hand.

8 February 1961

To Pibor today, 155 miles, mostly through grass plain and very light bush which has been burnt to death. One of our trucks broke a front spring as we left Akobo on the Abyssinian border. Got going again in little over an hour. We seemed to be in an endless grass plain and *Balanites* savannah. There was thicker bush near Pibor itself, which is an administrative post with District Commissioner and Police. Of course there is a jail as well and lots of prisoners doing garden work. The people round Pibor were almost wholly Murle. We stayed in the rest-house, having our beds on the lawn which had been watered by the prisoners. Lots of swallows roosted on a wire under the low thatched eaves of the rest-house and by day bats hung under the eaves.

9 February 1961

Bushir introduced us last night to Mahmoud Abu Sereina, his superior, and second in command of the Game Department. Mahmoud was sitting in the white gown of his people, relaxed and waiting for us. He rose with an easy grace and met us with exquisite charm of manner. He is the age I would have guessed — 37 — very dark brown, medium height, thickset and powerful. I knew immediately this was the man I had been waiting for, though I could not then say why, except that I recognized a man who was powerful in mind and spirit as well as in body. It has been only later, gradually and fragmentarily, that I have discovered the key position he must hold. First, his integrity, which is that of the selfless man doing work which has drawn him as an inspiration. He is of the upper class of Sudanese gentlefolk, well educated, well connected, married to the daughter of the Minister of Finance, a personal friend of the President and of several other members of the Council of Ministers. His complete poise extends to situations of quick decision in the field when he is sent to accompany distinguished but silly guests of the Government. A Yugoslav Minister on safari shot a lion by jumping out of the car and being melodramatic. He shattered one of the lion's fore paws and it came headlong and jumped into the back of the lorry following. Mahmoud killed it there. He also killed the elephant the Yugoslav wounded with a head shot at $6\frac{1}{2}$ yards. These things I heard from Dr Ibrahim. Mahmoud, in fact, is a crack shot. Being born a gentleman he has no ambition and has declined diplomatic posts in order to work for game conservation. He was trained as a surveyor but gave up that work nine years ago to take a junior position the Game Department. He could have been head of it now but declined in favour of Medain, who is his superior in years and in service. Also he wanted to get more experience in the field. This is undoubtedly the man on whom the future must be built in wildlife conservation in the Sudan, and he must be important in Africa as a whole before long. I must get him sent to the new Nature Conservancy course at London University and must help the Sudan Government frame the application for a senior man in conservation to come to the Sudan under David Owen's United Nations Fund. Meantime, Mahmoud is to be with us till March 15, and we say goodbye happily to our cheerful little rascal of a Bushir (who we hear has two wives and twelve children). We got away by 5.30 this morning on our 135 mile run to Boma, which is near the Ethiopian border south-east of Pibor. Again more miles and miles of grass plain and fire-degraded bush. But we passed through a Reserve for White-Eared Cob and saw many hundreds of these interesting antelopes. They are essentially the same animal as the puku of Northern Rhodesia and the Uganda Cob, but in this variety the males turn black when mature, with white ears, throat and muzzle – very handsome. Another difference is that the white-eared cob is a migratory species following *toich* country (temporarily inundated grasslands) from Ethiopia into Northern Kenya by way of south-eastern Sudan. They are to be seen in tens of thousands and on migration are rather like the caribou of the far North, indifferent to man. Another comparison with the caribou is that the lions move along with the cob much as the wolves with the caribou. We came upon sandstone kopjes a few hundred feet high west of Boma, and then to the mountain itself which reaches 7,000 feet and is basaltic over the sandstone. Many small volcanic cones are in the neighbourhood. The massif of Boma mountain should of course be heavily forested but I was sorry to see it ravaged by fire. The forests persisted only in cups and hollows where the rising fire from the plains could not reach them. The summit was fringed by thin trees and the sides light green with grass, acacias and tamarinds, and oddments of *Terminalia* scrub. The dark green raised surface texture of montane high forest appeared only in the little patches instead of clothing the whole. Rainfall is well distributed here throughout most of the year, so that some of the grassy slopes were brilliant emerald green. It looked so good from below and afar, but in fact there were no herds of game grazing because there was no water on the slopes. We reached what is called Bottom Camp, below the mountain, where there was a large grass *chitenji* in which George and I were to be housed for two nights, and two or three smaller ones. These buildings of rough poles, thatch, and grass walls are quite delightful to live in. The texture of the walls and roof absorbs all the rough edges from one's personality and you feel at rest. Two years ago we held the Luangwa Valley Conference in a large *chitenji* open on one side to a lagoon beyond which game wandered, crocodiles basked and birds played. No wonder it was so successful and amicable even though difficult matters were discussed. We went out in the late afternoon into the plain of this *Balanites* and *Acacia* savannah for perhaps 15 miles. There was plenty of grass but it was not being eaten. In short, the game simply wasn't there. We saw perhaps 20 hartebeeste in small groups, a few tiang (topi), 17 zebra, probably a score or more of oribi and a few reedbuck. I think there were three giraffe. Mahmoud was obviously disappointed and depressed. He had been here a year ago and saw large numbers of animals, herds of hundreds, and he had asked one of his assistants only three months ago and he had said all was well. Mahmoud says he will now put on a squad to find out to where the herds have moved. For myself, I just feel that the country has burnt to death and that game is scarce. I heard no hyaenas at night and we have seen none. The only carnivore seen as yet has been a dark-coloured jackal. All the animals we saw this afternoon were shy of the lorry. Some safari parties have been in and I am just wondering whether, for shooting or photography, the animals have been chased around.

10 February 1961

Out in one lorry at 5.30 this morning over the plain below Boma Hill. We saw pretty well the same animals we saw last night except for a few more giraffe. This damn lorry business irks me. You never get out on your feet and the bumping in a confined space gives you bruises on shoulder, thigh and knee, and you get more knocks in precisely the same places as you bump along next day. But I admit there is no point in walking around a bare plain empty of animals. Anyway, back to breakfast and then away up the hill on a spectacular track. Up there the air was cool and delightful, and the sunshine made everything happy. There is an Arab merchant living at the top of the mountain and numbers of Murle live in the vicinity, buying some things from him in exchange for wild coffee beans which they gather. This man was tall, young and fine-looking, really a striking character whom one would not doubt in any way. His wife died two years ago and he seems content to live up here in this delightful climate like an English spring. He speaks the Murle language fluently and it was obvious he had the confidence of the people. These Murle are really quite delightful people, like children, full of fun and ready to communicate. Hassan the Arab brought chairs and a table and offered us mint tea. The Murles gathered round with interest. We are near Ethiopia where there is a small tribe whose women cut a lateral aperture below the

'These Murle are really quite delightful people . . . full of fun . . .' The Arab trader, Hassan, and fine Murle men and women at Boma Hill on 10 February 1961.

lower lip and gradually stretch the lower lip itself till they can put in a round tin lid. A few of these women were here and allowed us to take photographs. The lid from a Lyle's Golden Syrup tin is the perfect one, being smooothly finished on the edge. The lower incisor teeth are also knocked out, and when these women eat they put the food on the tin lid, open the mouth and raise the lid to empty the food into the mouth. There was to be a dance today and the best part of 100 Murles had gathered. As we sat drinking tea a naked man came along with two tiny girls, perhaps five years old. The little girls were evidently dressed for the party with their little decorated fringe fore and aft and their hair done in intricate pattern. The man greeted us with charm and polished manner, and the little girls clung each to a leg of their father in proper momentary shyness. I bent down to shake hands with them also: one was obviously thrilled to bits and ready to try anything once. She smiled, raised and lowered her eyes, and chucked her shoulder and beamed at us. The other, being shyer, clung to her father, one arm round his leg and the other hand hanging on to his penis. However, with the example of her sister, her shyness was overcome, she shook hands with graver face and thereafter decided things were not so bad, because the two of them sat cross-legged in our circle and were full of fun. The father stood behind them smiling and, as I think, proud. Then a witch doctoress was introduced to us. She was of the usual small Murle non-negroid type, solemn-faced and appeared middle-aged, though I don't think she was because her breasts were firm and young and well carried. She had such poise as any society woman might envy. Looking at us with complete

The Murle dance at Boma Hill on 10 February 1961 – 'They formed into a ring, singing, with a man running within the ring and acting as precentor.'

frank appraisal, she was gracious enough to shake hands, the palm held upwards and at shoulder level. As we placed a hand on hers, she bent her head and kissed the wrist of her own hand. We bowed with proper gravity. She then informed us that she was all-powerful in four clans of the Murle and that she had much greater power than any government, because the people did exactly what she said they should. This was a plain statement of truth, I think. The translation of her Murle speech came from Hassan the Arab merchant to Mahmoud in Arabic and then from him to us in English. Ease of communication was quite striking. She said she would watch the dance if she felt like doing so, but if she felt she didn't she would not. No society woman ever left the play after the first act or came in at the second act with more detached insouciance. Her ears were heavy with multiples of blue rings and numerous charms hung about her neck. I wonder if absolute power corrupted? At this moment the young men appeared behind us in a running single file, each with a red crane feather on his head. They formed into a ring, singing, with a man running within the ring and acting as precentor. Now the men began to leap and come down flat on their feet on the earth. The sound was impressive and their voices came low and cavernous. I saw the witch doctoress also jumping up and down in time at the side of me, so I gathered she would probably condescend to watch the dance. The men ran singing to a ground 150 yards away where 30 or 40 young girls of 10–12 had gathered. The men formed a line, sang, leapt up and down and clapped their hands in time. The little girls ran forward and back. A few men in turn would run a few steps forward and all the little girls would run to them so that three girls jumped with each man, their hands to his sides and chest. Then all back again. The precentor was tremendously active, running along his line of men. We watched this idyllic scene on what might be called the village green for an hour or so. George and I then set out to walk down the mountain in order to enjoy the country and take some photographs. It was not more than four miles and I had not had enough exercise, so in the later afternoon I took a Game Scout and climbed into three different pockets of original forest half way up the mountain. The ground was boulder strewn, easy enough where the grass was short but harder where it was long, and in those places where the fire had newly invaded a bit of old forest there was that dense herbaceous growth to six or seven feet high that is difficult to get through. Once in the forest the ground itself is relatively clear and it is possible to walk easily. It suddenly struck me that in xeric, fire-induced plant associations so many plants have spines, hooked thorns, burrs and so on, whereas in the deep forest there are few thorny plants. I saw a troop of blue monkeys in the forest, and in a damp grove of long grass on the way home we flushed a roan antelope. Otherwise no game.

11 February 1961

We should have left at 4.30 this morning but the wind rose in the night and there was a heavy shower of rain. So we stayed where we were till 6 o'clock and then had our tea and biscuits at ease beneath the large fig tree by our *chitenji*. Away by seven and the ground was not really bad. The roads, such as they are in this part of the world – indeed a large part of the Sudan – are just the hardened black cotton soil which occurs on the lower ground in many parts of Africa. It is an excellent surface when it is dry and flattened, but if there is a shower of rain the soil becomes colloidal. The tyres pick up the top couple of inches of soil and throw it around, and the vehicle itself slides this way and that. The road is ruined as a motoring surface and thereafter every bumping vehicle makes it worse. All the roads in the Southern Sudan close by law in the rainy season and any gentleman who thinks he will have a bash at it anyway finds himself arrested and in jug for ruining the surface of the roads. I do so admire the Sudanese capacity for action at the top to stop individuals from being anti-social. They do not profess to be democratic, thank God. Equally, Mahmoud has very great power in making and unmaking game reserves. If he thinks the animals of a certain area need special protection he establishes a Reserve for as long as he thinks fit. In our so-called democracies it is the devil and all to get a Reserve gazetted. Sudan, anyway, has got rid of tin-pot politicians reared and indoctrinated in the Christian missions. The country was fast going to a Communist minority from Equatoria. The mutiny of 1955 showed up all this and the military government which followed has cut out the foolish inefficiency of democracy in a country unready for it, and my impression is that the present style of government by a president and council of ministers is benign and just. We are struck by the honesty as compared with a place like Kenya or Northern Rhodesia. In short, I am surer of what I have felt for some years, that Islam has had a greater moral influence in Africa than has Christianity. The political activities of

Christian missions and their anti-Arab bias puts them in no position to grumble if they are kicked out.

We had waited till the sun had dried out the surface and we had no trouble. By 30 miles along there had been no rain, so all was well. We stopped at perhaps 50 miles near a dribble of a river where were some shady trees. There we found Murles encamped, and one man was running round with a spear, singing. He passed us unseeing and went behind the trees. Soon two of them were running up from the river bed, crying loudly and spears raised. Again we were passed as if we were not there, and I sensed the man was in some sort of ecstasy. Mahmoud then came to tell us the first man had speared a bull in some ceremony and that he was now highly excited. The technique is that a family must not kill one of their own bulls but someone else's. Having done that, that someone else has the right to come and choose a bull from the other man's stock in recompense. The bull was somewhere beyond the bushes but I had no curiosity to find it. Soon, I supposed, the Murles would be having one of their few feasts of meat.

They were not quite alone in this, for at perhaps 80 miles Mahmoud got down among a big herd of white-eared cob and shot two black males. Our party was now in high spirits: a murle we had taken aboard at Boma to give him a lift to Bor began skinning one cob with a longish very sharp knife at great speed. I never saw such speed and co-ordination of two hands and one knife. One of our own Game Guards was busy at the other beast and both were finished at about the same moment. Cob skins are much desired as prayer rugs and one of our drivers, Jiger, is to have one, and Hassan, the cook, is to have the other. Jiger is a black Arab, taciturn and devout. Hassan is also devout, but likes his beer and is an extremely humorous though short-tempered character. At this moment Hassan was in high spirits, for the camp had meat.

The bellies were slit open and the gralloch removed. The livers were seized and cut up into walnut-sized collops. Hassan produced lemons, salt and chillies. The raw liver was smothered in this sauce and our party of eleven dipped their hands into the dish and soon the liver was gone. George and I agreed that raw liver so eaten, warm from the animal, was very good indeed. A fire had already been made in the road; the testicles were split, roasted for two minutes and eaten. Then the ribs. Good fellowship was now abundantly manifest and we continued our way. Back at Pibor in the rest house again for the night. Roast cob for supper and then bed under the stars.

*   *   *

Upper Nile is varied in character with vast *toich* pastures and ample, though fire-ridden bush. This province holds some of the largest reserve stocks of game and is the most exploited part of the country by safari hunting. The Shilluk and the Nuer are cattle symbiotes, and pastoralism is present among some Murle, but game and cattle are in no conflict.

CHAPTER ELEVEN

Equatoria

February 1961

Retracing our road to Pibor we moved to Bor across great plains of *Hyparrhenia*. Here we saw herds of Mongalla (red-fronted) gazelles with numerous reedbuck, a new and very pleasing grouping. Around Mongalla we were in low bush with savannah of tall trees, *Acacia*, *Kigelia*, and *Lannea* which held numerous oribi, roan antelope, buffalo, and elephant. We saw numerous elephant gathering in a swamp in the evening light, and detoured to Juba through a reserve of good bush and a great damp meadow of grass holding a few waterbuck, and the spoor of buffalo and elephant.

From Juba to Nimule near the Uganda border we were in the province of Equatoria, quite different from the colluvial Upper Nile which we had left. Here we were on rock of the Basement Complex, schists and quartzites with a relatively thin soil cover. The Lotuka were practising shifting cultivation of almost *chitemene* type, pollarding the bush trees and burning the lop and top to provide a stimulating dressing of potash. The country has been burned to a sad state. The soil was red. Short-staple cotton was being grown in small plots as a cash crop and there was too much evidence of cassava as the food staple. This crop will assuredly bring the people to a low standard of living as the potato did when it became the staple of the Chilean Indians, the Irish, and the West Highlanders of Scotland.

In the Nimule National Park (eighty square miles) it was exhilarating to see eleven square-lipped rhinoceros at one time and relatively tame. Elephants are both tame and numerous along the Nile with herds of buffalo, many water-buck, and a growing herd of Uganda cob. The Park showed signs of over-use: the grass cover was thin in places and the *Combretum* bush fragmented.

The way to Torit is through poor over-burnt country, much coppiced. Beyond Torit, the Imatong Mountains are of gneiss and still retain some of their rich montane forest of mahogany and olive — *Khaia*, *Mysopsis*, and *Cedrella*, but the fires have made dreadful inroads into the natural wealth of the Sudan. The work of the Forest Service is inspiring. We walked into the largest patch of the original woodland, the Letanga Forest (4,000 ha.), and saw colobus monkeys in the high trees. Higher still, at Gilo, in a perfection of atmosphere and scenery, there are extensive plantings of *Cupressus* on forest lands ravaged by earlier fires. On return to Torit we entered a flat colluvial plain of long grass and degraded ant-hill type bush used by roan antelope and elephant.

12 February 1961

Did not sleep too well, as a few mosquitoes buzzed in my ear and woke me. Then I noticed a clouding of the sky and a great deal of lightning and some thunder. We were to leave at 4.30 and I got up early, telling Mahmoud and George we would be lucky if we got our beds packed while it was dry. Sure enough the rain came but we had beaten it. It was not much rain, but too much possibly. We left at 5 o'clock and crawled away at five miles an hour, going crabwise most of the time. We reached a police and veterinary post at 18 miles, by which time it was truly raining and we must give it a chance. A veterinary friend of Mahmoud's offered us tea. An hour later, the rain over, we set forth again and had no trouble. Now we passed over a great grass plain, passing several good-natured groups of Murles walking along with their minimum belongings carried on their heads, their bodies unfettered by clothes. Soon we came into a country of Mongalla gazelles and reedbuck, thousands of the former and hundreds of the latter. This is a new grouping to me, and I have never seen so many reedbuck together as among these large flocks of gazelles. And so to Bor, outside of which town we saw a group of 27 giraffe. They are always lovely in numbers and George was particularly thrilled. Coming into Bor we had seen the last of the Murles, for which tribe I shall always have an affection, and were among Dinkas, much bigger people and more negroid in character. The missions had obviously been busy in Bor, because many people wore rags and cast-offs. Two women tried to beg from me unsuccessfully. This is a rather degraded Dinka area which should not be taken as representative. The huts are thatched down to the ground with an added thatched porch. The huts are dispersed like West Highland crofts over the settlement and I was sorry to see the countryside was taking a beating. I cannot admire the land-use pattern here; too large an area is devastated before passing on to a new one. Over time the whole bush picture looks sick. We were entertained to luncheon Sudanese style by the District Commissioner, a friend of Mahmoud's, and smart in his starched khaki drill bush jacket and shorts and plumed topee. The topee is a sign of office in this country. The D.C.'s house was on the banks of the Nile, surrounded by trees and very pleasant. Wire gauze pretty well covered in the house, for mosquitoes are bad. We left at 3 p.m. for Mongalla, over 80 miles away. The track was awful and steadily got worse. The journey took five hours and for the last two I was so tired I was scarcely conscious of much, though the continual bumping prevented one from being able to sleep. It was dark, anyway. We intended to have the rest-house at Mongalla, but Prince Bernhard of the Netherlands had got there before us and was occupying it all. So we made camp alongside. We were to stay two nights and Mahmoud wished to show he was conducting VIPs, so a grand tent was erected for George and me, a splendid marquee lined with red material, the sort of thing you erect at the agricultural show when the Queen is attending. Bushir was here conducting Prince Bernhard, also the governor of Khartoum, who made himself very agreeable and brought over little bottles of whisky. I was so tired I took some as a restorative and felt the better of it, though it made me pull a face. Then the cook and waiters from the Prince's outfit brought our dinner, at the orders of the Governor of Khartoum. Finally, Mahmoud, in virtue of his position, went to pay his respects to the Prince and his party, and came back saying the Prince had invited us round for the following evening. He had also hinted that he would like Mahmoud to accompany them and we could have Bushir. I had already divined this might happen but fortunately Mahmoud was adamant. I breathed again and slept comfortably.

13 February 1961

Away at sunrise to a Reserve 20 miles away where there were buffalo and elephant and too many Dinka. We saw plenty of elephant signs but no elephant. This was low-lying bush with savannah of tall trees, *Acacia, Kigelia* and *Lannea*. We searched long and used our glasses well, but no game except for a roan antelope I flushed. This landscape should have had a lot of game in evidence. Coming to a favourite watering place there were plenty of birds and one old bull buffalo, which greatly pleased George, for we stalked near enough for a good photo. Back again to lunch and away at 3 p.m. to another Reserve which involved 60 miles of devious motoring in the lorry. The bush was fairly good before coming through over-fired country on to a grass plain. No game except oribi and a few giraffe. Then as we approached a marsh at the turning point of our trip we saw elephants coming to water, eight of them in all. We got up to the other side of the marsh and could go no further without getting wet to the middle. So we watched the great and lovely beasts eating the marsh grasses until darkness fell. Back along awful tracks reaching camp at 8 p.m. Dinner immediately and over to talk with the Prince, his doctor, his aide-de-camp and someone else, perhaps another aide. He made himself most agreeable, producing brandy and cigars, and talked knowledgeably and with good questions. An entirely happy and informal evening. Added as a last word to me that if ever he could be any use, to write and let him know. We returned to our canvas palace for the night and slept extremely well.

14 February 1961

Away finally by 8.30 a.m. bound for Juba by way of a Reserve in which we saw absolutely nothing, though during a short walk I took in the bush I found plenty of sign of elephant and buffalo. Am wrong, because in one dampish great meadow of grass we did see a few waterbuck. Coming to a ford we found a decrepit merchant's lorry sitting plumb in the middle of it with a broken wheel. This necessitated some rearrangement of rocks and a perilous crossing for our three lorries. Our chaps cheered and yelled as each got clear on the other side. Coming into Juba after 35 miles we had

Passengers of the ferry boat at Juba on 15 February 1961. 'All was activity . . . every man wore shorts and there wasn't a bare bosom on the boat.'

to cross the Nile by ferry boat. All was activity and so active have been the Catholics and the Presbyterians and the God knows whats, that every man wore shorts and there wasn't a bare bosom on the boat. Yet so thin is the veneer that when the women rearranged their saris before putting the loads on their heads, they still did so without shame and it was apparent that beneath the cotton saris they still kept to their beautifully decorated little leather aprons fore and aft. Looking into the baskets from the bridge of the boat I saw little bundles of tamarind pods, husked, and quantities of durra, which is like giant millet. Juba is now quite a modern little town, the administrative centre of Equatoria, and with a very nice hotel in which we were to be housed for three nights while Mahmoud gets his office work done and I my writing. Received mail here and had quite a lot to do. We have also bathed well and got ourselves cleaned up. Out at night to a dinner party, buffet style, at the house of the Deputy Governor of the Sudan. Fully a score of people, including Sudanese, an Italian eye surgeon from Asmara, and Baron Krupp von Bohlen, the present head of the German arms firm and steel empire and his friend, who are on safari. Krupp is a pleasant quiet fellow, absolutely devoid of ostentation or Germanness. I have followed this man's career with some interest. He could not be convicted of any war crimes for merely producing and selling arms; he has kept well out of politics, and from having lost everything he has now got pretty well everything back and is very rich and very clever. I had a long talk with him and liked him and we met again on:

15 February

at Mahmoud's house where we had another party.

16 February 1961

Writing all day and got a hair-cut in the town in an open booth. Four weeks out now and tomorrow four weeks to go. Am lasting pretty well and have lost quite a bit of weight. All to the good.

17 February 1961

Supposed to leave Juba by 7 a.m. but in fact it was later. I went with Mahmoud to his house to see his little pet square-lipped rhinoceros. Its mother wandered into a cattle camp somewhere soon after it was born. The dogs created a commotion and drove the mother rhinoceros away. The newly-born baby lay down with the cattle and was thus orphaned in effect. The men let Mahmoud know and he fetched it. The little rhino was then pink-fleshed and not very rhinoesque. One has never been reared artificially before but Mahmoud persevered with cow's milk, sugar and maize meal. The little rhino is now seven weeks old and drinks a gallon of milk a day from the bottle. The poor little thing seems happy enough in a scrupulously clean pen complete with mud bath. It skips about very playfully and is of course completely tame. It has a fringe of absolutely black hair along the edges of the ears and along the eyelids. What a pet it would be in a children's zoo! I suppose it is now past the danger point, though it has not yet begun grazing or eating grass. Just as we were leaving again, an African girl brought in a baby white colobus monkey which had been found with its fingers damaged. One or two fingers had gone and one or two more would have to be amputated. Its face was distorted with pain and the creature was crying piteously. Mahmoud sent it over to the veterinary department for treatment but he is not optimistic about rearing it, for this species is delicate.

First we had to cross the Nile ferry. We went on the bridge again and were introduced to the merry-eyed old Arab skipper sitting cross-legged on a board across the rails on the port side of the bridge. This seems to be the traditional perch of Nile skippers.

On then for 110 miles to Nimule, which is a small National Park on the banks of the Nile on the Uganda border. We passed through a lot of undulating poorish to middling bush, all grossly overburnt and much of it pollarded as it would be in Northern Rhodesia for *chitemene* gardening, i.e., shifting cultivation. Indeed we are now on red soil and gravel roads. The people in the villages along this road seem a poorish tribe in ragged European clothes. They are Roman Catholics and are now growing American short-staple cotton in small plots of a quarter acre or so. One village was having a weekly market of this stuff.

We are staying in a new rest house at Nimule, one rushed up to take the President and Abd el Nasser when the latter came to see the little National Park of 80 square miles. The Nile is a swift running river of perhaps 5 knots and the depth here is 66 feet. The scenery is quite lovely with hills in the background, green bush and wild green meadows of *Vossia* and *Papyrus*. We crossed by dinghy and the jeep on a pontoon. The range of Nimule struck me immediately as being hit pretty hard. Even the grass is getting thin and the *Combretum* is getting fragmentary. Eighty square miles is nothing for a National Park and this one is carrying a good stock of elephants, a spectacular number of square-lipped or white rhinoceros and several good herds of buffalo. There is a growing population of Uganda cob. The cob bucks seem to be in breeding condition and have definitely swollen necks, rather like the entasis of a Greek pillar. We soon found four rhinoceros and walked to about 120 yards, possibly less. But they weren't in a good position. A little later, however, we saw a group of five or six and in a position to be stalked close. Imagine the thrill when we found ten together, and another was in sight. We were in a tumble of rocks and the animals actually grazed towards us. I took photos of several with just my normal 5 cm. lens. We had little luck with elephant, seeing only three on the other side of a swamp.

18 February 1961

Off to Uganda 5.45 this morning with Mahmoud, who was going the 25 miles for maize meal for his white rhinoceros calf. The two customs blocks let us through without trouble and we went to an Indian duker for the meal. The Indian entertained us to sweet spiced tea and chapatties and I got talking to him about conditions. He said he would stay as long as ever he was able, but if he had to get out he would then go to England, not India. Mahmoud and I called at the Traveller's Tree, near Nimule, on the way back. It is an old tamarind and there is now a marker – an Italian explorer 1857, Samuel Baker 1865, Emain Pasha 1875, Winston Churchill 1907. Back to breakfast 8.15 and George had had a lovely time in the park watching elephants at close quarters. Away again 9.30 to Torit, 90 odd miles, passing through a lot of burnt-to-death bush and coppiced stuff. Such people as we saw we did not think much of – scruffy villages, scrappy agriculture, and no cattle – typical missionaries' Africans. Most of these were wearing a cross round their necks. It suddenly struck me that it is the lowliest and least good tribes that are set upon by the missionaries: the warrior tribes and more powerful and proud sorts are left pagan. Having reached Torit, a typical administrative boma of the British sort, taken over complete by the Sudanese. I was introduced to the District Commissioner, a polished and charming Arab, dark, and dressed in exactly the same clothes the British D.C. would have been – starched khaki drill bush jacket and shorts and cotton stockings and well-polished brown shoes. President Abood's photo was hung in his office instead of the Queen's. That is the only difference. The Sudanese have taken over the British way almost in entirety. The same in the Forest Veterinary Services. Mahmoud was staying overnight in Torit to sort and despatch ivory. George and I came on with two lorries a further 36 miles to Katire in the Immatong Mountains. This is a Forestry Station, and Shawki in Khartoum had been most anxious that we should see this forest. These hills are spectacular, lots of bare rocky peaks in the style the old gneiss makes whether here or in Scotland. There is also a very coarse diorite which has not been pressed hard enough to be granite. It is so coarse as to be rather crumbly. The highest mountain of the Immatongs, Mount Kenyeti, is nearly 10,000 feet. The rainfall is 220 cm. or rather less, about 80 inches. Bamboo forests high up are nothing like so extensive as on the high ground of Kenya; possibly the rainfall is not high enough. I was disappointed coming into these mountains to see such dreadful evidence of fire. The slopes have been bared and the ridges are crowned with a silly singed fringe of *Acacia abyssinica*. Katire is at about 5,000 feet and very pleasant. One of my old students called Khamal met me. He is now assistant conservator of this Reserve. We also met the Oxford-trained mill manager and a highly intelligent forester called Jafar. We were to be settled for the night in the wooden rest house, a nice rambling panelled place. We then went to see the sawmill, where a good deal of African mahogany is sawn. Water powers the turbine and creates electricity for all the machinery and lighting. Most of the labour for all this forest comes from the neighbouring Lotuka tribe, among whom the mutiny broke out 6 years ago, when 485 northern Sudanese were massacred – men, women and children.

19 February 1961

Away by 6.30 to walk into Letanga Forest where there is almost pristine montane high forest. Quite lovely – mahogany, *Mysopsis*, *Cedrella*, and other fine species of 120-foot trees. Some particular trees are being used as seed trees and the bush is cleared beneath them for ease in gathering the seeds. You could see everything here, from burnt-out *Terminalia* and *Erythrina* bush, through broken canopy and heavy herbaceous layer (when fire has been through once) to unbroken forest with a floor clear of herbs or grass. There are 10,000 acres of this forest here and several more good patches higher up. We saw the black and white colobus monkeys dashing about in the high trees, making some spectacular leaps. We also looked at nurseries of teak and some of the forest species. The teak does well, better than in India, and comes into production at 60 years. We saw several 20-year old plantations of dead straight poles going to 40 feet. Back to breakfast at the mill manager's. Most hospitable, but rather trying in a way because Ramadan started two days ago and a Muslim must not break his fast till 6 p.m. each day for the coming month. Although the Prophet expressly stated that those travelling need not be strict in fasting, all our drivers insist on doing so, and they are already half dead by lunch time. Just hope they don't fall asleep from exhaustion while driving. After breakfast we climbed up the hill to Gilo, another rest house and sub-station at 7,000 feet. Quite lovely in every way. Where fire has burned the ground, cypresses and various exotic conifers have been planted. Even the African *Juniperus procera* has been planted, though it takes 120 years as against the 40 for *Cupressus*. We stopped at an exceptionally carefully tended nursery for all these exotic conifers. Three years and the nursery must move on to new ground. *Pinus radiata* seems a favourite. We climbed to a 7,500-foot hill after lunch to look over the hill country, observe the fire pattern and the planting pattern and the sanctuary areas of indigenous forest. Elephants come into this high ground from Uganda, and there are a few buffalo. Bushbuck are pretty well the only other game, though I would expect to find klipspringer on some of the rocky peaks.

20 February 1961

Rather a frustrating day. If I am in charge of any operation to do with time, it runs to time. I have almost a neurosis about time and punctuality. So we left Gilo on time – 6.30 a.m., though Mahmoud would have expected us to leave Katire at 6 o'clock. But we could not leave Gilo till full light because of the precipitous road. Down at Katire by 7.15 to say goodbye to our hosts but they were not around. Had to wait for them and it was 7.45 before we left for Torit. We reached this boma by 9.15 and found Mahmoud waiting. We were now supposed to go to Kapoeti and Lelewi near the Kenya border, but Mahmoud has had information that the country is now empty of game and that it has moved into Upper Nile for water. So we said it is no good bumping 300 miles there and 300 back to see nothing and that anyway we couldn't see the good of making such an itinerary that all our time is taken in getting there and no time left for getting into the bush. In any case we are not really getting into the bush and the outfit is not geared to this. Mahmoud is upset because wherever we go the game isn't there in any quantity and I am always unconvinced by the signs that there is much game anyway. Mahmoud says, then let us give up the prepared itinerary and get into some uncomfortable places where he is sure there is game. I say, fine. He is anxious this morning to take us a few miles into the bush to where there are many buffalo and roan antelope. So away we go bushwhacking in his jeep. The jeep is a crime against nature – human nature – in so far as it rests between coccyx and occiput. We went into bush of long grass, *Combretum* and *Terminalia* and degraded anthill community. Elephants use it in the wet season. I was sitting in front of the jeep, George in the back. We saw nothing but oribi, which I associate with poor conditions. Whenever we came near a waterhole or good conditions there was a herd of cattle. Half way I persuaded George to have the front seat for the return journey. Being the right sort of chap he said, oh no, he was alright, but he didn't look it, so I put on a little more pressure and we changed seats. The next three-quarters of an hour was hell for me. We had returned to the road by the shortest route being as we were seeing nothing, but I was so badly knocked about and shaken as to be scarcely able to stand up – nausea and general disorientation, so for the six miles back to Torit along the road George went in the back and I gradually recovered. Thermos tea at Torit and then back into a lorry alongside Jiger for a normal 90-mile bump back to Juba, where we are to stay in the hotel for the night. The road back to Juba is through hills and kopjes set in the plain. One sheer face of several hundred feet had some ledges of marabou storks. The country is plentifully sprinkled with land-spoiling communities of the Lotuka tribe. Candidly, I don't take to these people. They are too much like the Bantu farther south. These are the people who helped the mutineers in 1955. There is a memorial in Torit to the 76 northern Sudanese who were massacred in the town – men, women and children. The full total in Equatoria was 480. People in Britain, the

United States and the United Nations go on blethering about democracy, one man one vote and all that. The Sudan soon gave up that notion, and the Congo situation – so amusing to a cynic like myself – may soon persuade the pseudo-Socialists (Conservatives) in Britain and the fundamentalists (anti-colonials) in USA that Africa is Africa and must come along at its own pace. I saw several stretches of sheet erosion along this road. Having arrived at Juba and had a wash and change in time for tea on the terrace of the hotel, we found the same old gang had got there also – Prince Bernhard, Krupp von Bohlen, and their retinues. Rather amusing that these two are cousins. George and I were now greeted almost as old acquaintances. Annoys me like hell that try as we did to get the Sudanese to take us into the remote places where there must be a good game population, here we are wandering around in some kind of a social circus. Anyway, this morning has settled Mahmoud and we should get into the back blocks. This is the kind of thing that happens when Khartoum (= London or Washington) arranges things instead of the man on the spot.

* * *

Equatoria is least characteristic of the Sudan and more like many poor lateritic areas farther south in Africa. It is not a protein producing area in any sense demanding clearing and planting of the ground or establishing herds of cattle. Cattle are present but sparingly and some areas will never carry cattle. Nor will it be a bread basket. The acid rocks indicate to the seeing eye that the natural vegetation should be conserved and protected from too frequent fires, and that Equatoria should be the Sudan's forest. A broadminded economic survey should assess the forest and game potential *vis-à-vis* the small areas of short-staple cotton and the long term development of the province. The wildlife spectacle of Equatoria is impressive and the all-season gravel roads giving access to Uganda, together with Nile tourist traffic, give a naturally less well endowed province a fair prospect of an ultimate prosperity. The steamer journey from Juba to Khartoum is one which every African traveller hopes to take.

CHAPTER TWELVE

Bahr-el-Ghazal and Dar-Fur

February and March 1961

THE BAHR-EL-GHAZAL is entirely different from Equatoria. We were now in a country of rivers, one of the potentially richest reserve areas of the world – colluvial plains and water in vast expanse, a further development, as it were, of what we had grown to know in the province of Upper Nile. We entered the Bahr-el-Ghazal north of Terakeka, struck westward, and spent several days along the Aliab River and its lagoons, going fishing in dug-out canoes with the fishing Dinkas, and striking back into the bush among the pastoral groups who also treated us with a fine welcoming courtesy.

Game was prolific and the country so rich in water and grass that the Dinka are not jealous of the game. Through Yirol we camped beside Lake Nyubor and saw that distinctive bird of the Bahr-el-Ghazal, the shoebill stork. Then through Rumbek and Tonj in excellent forest bush on sandy soil. In the Southern National Park (7,700 square miles of tsetse-fly bush of excellent sort) there are no roads or rest-houses, just the river, a few water holes, the beautiful forest bush and the wide array of hoofed game including Lord Derby's (giant) eland (not seen).

The route was to Wau, a pastoral town by the river, and on through Aweil in good forest land to Nyamlell, where the rest-house is on a bluff overlooking a U-bend in the Bahr Lol. Fishing with cast nets went on all night and sounds of singing in unison came from camp fires until past midnight. Northward from Nyamlell we came to the river, Bahr-el-Arab, the boundary between the Bahr-el-Ghazal and Dar-Fur, which was ethnic as well (as geographic) in that we were leaving the Nilotic lands and entering the nomadic Arab country. The Rizeigat were camped on the Arab side of the Bahr-el-Arab, and their Chief entertained us with traditional hospitality and kindness. In Sofara we saw smiths at work on small artefacts, knives and tweezers, the men sitting under the lightest of grass shelters by the kind of small anvil suitable for the nomadic life.

The bush was now light with short grass curing on the stalk, and quite different from that of the Bahr-el-Ghazal. Beyond the day's cattle trek from the river there was much more grass and we thought it excellent country – of the nomadic Bagarra as far as Nyala, where we met many families on the move, always cheerful in their greeting; the man might be riding a camel, the woman on a bull, and two children perhaps on an ass, and their cattle would be moving gently in the bush.

On through Sibdu and Ed Da'ein to Nyala through many miles of *quos*, stablilized dunes carrying light bush and resting on clay. Motor traffic crossing the *quos* is a threat to this valuable grazing, by choosing (and disrupting) a way through where the binding vegetation is not broken. We were now seeing many baobab trees, *Adansonia.* Then northwestwards past Kass into the green, climatic penumbra of the Jebel Mara (3,071 m). Zalingei at an altitude of around 1,000 metres must be one of the favoured situations on the planet, with its great groves of *Acacia albida* trees, which lightly shade the whole town, its less than semi-arid climate, and its happy remoteness. The bush to the south-west of Zalingei still holds good stocks of game, including elephants, the loss of which from this central point of Africa would be tragic. The Jebel itself as a shapely mountain gives the impression of treelessness, but the foothills are well wooded in places with *Cordia* and *Khaia*, and terraced, an amazing relic of human industry in some long past pluvial epoch when cultivation was possible. The descendants of the ancient Fur are presumably the rather short, slightly negroid Fur of today who live in the mountain and are expert gardeners, their gardens are green, well tended, and a delight to the eye. We struck into the mountain from Murgatello, Baldong and Galoll. There is very little game but greater kudu are fairly common.

To return to Nyala and from thence to fly to El Fasher and Khartoum was almost an anticlimax, but even the air journey held lessons for us in observing patterns of landuse and the distribution of cultivation. We had covered nearly 5,000 miles, and yet such a small area of this great country.

21 February 1961

Enjoyed the clean-up: away by 11 o'clock ahead of Mahmoud to Terrekake, 57 miles, on the west bank of the Nile, through *Balanites* savannah and *Acacia* scrub. Waiting here for Mahmoud, in the shade of a huge fig tree, enjoying the wide view of the Nile and the sudd. Beautiful green-headed, scarlet-breasted sunbirds in the frangipani trees now in blossom. One must enjoy the moments and endure much else. Mahmoud reached us by 4 o'clock, after I had climbed into the fig tree and had a sleep stretched on one of the great boughs which are almost horizontal. We drove westwards through a country of much better bush than heretofore till we reached a considerable lake 21 miles from Terrekake. The lake was beautiful, palm-fringed and fairly densely bushed. There was even a small sandy beach where we were able to have a bath, keeping an eye open for crocodiles. The moon is on the rise now, so this was a particularly lovely camp.

22 February 1961

Away by 5 o'clock this morning driving through this good bush. Not that we saw much game except a few Tiang, which are here very dark. But there was plenty of sign of elephants. We came into the open river, sudd and *toich* country from time to time where the Dinkas have their cattle camps, and along the road we would meet these ash-coloured people, like grey naked ghosts in the half light. We stopped at one camp to see the large herd of almost white cattle with enormous horns. The creatures have quiet eyes and beautiful faces and are evidently at one with the Dinkas. One beast was basically black but with patching and marbling of white over the head and body. My old friend, Joe Broadhurst, was one of the earliest white men among the Dinkas, back in 1925-30; he gave me a photograph of one of these cattle and told me its special name. It has taken me a good while to get here, but the Dinkas are not much changed. They have a self sufficiency with their cattle and few outside desires, but unlike the Masai they live in the great river-strewn area of Bahr-el-Ghazal, where the seasonally inundated grassland seems limitless and is self-sustaining, not susceptible and sensitive to erosion like Masailand. The Dinka cattle are so much better than those of the Masai. The Dinka eat some durra (millet) and mainly subsist on milk and some butter. Whereas the Masai bleed their cattle to mix blood with the milk, the Dinka mix cattle urine with the milk and are a little contemptuous of our not doing so. We are losing much of the goodness, they say. We came on to the edge of bush and *toich*, thinking to camp. Two Dinka women soon came to inspect and talk. The young one was

Dinka herdsmen in the early morning of 22 February dusted with ash against mosquitoes, west of Terrekake, Equatoria.

astonished to see George and me, for she had seen no white people before. Mahmoud can understand some of their language fortunately. I gave them some snuff, which they put in their lower lip, and the older woman remarked that we were likely to have many visitors. The Dinka are essentially sociable people and the women do not seem to be depressed, for they are as much to the fore in the incessant conversation as are the men. No Masai woman would speak or smile to a white man, but here among the Dinkas the women are completely easy and poised and with charming manners in their open and child-like way. But all Dinkas spit a great deal and I wish they didn't. Many Dinka huts are set up four feet above the ground, with a platform outside the door. The place beneath is one of shade in the daytime and a pleasant spot for sleep and social intercourse. If we stop to talk we must shake hands with everyone, the Dinkas coming forward with high, formal smile, the hand raised almost to the shoulder and the fingers given in a soft, formal touch, rather than the heavy bear hug of the United States. The visitor is then asked to join the family in the shade under the hut. Mahmoud decided this place was not propitious and we continued another mile to a fishing camp for which he was in part responsible. There must be two acres or more of cleared ground at the edge of a lagoon of a larger river. Fishing with a mile long seine net is one of the things the Dinkas have been taught, though the poorer members of the tribe have long been fishermen. Lots of split and salted fish hang from strings and poles in one part of the cleared ground. On top of each pole a kite perches, but the salted fish is immune from its attentions. Nearer the water marabou storks walk

Dinka fishermen by Lake Aliab near Yirol on 22 February 1961. The Dinka have been taught to fish with a mile-long seine net.

A magnificent Dinka fisherman, Bahr-el-Ghazal, 22 February 1961.

around solemnly, scavenging and keeping the place clean. The camp is a busy place, full of talk and laughter. Whatever interest we may have in the Dinka is reciprocated by theirs in us. Our gear is commented upon, and if I wash from a basin it is distinctly funny. When I shaved I had a ring of women watching intently and remarking on details of the operation. We have been given a little *chitenji* in the middle of the camp, and as I sit here writing, the women and children come peeping in to see how it is done, always with a smile and evident good fellowship.

We went off in the jeep during the afternoon to stab into the bush in search of elephants and other game. The bush was fairly dense, on black cotton soil trodden by elephants in the wet season and now set rock hard. Sometimes we went almost as fast as walking, to the accompaniment of a dreadful row of metallic bumps and thorns screeching against metal. Most people were scratched, and my left ear projecting so far from my head, got nicely slit. Unlike the elephant I have not developed the musculature for neatly folding my ears back to the head when going through dense bush. At one point I could suffer the jeep no longer and like the man who rode an ass, got off sometimes for a rest. A Game Guard and I found ourselves a mile ahead of the jeep and then there was a whistling and a rifle shot to recall us. Jeeping in this country really iritates me. We saw one elephant, a few tiang, one roan antelope, some giraffe and two oribi, and endured a row and a bumping enough to ruin any enjoyment of the bush. I would sooner stay at home and read about animals.

23 February 1961

Away at 6 o'clock to find where elephants crossed the track going to water and had returned again into the bush. We found the place near where we went in yesterday and the animals had crossed the track probably an hour beforehand. I had thought we were late in starting: instead of a cup of tea from the mudguard and away without food, we had tablecloth on the table and George wanted coddled eggs. So now we set forth into the bush in the jeep, the idea being that we should catch up with the elephants. We must have been doing two miles an hour, with my temper getting shorter, when the driver got a stab in the eye from a thorn. I got out and said men should have legs as well as testicles. I knew this would get Mahmoud on the raw, for he is a great walker and hard goer. But I felt he was doing this to save our legs and I didn't want them saving. He got out in a huff and we began walking. It was absolutely lovely in this early light and coolness, hearing the birds and smelling the sweet air. We walked fast and for a time much faster than the jeep could have gone. Then, I suppose, the ground was easy enough that the jeep could have gone faster than we did. The elephants had at first been feeding and amusing themselves as they progressed, but quite suddenly it was as if the head cow had said, 'Come on; it's time we got moving.' And they proceeded in single file along a good path of their own making. It was an easy path for us and we walked hard, the sun rising and our shadows shortening, but I knew the elephants would be outpacing us by one or one and a half miles an hour. We walked seven miles. Mahmoud said he thought the elephants would not stop before noon and that we should not catch up with them till 1 o'clock, which would leave us only five hours before dark. We had started walking at 7 a.m. and it was now half-past nine. Our lateness in leaving camp and then my making us walk had put us into an impossible schedule and I knew the right thing was to walk back. The Game Guards had carried water from the jeep; we stopped each hour for water because it was now very hot, but of course I declined it all the time because I had caused it to be carried. Anyway, I don't go much on suckling a water bottle every two or three miles. Go thirsty and drink when you get back. George says he doesn't go much on walking for walking's sake, and admittedly we did not see much game. Mahmoud has said he wants to lose 10 kilos of weight on this safari. I reckon I have taken off one of them today. But he is an excellent fellow and his only grumble is that I did not stay in the jeep for another hour, he being convinced we could have beaten walking pace. But he admitted to me that he was trying to save our legs, and I still hold we are not here to save our legs but to get moving early enough. Eating is largely a habit and for nutrition is not necessary more than once a day in the tropics. When I first came to Africa I did not like starting the day on nothing and doing the day's march before eating, but I soon got into the way of it and now prefer it. Tea as often as you like, but a damn good feed in the evening is splendid: one gets rid of all belly fat and the body tones up into a tireless co-ordinating whole.

We got back to a Dinka settlement near the track and from where we had taken a man for a guide yesterday. The group of a dozen or so men, women and children were delighted to see us and asked us into the shade beneath their hut. So we were well packed in. We were offered water, but we had our own, and in no time the runner had contacted the jeep and the Thermos flasks of tea now came along. I sat down now to drink tea and went on drinking. The Dinkas watched with profound interest and one woman asked if any other white man could drink as much tea as I could. Mahmoud told her I was the greatest tea drinker of all men. When I had passed round the snuff in the old leather box Major Seymour gave me, it was suggested my snuff must have come from the Nuer, as it was different from their own. Must tell that to Freiburg and Treyer of the Haymarket. By this time the old Headman had relapsed once more into sleep and did not wake when we left after a very successful party. The afternoon was very hot, but as a fishing crew was about to go three and a half miles to fish a lake with the seine, George and I joined them in the big plank boat. The seine was neatly laid in the stern and four naked Dinkas stood on the gunwales with long bamboos, poling us through the lagoon. As many more were in the boat. One of the men poling was one of the most magnificently built men I have ever seen, like the Zeus that was brought up by the fishermen's nets in the Aegean in 1925. He revelled in this beautiful work and is known as one of the most goodnatured of men, always willing to help anybody. Nobody can knock him down, the foreman told me, but he can knock everybody down and take on two or three at a time. The *toich* fires were burning with dramatic great clouds of smoke rising to the empyrean, and as we passed for a mile through a channel little wider than the boat, a fire more smoke than flame came to the very edge of the channel. The laughter and talk were continuous and our white skins were apparently the funniest thing of all. Now we cleared into an extensive lake where the imported water hyacinth is growing apace and threatening the fishery and the stability of the water system. It is so beautiful, but how it got from South America into the head waters of Nile and Congo no one seems to know. Three fishermen jumped overboard into the waist-deep water to take the end of the seine and the strong boy jumped over to haul the boat for its circular mile run back to the beginning of the seine. I threw off my own clothes and had a most glorious half hour helping to haul the boat as the seine was paid out from the stern. Two more Dinkas were soon overboard because as the net was paid out the weight was greater. How lovely that water was on this hot afternoon, and I just hoped I would not get a bad burn from the sun. When we were within a quarter of a mile of closing the circle, three men began to thrash the water with the bamboo poles, to drive the fish into the net. Jubilation and water play as we joined up, and then the hauling aboard of the net and the score of large fish we had caught – Nile perch and of about ten pounds each. Not a good haul but no one seemed the less happy. Back again into the home lagoon and sharp lookout kept for a hippo bull which is tending to dispute possession of the lagoon.

24 February 1961

Mahmoud went to hunt meat this morning and George has gone along with him. As Mahmoud would be cruising the bush in the jeep I thought there was no point in going along, but would take the chance of doing some writing. So here I am sitting in the *chitenji*, the women looking in and the boys asking to use my binoculars, which are a never-failing fascination. Blue of water, green of *toich*, the vast dome of sky, and the happiness and leisure of these childlike people. I am enjoying my rest. Forgot to mention that at the Dinka party yesterday we saw some little snakes coiled in the roof above us, which was the hut floor. They are poisonous snakes but the Dinkas say they have a regard for them and as the snakes do them no harm they see no reason for disturbing the snakes. They are a curiously tolerant people, not utterly ridden with the deadening beliefs of farther south in Africa. The Dinka also respect the hippopotamus, saying if he is killed the lakes and lagoons will dry up and the fishing will go. This is not superstition but good ecology, recognizing the hippo as an amphibious bulldozer and eater of surface vegetation along the waterways. They have a superstition that they are descended from the giraffe and these animals often walk around the settlements. If a giraffe calf is born among the cattle it is rubbed with cattle fat and the mother seems not to object. Late in the afternoon I went to a cattle camp and watched 500 of the the great-horned cattle come to the clearing. Each went to its place and was tethered by a thong to an inconsiderable peg. The animals purposely kicked up dust to create a fine cloud which keeps off the flies. Dinkas hospitable as ever. Mahmoud and George came back with two tiang this evening. They had had to work to get them. Game is not plentiful and the Dinkas are remarkably well sprinkled through the country. The Dinka cattle culture of the temporarily inundated grasslands is one of the few examples I know of cattle being kept without erosion of the land, but even this may be an over-statement because the bush around the *toich* gets rather a beating, and though the carrying capacity of the *toich* is much greater than its present use, it is probable that the surrounding bush is being over-used. Coming back from the cattle camp past many very simple fishermen's dwellings, the children ran away from me – their first horrible vision of a white man. The Dinka call us pink men, which phrase, foolishly enough, does not make us feel quite the chaps we think ourselves. The Dinkas at the cattle camp were outspoken about my appearance. They say I am like a hippopotamus – pink in places. And this pink skin couldn't be good – if you poked it hard the blood would run out; there is no strength in it. So the Dinka tell me, without the slightest hint of wishing to offend. What a pity there is confusion of tongues! I had to have an interpreter.

25 February 1961

Away by 6 o'clock this morning leaving this pleasant Lake Aliab and heading for Yirol and Lake Nyebor. The road was bad almost all the way, but we passed through some good bush on sandy soil. Here, I think, the cattle do not come because it is waterless country, and the game is out of it now for the same reason. But in the wet season the game runs through it, but it is too remote for the cattle. So the bush remains good. Plenty of *Combretum*, tamarind, *Isoberlinnia* and other trees I did not know. Yirol is an administrative boma and far from pretty. Lots of convicts, lots of halt and lame, and a strong posse of police. On through more sandbush and inundated land with a terrible raised, single-track road surface, and eventually a gradual descent to the lake. There is a collection of Dinka here, the young men sitting around and even the little girls pounding the durra and fetching water. Across the lake we can make out two great cattle camps. The game appears in the early morning before the cattle leave their tethers.

26 February 1961

Away soon after 6 o'clock in two dug-out canoes poled along easily while George and I, Mahmoud and two Game Guards sit like nabobs on bits of ambatch trunk on the floor of the canoe. All the same, after three and a half hours George remarked that it wasn't so frightfully comfortable, nor was it. We saw no game at all, because some Dinka from a different angle had set up another cattle camp at the north end of the lake. There was also a bunch of Nuers up there who had come for the fishing, using canoes and throwing spears. It was really a fine sight to see four canoes being poled along from the stern, with the spearman in the bow, casting a spear on chance. The distal end of the spear was attached to a line so that the spear could be pulled in again. Two more canoes went in among the hippopotamuses, using a very long throwing spear which was whippy, went into the water at a low angle and would explore the mud. The hippos did not seem to mind and the men did not interfere with the hippos. This kind of fishing is pure chance – as George says, just the same as the pelican which seems to take pot shots with its big beak and gets a fish sometimes. The high spot of the morning was to see a shoebill stork, a species peculiar to the Sudan. It is slate blue in colour and appears to have a distinct poll at the back of its head, like a hammer head. The bill is 7–8 inches long, convex laterally and about 5 inches wide. Spent the middle of the day writing while Mahmoud went to see the Chief 20 miles away. Our Dinka neighbours seem to do nothing but pound a little durra, talk, laugh, lie around, and talk to us. Surprising how much communication there can be without a word of language. Across the lake are the great cattle camps of many hundreds of quiet cattle. The commensalism of man, cattle, and other oddments of domesticated animals, sheep, goats and chickens, is so peaceful and accepted. The lizards and skinks even are tame, the kites and hooded vultures wander through camp, weaver birds throng round the durra pounding, and yellow and white wagtails walk around our chairs. This is a pleasant rest. Out for a couple of hours in the evening, bush-whacking in the jeep. We saw nothing but a few tiang and oribi, but it is obvious this country is well used by elephant and buffalo in the wet season.

27 February 1961

Away at 6 a.m. over some rough road to Rumbek, which is a boma. I sent a cable of birthday greetings to Richard at school. The final word 'Dad' bothered the post office clerk a good deal. On again, 156 miles for the day to Tonj, another boma and looking almost precisely the same as Rumbek, the latrines being the most prominent feature of the architecture. The door of one of these was open as we passed and I felt that as far as hygiene is concerned it would be better to do without these structures at all. We have passed through an area of really first-class bush, almost forest, on sandy soil and without water. This was preceded by miles and miles of Dinka villages clustered along the road. They were not the same class of folk we have come to know. Both George and I could smell missionaries and their terribly depressive influence. We are encamped in the rest-house and are having a bath in a canvas receptacle, and our laundry is being done.

28 February 1961

Away at 6.30 into the bush, first crossing the *toich* of Tonj, where a hundred or more cob were grazing. Forgot to say we went out last night for two hours along the edge of the bush and the *toich*. We saw no game at all because there are cattle camps all the way. Yet this is reckoned good game country. The Dinkas are on a good thing here – the bush, the river, the *toich*, everything right for a cattle people. Hope I have got good photographs of Dinka cattle. They are so good-looking and the horns are really enormous on old cows and bullocks. Today we bushwacked 40 miles and did 70 miles on the road, getting into what is called the Southern National Park. This is an area of 6,000 square miles of tetse bush of excellent sort. May it long remain so. There are no roads, no rest-houses, no nothing except the bush, a river, a few waterholes, and the game. Elephants, buffalo, Lord Derby's eland, roan antelope and most of the smaller things like hartebeest and so on.

Just to finish off the story of the Chief's policeman of two or three days ago, it wasn't his good day. When Mahmoud went to see the Chief he told him the charge the policeman had made for the hire of the canoes. The Chief was furious and had his

A Dinka cattle camp near Yirol, 28 February 1961.

secretary write out a chit in English reducing the charge to a quarter of what it had been. Mahmoud brought back to Lake Nyubor another of the Chief's wives (he has 16) and her daughter, a dashing young creature of about 17. This second wife in our little party is obviously a gentlewoman of considerable distinction. Black as ebony, tall, slim, no negroid features, and a poise, manner and graciousness which any society woman would envy. She was wearing a long double necklace of opaque small blue-green stones like beads. These are hereditary treasures of great age and most Dinka women are glad to be able to display a few, but here were hundreds. Mahmoud tells us she is a very wealthy woman in her own right.

After our bath and shave at sunset, we were doping up with the German mosquito dope we were given by Krupp's friend in Juba. It is excellent. The second Chief's wife indicated that she had not brought her mosquito net, so I held out the flexible bottle labelled 'Nurse Harvey's Baby Powder' in which Johanssen had put the stuff in Khartoum. I poured a little into the first wife's large practical hand, then into the second wife's beautiful slim hand. The daughter was obviously tingling to have some. The party was now busy daubing the dope over hands, arms and legs, and the daughter, who was dressed in a tiny bit of check cotton about the size of a towel, carefully lifted this and sat down on her bare behind. They all do this because it is a way of keeping the cotton clean, but this lass, having one hand full of dope and being young and careless anyway, still displayed a lot of behind. The Chief's second wife pulled her cotton down for her very promptly, and the girl was embarrassed at a sharp word from mother. Mahmoud said 'Tut-tut; what does it matter when the girl need not be wearing anything at all.' The mother said, 'Exactly, that is the point; if she wears nothing she wears nothing and that's all right, but if she wears a cotton she wears a cotton and must keep her behind covered while she is wearing it.' This seems to me an absolute logic of propriety which Emily Post could not improve. And then the Chief's second wife graciously gave the policeman the privilege of lending her his mosquito net for the night. Things always come in threes! Finally the two wives slept together under the mosquito net and that should show how amicable an affair polygamy can be in its right setting.

1 March 1961

Back to the bed sores. In other words, a long day (200 miles) of travel in the trucks finding the sore places again on knees, elbows and shoulders. We were away by 6.30 and stopped in Wau, which is the largest town in Bahr-el-Ghazal. We passed through a lot of good forest bush and then came to a wide flood plain of one of the rivers which ultimately augment the White Nile. It was now 9 o'clock and large numbers of people were crossing the river going to market. Many carried bottles of milk for sale. On the over-side was a big concrete slab for slaughtering cattle and there were high frames for stretching the patterned skins. Women were cleaning some green hides. The market place itself was all activity, the ground being the place where everything was laid. The many trees showed a good 18 inches of their roots, which shows the wear and tear of such busyness. We were taken to the Veterinary Department's office while Mahmoud got ahead with his considerable lot of business. We felt imprisoned but had to sit it out. Eventually we got some mail, the second since coming to the Sudan. We thought the town awful, especially the system of building the houses so that the latrines backed on to the street, for municipal efficiency in emptying the buckets. The streets were therefore noisome and unpleasant of passage. Was it the British regime, I wonder, that dreamt up this idea, or is it some Moslem notion of practicality? We got away by 2 o'clock and shot along at too fast a pace for more than 90 miles to Aweil, a town which is beginning to show more Arab influence. Then on again, on a frightful road for nearly 50 miles to Nyamlell; at first the excellent forest bush and signs of the Forest Department's activity, but between Aweil and Nyamlell the bush began to thin out. Nyamlell is on the banks of the Bahr Lol. The Lol flows into the sudd of the White Nile (Bahr el Jebel). The rest house is set on a bluff above a sharp bend in the river and is quite the best of its kind we have seen. There is a terrace right to the edge of the bluff. We reached here at 6.30 o'clock, just with enough light to see the tremendous primitive fishing activity in the river below. Men in dug-out canoes were flinging cast-nets over the water, singing, laughing and obviously enjoying life. Song in unison was coming from the banks where little fires were twinkling in the short twilight. Full moon tonight, so the fishing and the fun has continued, and we, dining and sleeping on the terrace above the river, have shared vicariously. Just after we came in tonight we saw a cavalcade of horsemen coming into the village. Soon we learnt that the police had captured 53 Arabs of a wild tribe to the west and 59 horses. They had been on an ivory and giraffe raid into Dinka territory.

2 March 1961

A cold night with wind from the north. George and I had our beds out on the terrace, as is our habit, and we found two blankets insufficient last night. Along with Mahmoud to the compound where the 53 Arabs and 59 ponies were. What a crew! Some of the wildest men you ever saw, who are not content with slitting someone's throat with whom they are angry; they decapitate them as well. Mahmoud began a parley and I was glad to see that he shook hands with their Chief. This is as it should be; a captive is a specially unfortunate individual who is in your power; therefore you do not display power. Piles of primitive saddlery lay around and the abject horses were tied to the fence. How abject they were! I have never seen a worse bunch – narrow-chested, cow-hocked, roach-backed, hammer-headed (for they are Barbs), and carrying many sores. These ponies have been used hard these days and will now have a rest. We went to have a look at their weapons – very long spears like lances for elephants and throwing spears. One man is in hospital already in Aweil with a tusk through his chest. And then the arm knives, with which they are so likely to come at you if they get angry. Also lots of little charms sewn in leather, including a little book in Arabic of verses of the Koran chosen for daily use and thought by the Mahdi. My personal weakness of pity for anyone captive, man or animal, was already at work within me, though I would cheerfully shoot to kill if they were at large. Mahmoud went off with a lorry load of the prisoners to the gaol at Aweil.

George and I headed southward with a Dinka guide and a Game Scout in the hope of finding game. We passed through good bush with a lot of *Pterocarpus* trees, but the Dinka villages strewn along the way have cleared an awful lot of ground that will take a century to recover. We crossed the River Lol once at a ford but at 32 miles we could not cross the next ford. So we walked for 6–7 miles and waded over the river twice. Plenty of sign of elephant, buffalo and giraffe, but we saw only hartebeest, tiang and reedbuck. The elephants were too far in. The Dinka thatching we have seen today is very good. These Dinkas do not seem as tall as our friends farther east and I wonder if there is admixture with a tribe we struck two days ago – the Jur, who are shorter, fuller-bodied and pronouncedly negroid, not Nilotics at all. Back soon after 4 o'clock and had a marvellous shower bath (the first encountered in the Sudan), enjoyed the fishing activity observed from our terrace, and had a marvellous dinner cooked by Uncle Hassan, our beloved cook. We are a very happy crowd.

3 March 1961

What a day! We left Nyamlell at 7 o'clock and came through over sixty miles of Dinka country, much of it cleared in the way the Dinka do – space no object, and I think agriculturally they are land spoilers, but not so as cattle keepers on the *toich*. There were a few patches of pleasant forest bush, but on the whole there were distinct signs of getting into drier country. Then the river, Bahr-el-Arab, which is the boundary

The Bahr-el-Arab river which is the boundary between Bahr-el-Ghazal and Dar-Fur. Fraser Darling and Merck crossing on 3 March 1961.

between Dar-Fur and Bahr-el-Ghazal. Here we came to a busy scene of comings and goings across the river in dug-out canoes, and a meeting of cultures, the Nilotic and the Arab. There was a village on the other side of the river called Sofara, which means misbehaviour. No structure in the whole village was more permanent than a flimsy grass hut. We crossed over in the dug-outs though the lorries had to take a one and a half hour journey to get across. Mahmoud, George and I were taken to see the Chief, the tribe being the wild, knifing crowd of which we had seen 53 members and their mounts at Nyamlell. We were taken to a cool grass *chitenji* where deck chairs were set out for us and grass mats on the floor. A few Arab gentlemen in white robes shook hands with us and entertained us until the Chief arrived, an elderly man in white, with his small embroidered Arab cap but no turban. He is extremely gentle in manner and waves us courteously to our seats once more. Drinks of lemon are brought for us, the Chief sits on a short bed-like structure and the gentlemen sit on the palm matting. (The tribe is the Raziegat.) Then starts a battle of wits between Mahmoud and the Chief. The Chief wants his men back and knows he has no cards. One of the elders of the tribe sitting at his feet is a fine-boned, exquisitely featured little man with the mildest eye you ever encountered, so very gentle and with a whimsical half smile. This little gentleman had five sons in the wild crowd we saw in the jail. The Chief and Mahmoud are so courteous to each other, and an amused, rather gracious smile, seems always playing round the Chief's face. You and I have not met before, says the Chief, and it is unfortunate we must meet in a fight. Not a fight but an argument between brothers,

The Raziegat chieftain (first left with white skull cap) and elders at Sofara on 3 March 1961.

The Raziegat chieftain's stallion offered in payment for the freedom of his tribesmen at Sofara on 3 March 1961.

Jiger, our driver, is a black Arab from Dar-Fur and a fine character, 3 March 1961.

says Mahmoud. The Chief accepts the gesture. Could the men be returned to Dar-Fur for trial in the Chief's district? No, said Mahmoud, because it is a principle that trial is in the province where the crime is committed. On and on they go, the Chief never losing his gentle expression, and Mahmoud then making the gesture that he will do his best to get the men sent back into Dar-Fur to serve their sentences, so that the tribe will have access to the men. This is well taken. The Chief tells Mahmoud he likes him and that he feels he has met a man of honour. Mahmoud is modestly grateful. One of the gentlemen is sent away and shortly tea is served for us, really lovely strong tea for which I was thankful. Then the Chief's stud horse is led before the *chitenji* for our inspection. It was a good-looking chestnut pony of good bone, hard well-shaped unshod feet, a really lovely shoulder, and a strong loin. It was a good pony altogether and the Chief was gratified that I wished to photograph it. Mahmoud is of a horse-loving family and his eye lit up at the pony. The Chief was not unobservant and said he would like to give this stallion to Mahmoud. (Had Mahmoud accepted such a graciously offered bribe, I imagine the Chief would have taken his full price out of those 53 men when he got them back.) Mahmoud declined with gentleness but said he so admired the horse that he would like to pay £100 for him. The Chief said he would now take care of the horse as Mahmoud's until such time as Mahmoud should fetch it. The Chief now indicated that the heat of the day was approaching and he had had prepared for us a place to rest. He himself led us to a cool grass *chitenji* in which were three beds laid for us. What could have been more perfect human behaviour? Eventually we rose to go when we heard our lorries approach, but by this time our cook Uncle Hassan had been off with a Nyamlell trader called Hassan and had been drinking the spirit arake. Our Hassan gets abusive when drunk and the object of his abuse was Jiger, the black Arab driver with whom I sit. Jiger is a very fine man and very devout. Hassan rides in the back of our lorry and I have often felt acutely for him when the lorry goes over bump and hollow and chucks everything in the back up against the roof. Hassan now began to pour the whole of his pent-up sufferings as abuse on Jiger. I knew exactly the kind of things he was saying as his voice got louder and higher and he began jumping about. I went over to try to lead Hassan away but I could tell he was crazed and did not see me, or anyone other than Jiger who, stern and rugged-faced, leaned as if unheeding against his lorry. It was very embarrassing for Mahmoud, and the Chief, seeing that, eased the situation by saying that we had all seen things like this happen and one must remember the name of this village, which he thought well-earned. We got away in due course, but 10 or 12 miles along we had seemed to lose the second lorry – Ahmed's. Jiger and I waited and then went back, and met Ahmed and George coming along. George said that as they were about to leave Sofara, Mahomed had disappeared and he had to be found in the less reputable grass huts. No sooner had they got going than Mahomed yelled that the lady had lifted his wrist watch and he had to go back after that. Mahomed came now into the back of our lorry, half drunk. On we went again for an hour and then at a particularly bad bump there were yells in the back and Jiger stopped. Mohamed came to the cab, his face contorted with rage as to be unrecognizable; this our mild handsome waiter Mahomed. The drink had got him. I told Jiger to drive on and leave him in the road, for I could hear Ahmed's lorry draw up behind and I thought the soldiers in the back of that could deal with Mahomed. So on we went through miles and miles of dry thinning acacia bush, the long grass being replaced by short. No more fire trouble but intense overgrazing. We now came into the country of the Bagharra nomads, small people who lived in little oval tents of palm matting. Their cattle were reds, roans, blacks and a few white, and I thought surprisingly good. The people are small and brown and happy and wild. Many of the women go bare to the waist in day time and wear their indigo veil at night. The women wear their hair in a multitude of little plaits. The family on trek may be a man on a camel, a woman riding a bull with the rein through its nose and a couple of children on an ass. Some of the women and children are surprisingly goodlooking, but the older women get like old gypsies from their exposed life. You remember my mentioning the delicacy of features of a little girl baby of a Bagharra woman in Malakal. All seem to have the most wonderful teeth. We stayed the night at Sibdu, where there is a large rainwater pool and where there is a tin rest-house put up for the ceremonial visit of Abd el Nasser. A Chief's brother sent us a fat sheep as a present. The upper lot of Arabs around here are big strong-looking men with thick necks, dark brown to black in colour.

4 March 1961

On 40 miles to Ed Da'ein, which is where Jiger comes from. We thought it a clean and pleasant little town where everyone seemed to have lots of time. Mahmoud telephoned to Nyala, supposed to be four to five hours away, but it took us seven hours because of the sand. We passed through light short grass bush with cattle grazing in large herds wherever there was water to be had. The country was badly overgrazed wherever there was water but as soon as you get farther away from water than cattle can walk in a day, the country has short grass under orchard-type bush.

Then into cattle again and the camel bush comes back into the picture. This grass obviously cures on the stalk and provides excellent feed for cattle which are obviously doing very well. Here the cattle are sold willingly and are not just used as currency. We came through an area of *quoz*, or stabilized dunes on a clay bottom. The dunes probably formed in the quaternary period and later became stabilized by vegetation. They would remain so were it not for overgrazing and motor traffic. The 'road' is merely where you go over these large low dunes of bush, which means the 'road' may be half a mile wide or more. It was difficult for us to keep station in this journey and we had to stop and reform every few miles to be sure we were all together. All of us got stuck in the sand now and again and we had to help each other out. Everybody cheerful and happy. Into Nyala at 5 o'clock, the remote town of camels and cattle, where evidences of western civilization are fragmentary. Mahmoud's cousin is colonel-in-charge here, and he asked us to Ramadan breakfast at 6.30 p.m. where we drank large quantities of the durra Ramadan cold drink and lemon, Sudanese food, durra paste and chillie sauce.

5 March 1961

Down into the town this morning by myself. We are in another blasted rest-house, but I suppose it gives us what passes for a bath and some shade. The weather these days is really very hot. When we are using the metal treads to get the cars out of the sand they get so hot you can't pick them up without something in your hand. We get rather dehydrated, but we are certainly toughening – one meal a day for the past week. I am pulling the fat off Mahmoud in good style and I also have lost a lot of flesh. George says he didn't really think he needed to lose much, but I guess his features are now a little cleaner cut than they were. The market place in Nyala is immense, just people sitting around selling much the same things – baskets, camel hobbles, knives, spear heads, earth nuts, onions, vegetables, leather, durra grain, earthenware pots, brightly coloured enamel basins, and so on. In the shops proper – open booths – lots of cottons for sale, some grocery items, and more basketry and tins. As you look over the market place it is a haze of indigo with the hundreds of women in their veils, but when I say veil I don't mean over the face. No women here go veiled. I took my wildebeest-tail fly whisk into a booth to get a plaited leather handle put on it and managed to get the idea over. Then back in the market place, I saw two Nuer boys; they greeted each other in good English, so I talked to them and went back to the leather booth with one of them who also spoke Arabic – to get things plain with the old boy. Mahmoud has rather left us to it today, so we have spent the time cleaning up and writing. The Sudanese do not excel in organization. A telegram came this morning saying I would be lecturing in Omdurman the day after I leave the Sudan. And our coming week is likely to be nothing but driving trucks through sand if we are not careful.

6 March 1961

We were to leave Nyala at 9.30 o'clock but we didn't of course. Not only did we go into the market place or *suk*, but we went to see Colonel Mahomed's horses. He is a real horse lover and has won a lot of races. His better stallion was a beautiful dark bay with black points. It was remarkable for its strong bone and good hock. It was comparatively light in body with this tremendous propelling apparatus, as you might say. Another stallion of seven years looked a lot older. Showed too much white in the eye and was slightly over at the knees. But he was still going well. His best mare was a brown, exceptionally kind and feminine, but big and well-boned. She had an unbelievably good shoulder. Had been fired for a strained tendon and the treatment had been too drastic. Then one of the prettiest chestnut fillies you ever saw, only 19 months old, but would make up into a very considerable mare. Sweet in temperament. Then a chestnut roan filly, and finally a country-bred chestnut stallion bought from the Arabs. Had not been reared too well, hard as nails and very fast in his class. But not the horse Mahmoud bought from the Chief the other day. Away at last, for 133 miles of boneshaking to Zalingei, west of Jebel Mara and our farthest west point in the Sudan. We passed through a lot of short grass acacia bush, where families of Bagharra moved with their cattle, asses, occasional camels, and the riding bulls. These dry country cattle are mostly looking very well. The Bagharra range through a lot of country in small groups. The Raziegat we met at Sofara seem to move more as a community, spending the driest time on the banks of the Bahr-el-Arab at Sofara where their smiths keep busy under their little grass shelters, and in the rains they range back as far as Nyala. We thought the Bagharra people extremely good-natured, and some of the women particularly good-looking. Their place in the tribe is much more that of equality than in some more sedentary Arab cultures. After a busy market town called Kass we seemed to come into the climatic penumbra of the Jebel Mara, because the bush got greener and more forest like. (Note. Lot of baobab between Nyala and Kass and 40 miles east of Nyala. All mutilated for rope fibre.) We learned there were elephants in this bush, between here and the French Equatorial border – Tchad, as we should now call it. About 15 miles from Zalingei we had a puncture, so George and I began to walk on and had the pleasure of five miles alone walking along the track. Several small groups of Arabs and cattle passed us, a little astonished at two white men wandering along with no visible means of sustenance, but always the polite greeting, and some would come over on their ponies to touch hands ceremoniously. I suppose part of the Arab paradox is that here we were defenceless in one of the most dangerous places in the world (other than Central Park) and were perfectly safe. This is also one of the politest places in the world, and good manners are in part the evolutionary consequence of the quickly provoked nature of these people. A Highlander once put it that way to me, many years ago. 'If you weren't good-mannered in the old days, you would not have lasted long,' he said.

Coming into Zalingei at sunset we were constantly crossing wide sandy *wadis*, lined with beautiful groves of *Acacia albida*. I have never seen so many of these trees which, nevertheless, are so essentially part of the African scene for me. They make up into large spreading trees of broken shade and are characteristic of moist situations, even liking inundation of several feet, as in the Luangwa Valley. And coming into Zalingei you could smell the beauty of moisture, and in effect this lovely place is just a huge grove of splendid *Acacia albida*, thousands of them. The people build their straw houses under them, surrounded by a square of grass fencing, so that the streets are formed by the grass fences and are constantly shaded by the acacias. The dust is beneath your feet, silent to the tread, and you are aware of the quiet texture of the whole physical environment. I decided I could live in Zalingei.

7 March 1961

Did not leave Zalingei till noon, so George and I had chance of a pleasant stroll through this lovely place. Camels, sheep, goats, cattle and ponies here and there and everywhere, water-boys with asses going to the well and coming back into the town shouting 'Moia'. There are enclosed gardens where papaya and other fruits are grown, and a lovely plaza which is merely that because it is surrounded by trees and grass fences, and on one side there is a fairly important-looking building of whitewashed red brick. But as I said last night, most of Zalingei is grass huts under the *Acacia albidas*, leisured, polite, and my notion of a place to live out the rest of my life. We talked to the District Commissioner who was a pretty civilized type, speaking English entirely easily. When I said I wouldn't mind coming and living out my last few years here, he said, 'Right enough; it would be a good thing all round because initially man likes man, but,' he said, 'in our modern political world if an Englishman applied to reside in Zalingei just because he wanted to and was able to, the authorities would immediately react by asking why should he wish to do that? And the plain true reason would be too incredible to be accepted at its face value.' That is just about the way life is in our modern world, cynical and characterized by that dreadful expression 'Sez you', the epitome of unbelief and loss of all sensitive values. We came through good bush for 55 miles, some of it uphill, to Murtagello, which is at about 4,000 feet at the foot of a foothill range of the main Jebel Mara. A conservator of forests called Abdullah had joined us in Nyala and was to show us something of the forestry aspirations in the Jebel Mara area. We are now staying in a forest camp and village of the Fur tribe, the rather primitive people of non-Arab stock indigenous to the Jebel Mara. They are short, well-made people, brown in colour and probably negroid but not markedly so. I suppose the Fur really have been here a very long time and have been pushed upwards into the Jebel. They are great gardeners, creating these in the little open acres where the streams of the Jebel come out of ravines. They grow onions as a speciality. The women, who do most of the garden work, therefore make good people for forest nursery work. They are also making the road which is being extended precipitously into the Jebel by the Forest Department. George and I went up stream to a pool and waterfall for a most glorious bath. It is a pleasure beyond description after the heat and dust and bumping. We broke our fast at 6.30 p.m. with our Ramadan-keeping friends, the Forest Rangers, who invited us to their Sudanese meal.

8 March 1961

Away uphill to Baldong at 7 o'clock this morning. The lorry stopped after seven miles and we walked on upwards into the hill. We are now in one of the major archaeological wonders and puzzles of the world – the terracing of much of the mountain of Jebel Mara. The hill is volcanic and the stones and boulders of basalt are everywhere with the earth between. If you were going to cultivate at all you would have to shift the boulders and presumably you would gradually get the notion of terracing to create the flat areas varying from two feet to 40 feet in width. From the Baldong Forest Nursery, set in a hollow among terraces, we went up a steep hillside of terraces cleared as a fire-break. These terraces were only two feet wide but well made and may have done no more originally than check the run off of water, but later we were in little hollows and valleys where the terracing was wider and very prettily done. Presumably these terraces were made in some pluvial period more than 2,000 years ago. Are the onion-growing Fur the direct descendants of the terrace-making culture? The Arabs say these works are too considerable for man and must be the work of Satan. Oh dear, what wrongheadedness! The Arab culture has never really got down to major physical work demanding great application, and the sight of these terraces all over the mountain must have quite shocked the 7th–8th century folk-wandering which brought Islam across Africa. George made a point the other day when he said the Arab stopped where the horse could not go (tsetse flies and trypanosomiasis) and that where in Africa the Arab could hunt on horseback, the game was pretty well gone. Not entirely true, of course, but the point is worth keeping in mind when considering this chequer-board of Africa. Abdullah and I walked a mile further into the hill and put up a young female kuda, all on her own. Kudu seem to be the only game on the hill. The Forest Department are trying to get trees growing, cypresses and eucalyptus of various species. I am far from sure that they can be successful on these highly porous and permeable basalt slopes. Probably it was the change to dry in climate which knocked out the terrace culture, but I think archaeological research should be done here. I am quite fascinated. Back to camp for an afternoon of writing, and bathing at another pool and waterfall in the same stream. Wonderful.

9 March 1961

Bed outside as usual last night and enjoying the sky before falling asleep, and then the Fur drums started in the village 150 yards away – a main beat higher pitched, and deeper variant undertones. Very good to listen to and the drums seemed good. They were going a long time before the people seemed to gather and start their monotonous singing and dancing. I listened a while but fell asleep and woke up again. George went over to look while I was asleep and came back when I was again asleep. He was not greatly impressed with the dancing, but the drum was interesting – a single instrument less than three feet long and a foot wide at the larger end. A big man bent over double and played the deeper variant notes with two sticks, and a little boy bent over the other end playing the main beat with two sticks. The women were really the dancers and had considerable difficulty in getting the men going – just like social goings-on in our own societies. The dance stopped soon after midnight but not the dogs and the braying asses. The dogs will start up barking for a quarter of an hour on end and you wonder why it should ever stop, and the donkeys set up their own preposterous row in among it all. Did not have a good night.

10 March 1961

Away soon after 7 o'clock to Galoll, which lies under the south-western face of the mountain. Many of the Fur women and a few men were out to wave us away with wide smiles. One characteristic of these people's faces is the prominent forehead, almost vertical. Galoll was less than 20 miles from the Murtagello and took us more than an hour and a half. Another rushing stream and waterfall with a beautiful pool, which provides for the Forest Nursery and the Fur vegetable growers. Some of the Fur are beginning to grow more than their onions and bearded wheat, and are going in for a little fruit as well. The bush around here could easily be forest; *Cordia* and African mahogany. Fire prevents its full expression, I suppose, but if the Forestry Department is going to prevent fires, I get the impression they would do better sticking to the indigenous species or at least planting in the shade of them. The stream sides should be tackled first and work outward from them. This is not being done. The Jebel itself is of basalt and seen from Galoll reminds me very much of Ben More, Mull, the rock being the same, the shape of the hill and buttressings, and the saddle between two summits. I went for a five or six mile walk with Abdullah, the forest conservator, in the early evening. Got some very good views of the Jebel with the sun low, and was able to examine its architecture through my binoculars. Saw sign of kudu and duiker, but nothing else alive except guinea fowl, of which Abdullah shot five.

11 March 1961

Out soon after 7 o'clock and back at 10.30 for breakfast of roast guinea fowl. George remained at camp to photograph birds, but Abdullah and I went towards the Jebel into the terraced country. Up here I found village sites of systems of round huts round central halls and little cupboards of places off the round huts. All the building was of dry stone to 2 to 3, and occasionally 4 feet. One village became almost a town. It was on a flat ridge and the valleys were terraced and even roads were lined by dry-stone dykes. I began to look for artefacts and between us we found a score or more fragments of grinding querns, and one or two complete ones. They were of granite or foliated gneiss. We also found two rubbing stones for the grinding, and these were rectangular, the end being used on the lower stone and they were of the basement complex, i.e., quartzite. So these people were pretty definite about the stone they wanted for grinding and would have had to carry it some miles to this basalt country. I also found a basalt basin – or the half of it – about the size of a washbowl, and obviously a basin because the inside had not the polish of a grinding stone. I took a lot of colour photographs of all these things and of the walls and hut circles. There must be some good archaeological work to be done here. Again, plenty of sign of kudu

Ruined villages of the terraced lands below Jebel Mara – walls and hut circles on 11 March 1961.

and duiker, and we saw one of the latter. Left at 1 p.m. for the four-hour (sic) journey to Kass. Reached that town at 3 p.m. exactly. Time as an accurate measure does not exist for these people and it is no good getting cross about it. We at least had a glorious bathe in the waterfall pool before coming away. Rest-house again at Kass; these people seem to think camping is derogatory, but thank goodness, several other officials are in occupation and we are outside. Mahmoud is away finding the Jebel elephants – I wonder with what success?

12 March 1961

We left Kass at 7 o'clock this morning and had an uneventful drive through the dry bush and sandy desert to Nyala. We did our bit of shopping in the *suk* and got warrants from the District Commissioner for our return flight to Khartoum, a thousand miles away by road with a lot of deep sand. It would have taken us all of a week to get through. We fly by DC 3 tomorrow and the trucks and crews get aboard the new railroad at Nyala and take 3–4 days to reach Khartoum. Mahmoud and his Dar-Fur game officer turned up in the early afternoon. He had not found the elephants but plenty of sign, and a good sprinkling of other game not showing signs of fright. The journey had been much longer than he had been led to believe and they had been stuck a few times in deep sand. All pretty much as I had expected.

13 March 1961

This is the end of the active part of our safari, getting on the airplane after Mahmoud's cousin the Colonel had given us a delicious breakfast of fried liver, fresh bread, and salad, and had shown us a pretty little Arab pony mare, dark bay and black points, which he was keeping for a brother officer. I could have become a horse-thief for her. The airplane came over but we still talked horses. Then a wild dash to the airfield and away in five minutes. Very civilized and personal and more enjoyable than going to London Airport and Idlewild. I am sorry to be leaving the Dar-Fur. No wonder John Owen said he spent the five happiest years of his life in Zalingei. The airplane put down at El Fasher and El Obeid and I just survived. Surprised to see how much of the semi-desert was and had been cultivated. Medani was at Khartoum to meet us, but in his usual slapdash way had not reserved our rooms till this morning and we had to share one, which pleased neither George nor me. And when we went up the previous folk were still in possession, so we got cross and shook up the hotel into promising separate rooms tomorrow morning. The Cambridge geographer, Richard Grove, walked to the terrace at sundown, so we had a good talk.

* * *

Bahr-el-Ghazal, the country of the rivers, is one of the most inspiring in Africa – an area of alternating water and seasonally inundated plain or *toich*, with sound bush of the slightly higher ground. The *toich* can never be overgrazed. It is so rich and productive and the seasonal inundation gives it rest perforce. The annual burning does no harm either, for no forest is being cut back thereby and the dressing of potash has value. The water table is so high that the short period of only partial exposure of the soil after burning does not result in desiccation. The limiting factor in the grazing cycle is the bush which is grazed in the wet season, but again, the bush can best withstand grazing and browsing in the wet season and it gets a measure of rest in the dry season.

The Bahr-el-Ghazal is one of those few areas which can withstand the presence of a large population of cattle in addition to a varied ungulate game fauna and even productive forest. Not only is it a very rich series of habitats, but the Dinka cattle culture is in itself a considerable asset in that it is a vast repository of knowledge of land and cattle management. Admittedly, an easy self-sufficiency is reached in a habitat so favoured by nature, with little desire to trade in the creatures so dearly loved and understood. Each animal is named and its lineage known as well as that of any human member of the tribe; this is an attitude I understand entirely from early experience and I react immediately in the cattle camps to the depth and beauty of the symbiotic relationship – the accummulated dust and ash on the tethering grounds which the cattle immediately began to kick up with their fore feet as they come to their tethers; the token stakes to which each beast went for voluntary tethering; the smudge fires; the grass fence sheltering the calves from wind; the play of calves, children and puppies; the protection against carnivorous animals fully understood by the cattle, and the readiness of the cows to share their milk with man even if the calf was not present. The Dinka so fully realise how they share their country with the game; some of them so far accept the giraffe that these animals will enter the cattle camp to calve, and their young will be rubbed with cattle fat as being accepted into the tribal economy.

We look upon the Sudan as one of the most fortunate naturally endowed countries in Africa and the Bahr-el-Ghazal is a very important part of that endowment. Protein producing areas of the world are much scarcer than starch and cellulose-producing ones: the Bahr-el-Ghazal can produce a very large amount of animal protein and that should be its destiny, not to be dissipated in the degrading expedience of starch growing or vegetable oil for hydrogenation to margarine.

[Nine years later in his Reith Lectures Frank put it in a nutshell (*Wilderness and Plenty*, 1970): '. . . the Bahr-el-Ghazal . . . is potentially one of the richest protein-producing areas of the world, but the starch-growers have already got 8,000 acres as an experimental plot for rice. The Dinka are a fine race of tall Nilotics, living in symbiosis with their beautiful cattle and wild game. Will a larger number of rice-fed Dinka living in a deteriorating habitat be a more contented people? I doubt it. Starch and sedition go together.']

EPILOGUE

An Appreciation

CHAPTER THIRTEEN

Axioms in Human Ecology

By THE TIME Fraser Darling went to Africa, he had grasped the sense of human ecology, which yet escaped succinct definition in his writings. Starting from his childhood rambles with his shy mother to the 'great wood' in the early morning before the populace was out and about, through his agricultural training and years of field ecology and animal behaviour, to his later endeavours in the realm of human ecology, he had acquired a sharp eye for country, wildlife and people. What he had pondered in the lives of deer and caribou, he now applied to lechwe and wildebeest; what he saw of the human condition among crofters in Scotland, Hopi Indians in Arizona and Eskimos in Alaska prepared him for encounters with the Bantu, Masai, Dinkas and many other tribes. Without rehearsal he went to Africa to do a job *ex tempore* which to others would have needed a lengthy period of work up. His analytical mind disentangled salient facts and quickly reassembled them in a rounded diagnosis of environmental conditions. This, he claimed, he could only achieve by being a non-specialist willing to span many disciplines at a fairly shallow level to arrive at a judgement. This he saw as an expertise in its own right, the forte of the human ecologist, and led him to deplore the arrogance of the specialist as he made clear at a Nairobi conference in 1963:

> The Chairman [Fraser Darling] remarked on the arrogance and conceit of some so-called agricultural scientists who were so readily listened to by the vote-catching politicians, and described such men who *used* science rather than *discovered* principles as dangerous modern-day adventurers.

W.L. Astle, who went to Northern Rhodesia in 1959 as pasture research officer and later became the Chief Wildlife Research Officer in Zambia, might have been one of these 'agricultural specialists'. As a scientist in the field Bill Astle found Fraser Darling's writings 'extremely superficial and of very little value' to him. Unless great care is taken, the use of science in politics can bedevil both and, as we shall see later, this was a recurring risk to Fraser Darling's science. Be that as it may, when it came to sorting out the broad ecological rationale of a country, it was not the narrow, gifted specialist but the sagacious all-rounder whose judgement was rightly sought and trusted by Government and people.

Five human activities dominated Frank's assessment of country: distribution of water, deforestation, predation of wildlife, grazing of domesticated stocks, and burning of the vegetation. His assessment of status and trend of habitat for both man and wildlife would involve the separate examination of these impacts. The permutation and correlation of the five are unique in each watershed, are vastly complicated, and could employ an entire research institute over a space of years to elucidate fully. Yet Frank did such assessments as a one-man effort in a few weeks of personal observation on the ground and perusal of key papers.

Though not ignoring facts based on hard data, he was not a man of numerical inclination, and, following his investigations, he was guided as much by instinct as by logic and research. To travel with him in a strange country was to obtain an instant feeling for that country through his ecological insights on the response of the natural system to man's use of it. His words were often spiritually evocative of tragedy, triumph, gaiety and even comedy. His persona had a mystical dimension which 'hard' scientists mistrusted, but in the end his fearless utterances on crucial issues usually made good sense and have stood the test of time.

A typical example of how Fraser Darling's mind worked lies in his interpretation of the black rhinoceros in the whistling thorn (gall acacia), which I have chosen as title for this book. Without research and based largely on his own observations on the spot, he saw that this great beast had the ability to deal with the thorn bush which was shunned by other herbivores. The diligent action of the rhino in getting hold of and ripping out young thorn bushes (gall acacias) set him thinking of the role which it has played in the past in maintaining open space in bush country. The disappearance of the rhino he could foresee would result in the unchallenged advance of the thorn bush, and the occlusion of the habitats of many other herbivores. That seems logical, but it took someone with Frank's powers of observation and ecological turn of mind to articulate the problem, and set it in its management context.

One of the axioms which came to dominate his thinking was that, in nature, predation plays a secondary role in the fortunes of populations of wild creatures – the primary role is played by habitat change. Predation by man on wild stocks may greatly reduce or exterminate quarry species in special circumstances, but large scale habitat change will irreversibly transform or exterminate whole communities of species. He had plenty to say about the dangers of organised poaching by muzzle-loaders satisfying the appetite for meat of the Copper Belt in Northern Rhodesia, and of the trapping gangs serving the biltong trade from the Lamai Wedge on the Kenya-Tanzania border. From his earlier days in the Scottish deer forests, he saw local people hunting for the pot or in a properly regulated cull of wildlife, as a close and often necessary simulation of natural predation. Large scale poaching and overkill for profit and export was, to him, vandalism of the natural predatory mechanism upon which populations of wild creatures depend for their health and survival.

The man-induced agents of change are well described in *West Highland Survey*, *Wildlife in Alaska*, and *Wildlife in an African Territory*: deforestation, overgrazing, and fire. The depletion of the country of its trees is self-evident; but the effects of grazing and burning are more subtle, requiring a detailed knowledge of plant communities, and the ability of constituent species to sustain continuous cropping, drought, and fire. All three acting together in force can lead to physical and biological collapse with severe erosion, flooding, drought, mass migration, and die-off. Articulated to these primary factors in Frank's eyes were many other factors relating to the regeneration of the vegetation, and numbers, condition, and behaviour of animals seen against the known climate, hydrology, geology, topography, and soils.

There is little wonder, therefore, that he often recorded an interest in the wayside flora, since he was continuously making up his mind about the proportions of browse- and fire-resistant herbage. He had to learn as he went, becoming familiar, for example, with the nutrient-rich grasses and shrubs, and the signs of large herbivores and their predators. In this he was greatly helped by a succession of travelling companions whose knowledge of Africa he greatly respected, and whom he mentions in his Journal. I went on safari with several of them and can testify to their great knowledge of Africa, and of the effect which they must have had on Frank's thinking. Men like Desmond Vesey-Fitzgerald whose knowledge of African wildlife was encyclopaedic, spanning both flora and fauna, vertebrates and invertebrates, and elucidating the pulses, rhythms and cycles of the African ecosystem. Like Evelyn Temple-Boreham who knew, conversed with, and enjoyed the respect of the bush African in a way which I saw in very few other white men; his barazzas (council meetings) with the Masai liabon (chief priest) and elders were dignified, business-like, and charged with trusty good humour. I can picture Frank listening with rapt attention and admiration to these men, and others like them, their faces aglow in the flicker of the camp fire.

From time to time he went walking by himself, possibly at some considerable risk of attack and injury by lion, elephant or buffalo. Then, free from distractive conversation, he was led by his own curiosity, and moved from one key observation to another on a path of discovery towards a point of ecological truth rarely reached by his peers. Each walk was a revelation of local conditions which he was keen to describe to his colleagues on the spot, as confirming or detracting from the broad canvas of his knowledge and experience.

Frank, in discussions with Hugh Lamprey, expressed great interest in the ecological separation of species of ungulates – how, over vast undisturbed periods of evolutionary time, nature has elaborated in the ancient African environment an intricately stratified grazing regime. From the plains to the highlands, a diverse array of herbivores has evolved, each in its own graze-browse niche. The whole system constitutes one of the wonders of the world – on land, the greatest flow of solar energy through green vegetation into animal life of unbelievable variety and abundance. The processes of evolutionary adaptiveness and ecological integration held an enormous fascination for Frank. He had endless pleasure in, figuratively, fitting together the jig-saw, and his writings describe the interdependence of the pieces as he saw them, say, in the conversion cycle. Success in evolution, he believed, was to find an unoccupied stratum of forage, exploit it, and have an assured place in the conversion cycle.

In the report on his survey in the Sudan, he is axiomatic:

> that it is a rare phenomenon in nature to find two species making exactly the same demands on the environment or treating the environment in the same way in any one habitat, though it is equally remarkable that in different habitats an animal in each may be found adapted to do the same thing. For example, what the hyaena does in the tropics the wolverine does in Arctic lands. But in any one habitat, each animal occupies its own *niche* in the economy of nature. Thus of all the hoofed species, of which there are so many different in Africa, each is making a slightly different demand of the environment or doing something slightly different to it. This means that the habitat is never stocked up to the limit of what it could hold of any one species.

He held to Lotka's doctrine that each species is employed in the conversion of energy from one level to another. Some are concerned with birth and regeneration of life, others with death and decay, and within the system there dwells the evolutionary force towards a state of maximum energy flow, or climax. In his quest for areas of untouched natural tropical forest (*mushitu*) in Luangwa, there was the desire to stand within the biological climax of Planet Earth (see colour plates).

Frank had already a reputation for candid utterance when diplomatic expediency was likely to mask ecological truth. His altercation with the Scottish Office over the *West Highland Survey*, the ecological incisiveness and didacticism of *Wildlife in Alaska*, the great measure of American bonhomie in his newly published *Pelican in the Wilderness*, and the overt respect accorded to him in the realm of international conservation, all made him a person to be reckoned with. Those who had the courage to invite his advice ran the risk of having their world collapse around their ears, with government departments reaching out for disclaimers of his reports and commentaries. Perhaps it was inevitable that the colonial regimes, and the land degradation which accompanied them in Africa, would produce in him at least as great a sense of outrage as did the Scottish establishment.

Frank faced a dilemma between what he saw with his own eyes and garnered from trusted companions, and the environmental questions which he had been invited to address by his hosts. Was he to knee-jerk governments and peoples in bringing them face to face with their addle-pated (his word) way of thinking, or was he to adopt the non-political stance of the scientist in describing cause and effect, without social and political attribution?

One of the axioms he propounded in the *West Highland Survey* was that man's politics were a powerful ecological factor; a main tenet of human ecology which was not sufficiently appreciated by the senior administrators of the colonial establishment in Africa. Frank saw the land as a careering chariot drawn by three horses – social, political, and natural steeds – each affecting the other two, and all interdependent in the destiny of the environment. The clinical separation of the ecological from the political factors, as was hoped for in Lusaka and Nairobi, he would find nearly impossible to deliver. In diplomacy he was discredited in lacking the softly–softly touch. He was certainly a direct and spontaneous person, fierce when the chips were down. However, he was something of a prophet in his own time who included within his thinking imperatives of which diplomats would be unaware, or consider unimportant and irrelevant.

It was in the Sudan that Frank came face to face with the habitat which he espoused more than any other in his long and varied life as an ecologist – the desert. I knew him as one who felt himself denied a place in history for original scientific discovery. His pioneer work on red deer, sea birds and seals is still recognised, but it is for his interpretation of the North-West Highlands of Scotland as a 'man-made desert' that he is mainly remembered and often quoted. This was his round impression of a land, once rich and inhabitated, now devastated and desolate but yet serene and beautiful to behold. That impression in the *West Highland Survey* chimes with that on seeing Dar-Fur recorded in his report:

> There are some who would say much of desert is useless. I am not
> of that opinion: desert is rarely absolute desert; the herds of addax
> and oryx and the flocks of gazelle which roam the semi-desert north
> of El Fasher are an adornment to the area which, in a fast developing
> world, will have a wilderness value far beyond what is being
> realised at the moment. But again, there is little chance of multiple
> use: the desert will either be used sparingly for its recreational and

rehabilitative value to a jaded humanity or it will not be used at all. Destruction of wild herds of desert animals would be wanton and altogether foolish. Their conservation is a responsibility of world citizenship.

CHAPTER FOURTEEN

Finale

IN LOOKING AT the African scene Fraser Darling became simultaneously aware of the ecological structure of these vast and diverse tropical environments and the agencies of change at work within them. With a historical perspective, he saw the signs of man's intervention and his rapid acceleration of otherwise benign natural processes leading to widespread degradation and often local devastation. The reduction of the tropical forest by the native Africans and colonial Asians and Europeans long preceded his time. In passing through the Luangwa Valley in 1866 Livingstone described the country as follows:

> A good portion of the trees of the country have been cut down for charcoal, and those which now spring up are small; certain fruit trees alone are left. The long slopes of the undulating country, clothed with fresh foliage, look very beautiful. The young trees alternate with patches of yellow grass not yet burned; the hills are covered with a thick mantle of small green trees with as usual, large ones at intervals.

This description could be applied to much of the surviving African forest of today; it is characteristic not of primary tropical forest, but of secondary forest which has naturally regenerated after man's felling of the great primaeval trees.

Frank had studied and was possibly inspired by Livingstone's *Last Journals* edited by Horace Waller (1874). Though not written with the same style and content as his own Journal, he saw Livingstone's account of his travels as an ecological statement of Africa in the mid-nineteenth century. He knew that the habitats and ecological processes within them which he was observing in his time, bore a strong resemblance to those of the previous century. Indeed, it was obvious to him that Africa had been greatly changed by human beings over many centuries, and that the rate of change had simply accelerated in the last century, in parallel with human expansion.

As with Livingstone's Journal which possesses the poignancy of the separation of the great explorer from his wife and family, so in the same way Frank was in Africa separated from Averil and the children at their time of greatest need. Though he did not seem to know it at the time of his visit to Northern Rhodesia in 1956, Averil was dying of cancer. The following hand-written notes addressed to Averil accompany the manuscript of his Journal, and reveal the sacrifices in family life which separation brings, often with tragic results.

On 24th August 1956 from the Kenya-Tanganyika border:

> Having a job to keep up to date. 1,500 miles in Kenya last week. So glad to be coming home a week earlier. For next year's trip here we can have a house at Chilanga, and the boys could come for their summer holidays. We can also have a grass hut camp in the Luangwa Valley, all laid on, where you can watch elephants and rhino and so on from the verandah. This camp would be in the nature of a summer holiday. Health situation would be all right. Oh I do love you so and am thinking about you because I am not sleeping well.

On 7th September 1956 from the Kafue River, Northern Rhodesia:

> This will go out by runner tomorrow and get to Lusaka for posting sometime during the week. I am thinking so much about you and also worrying, because I shall hear no word of you possibly until Sept 24. I keep dreaming of your coming with the children next year and of that holiday in the grass camp in the Luangwa. A grass hut is such a pleasant thing to live in. Kiss my children for me please and accept so many for yourself.

On 15th September 1956 from Kasempa in the Kaonde country, Northern Rhodesia:

> Four weeks today I should be home, and four weeks yesterday since I left home . . . I am in splendid health and so enjoying the chance of getting out on my own feet. The natural history here is so thrilling and there is so much needing record and study. Nine days hence I should pick up my first mail and how much I am hoping that the news will be good. Richard and Jamie will be going to school this week, and I am wishing them much good fortune and love . . . I think so much about you, Sweetheart, mostly in the middle of the night, but always with love and longing to be with you.

On 30th September 1956 camped on the banks of the Zambezi near Mogambwe Falls: 'Last before I return I suppose. I think of you all, individually and collectively, so often, and for periods as I am bashing along, like today, at four miles an hour.' After his return home in October 1956 and after Averil's death in February 1957: 'This batch [of the journal] is from last year's second trip, not sent home because it would have taken longer than I would. And after getting home, my darling never had a chance to read it.'

Averil's death came as a great blow to Frank. He was a naturally selfish person with a passion for career which subordinated family affairs to the demands of his work. The life he led commuting from Berkshire to the United

States, Africa and a continuous stream of national and international meetings was beyond his control, and it is not surprising that it should suffer partial collapse. Nonetheless, he sustained grief with courage and resolution to continue immediately his international career, and he was back in Northern Rhodesia in July 1957, continuing where he left off. At home his two sons were already at boarding school, but his daughter Francesca was only seventeen months old when her mother died, and for her particular care was required. At this time Frank's good friend Douglas Grant of Oliver and Boyd, Publishers, Edinburgh, introduced him to Christina who joined the family as house mother and to offer affection and consistent long term care to all the children.

The period of Frank's life covered by his African Journal was traumatic. Yet, it was simply the latest crisis in a life which from childhood had been marked by painfully enforced changes. His Journals are testimony of how he could put his domestic worries aside to get on with the job in hand, but at home conditions were different. He had too many distractions and found it very difficult to settle to thinking and writing. His report on the Mara was a victim of his disturbed life. He seriously overshot the deadlines set by his sponsors, when it was urgently required by the Government in setting up the Masai-Mara Game Reserve.

The tour in Sudan in 1961 which concludes the Journal was yet another ecological reconnaissance of the Conservation Foundation with the support of the New York Zoological Society. Fairfield Osborn was again the moving force and by 1959 Frank was himself a Vice-President of that Foundation, and George Merck, who accompanied him in Sudan, was then the Secretary of that Society. Sudan he found in a more degenerate state than any other country he had yet seen. The report on the tour to the Conservation Foundation almost predictably covered the weaknesses of the game ordinance and licensing system, the need for a system of Parks and Reserves related to a flexible wildlife policy, bringing the benefits of a game conservation policy directly to the people in or near Game Reserves, a rationale for 'safari' tourism, and the organisation of the Game Department with research and training echelons.

Frank was now ecological clinician extraordinaire. Few others could by now match him for breadth of experience and knowledge of environmental ills caused by man's use of natural resources. At the CTCA/IUCN Conference in Arusha in 1961 Frank was one of the leading personalities, bearing credentials which were at once British and American and stretching from tropic to arctic. Throughout the 1960s, while expending most of his energies on the work of the Conservation Foundation in New York and Washington, he still maintained his personal connections in Britain, living in a fine house at Shefford

Woodlands in Berkshire and a cottage in the Scottish Highlands (which he always said was his original field laboratory). In this period he played a leading part in the Foundation's international research programme.

In 1965 he was chairman of an historic conference at Airlie House, Warrenton, Virginia on *Future Environments of North America* (ed. Darling and Milton, 1966), and also in the mid-1960s he carried out an enquiry into social, political and ecological policies of the National Parks of the United States with N.D. Eichhorn (*Man and Nature in the National Parks*, 1967).

In 1966 the Nature Conservancy in Britain changed its status and became a component body of the newly formed Natural Environment Research Council, and new personalities were introduced. Among these were the Chairman, Lord Howick of Glendale who, as Sir Evelyn Baring, was Governor of Kenya during Frank's ecological survey of the Mara, and Professor Duncan Poore who was Frank's friend and admirer from Edinburgh University days, and now Director of the Conservancy. It is perhaps not surprising, therefore, that in 1969 Frank regained his place in the British establishment. This he did with a bang, for in 1969–70 he was appointed to the council of the Nature Conservancy, and to the Royal Commission on Environmental Pollution. In 1969 he delivered the Reith Lectures for the BBC, entitled *Wilderness and Plenty* (1970). Honours followed: a knighthood in 1970, honorary degrees from Heriot-Watt, New University of Ulster, Williams College Massachussets (to add to the Hon. LLD Glasgow 1958), Centenary Medalist (US National Parks Service), Commander of the Order of the Golden Ark (Netherlands), John Phillips Medalist (International Union for the Conservation of Nature).

Frank and Christina retired in 1973 to a finely adapted farmhouse near Forres in the Laigh of Moray, which he always maintained had the most clement climate in the British Isles. As the years advanced, his family and friends from many countries came to see him. He became progressively afflicted by angina and by a loss of balance; his great frame would lurch alarmingly in advancing to offer a hand of welcome. His achievements were never strong enough to erase from his conscience the circumstances of his birth. The Victorian stigma of bastardy, so carefully shrouded by him from his peers, was indelible in his heart. In the end he became preoccupied with the trials as well as the triumphs of his long and varied life, and with death itself. He died at Forres Hospital on 22nd October 1979.

Today in Africa and America the name of Fraser Darling has passed into history and may be almost forgotten. In Scotland, however, Frank's name is graven in the collective memory of the people among whose forebears he lived half a century ago, when he wrote:

The art and science of wild-life conservation is that which brings stability

or regular rhythm into disturbed habitat, beauty and balance into the wilderness itself, and renders it productive of material and spiritual values.
There, in the Highlands and Islands, he is remembered as a man before his time – the pioneer of the ethos of conservation uniting nature and human nature in a single harmony.

Index